BOOK 2

3rd edition

STP *Caribbean* MATHEMATICS

CE Layne
FW Ali
L Bostock
S Chandler
A Shepherd
E Smith

OXFORD
UNIVERSITY PRESS

Great Clarendon Street, Oxford, OX2 6DP, United Kingdom

Oxford University Press is a department of the University of Oxford.
It furthers the University's objective of excellence in research, scholarship,
and education by publishing worldwide. Oxford is a registered trade mark of
Oxford University Press in the UK and in certain other countries

First published in 1987
Second edition published by Stanley Thornes (Publishers) Ltd in 1997
Third edition published by Nelson Thornes Ltd in 2005
This edition published by Oxford University Press in 2014

British Library Cataloguing in Publication Data
Data available

978-0-7487-9088-3

10 9 8 7 6 5 4 3 2

Printed in India

Acknowledgements

Illustrations: Peters and Zabransky, Rupert Besley, Steve Ballinger, A & R Nelson,
Guyana, Linda Jeffrey
Page make-up: Tech-Set Ltd

Although we have made every effort to trace and contact all
copyright holders before publication this has not been possible in all
cases. If notified, the publisher will rectify any errors or omissions at
the earliest opportunity.

CONTENTS

Contents

Contents

Contents

INTRODUCTION

This book attempts to satisfy your needs as you begin your study of mathematics in the secondary school. We are very conscious of the need for success together with the enjoyment everyone finds in getting things right. With this in mind we have divided most of the exercises into three types of question:

> The first type, identified by plain numbers, e.g. **12**, helps you to see if you understand the work. These questions are considered necessary for every chapter you attempt.

> The second type, identified by a single underline, e.g. **12**, are extra, but not harder, questions for quicker workers, for extra practice or for later revision.

> The third type, identified by a double underline, e.g. **12**, are for those of you who manage Type 1 questions fairly easily and therefore need to attempt questions that are a little harder.

Most chapters end with 'mixed exercises'. These will help you revise what you have done, either when you have finished the chapter or at a later date.

All of you should be able to use a calculator accurately by the time you leave school. It is wise, in your first and second years, to use it mainly to check your answers, unless you have great difficulty with 'tables'. Whether you use the calculator or do the working yourself, always estimate your answer and always ask yourself the question, 'Is my answer a sensible one?'

PREFACE

To the Teacher

The general aims of the series are:

(1) to help students to
  - attain solid mathematical skills
  - connect mathematics to their everyday lives and understand its role in the development of our contemporary society
  - see the importance of thinking skills in everyday problems
  - discover the fun of doing mathematics and reinforce their positive attitudes to it.

(2) to encourage teachers to include historical information about mathematics in their programme.

In writing this four book series the authors have attempted to present topics in such a way that students will understand the connections in Mathematics, and be encouraged to see and use mathematics as a means to help make sense in the real world.

Topics from the history of mathematics have been incorporated to ensure that mathematics is not dissociated from its past. This should lead to an increase in the levels of enthusiasm, interest and fascination for mathematics, and should also enrich the teaching of it.

Careful grading of exercises makes the books approachable.

Some suggestions

(1) Before each lesson give a brief outline of the topic to be covered in the lesson. As examples are given refer back to the outline to show how the example fits into it.

(2) List terms on the chalkboard that you consider new to the students.
Solicit additional words from the class and encourage students to read from the text and make their own vocabulary.
Remember that mathematics is a foreign language. The ability to communicate mathematically must involve the careful use of the correct terminology.

(3) When possible have students construct alternative ways to phrase questions. This ties in with seeing mathematics as a language. Students tend to concentrate on the numerical or 'maths' part of the question and pay little attention to the instructions which give information which is required to solve the problem.

(4) When solving problems have students identify their own problem-solving strategies and listen to others. This practice should create an atmosphere of discussion in the class centred around different approaches to the same problem.
As the students try to solve problems on their own they will make mistakes. This is healthy, as this was the experience of the inventors of mathematics: they tried, guessed, made many mistakes and worked for hours, days and sometimes years before reaching a solution.
There are enough problems in the exercises to allow the students to try and try again. The excitement, disappointment and struggle with a problem until a solution is found provide a healthy classroom atmosphere.

To the Student

These books are written for you. As you study:

Try to break up the material in a chapter into manageable bits.
Always have paper and pencil when you study mathematics.
When you meet a new word write it down together with its meaning.
Read your questions carefully and rephrase them in your own words.
The information which you need to solve your problem is given in the wording of the problem, not the number part only.
Your success in mathematics may be achieved through practice.
You are therefore advised to try to solve as many problems as you can.
Always try more problems than those set by your teacher for homework.
Remember that the greatest cricketer or netball player became great by practising for many hours.

We have provided enough problems in the books to allow you to practice. Above all do not be afraid to make mistakes as you are learning. The greatest mathematicians all made many mistakes as they tried to solve problems.

You are now on your way to success in mathematics. – GOOD LUCK!

Picture Credits

Pascal machine, page 128, Poyet/La Nature 1904/MEPL;

Klein bottle, page 254, Science Musuem/Science and Society.

1 DIRECTED NUMBERS

AT THE END OF THIS CHAPTER…

you should be able to:

1 Plot points with positive and negative coordinates.

2 Use positive or negative numbers to describe displacements on one side or the other of a given point on a line.

3 Apply positive and negative numbers, where appropriate, in a physical situation.

4 Perform operations of addition, subtraction, multiplication or division on positive and negative numbers.

5 Multiply expressions with directed numbers.

6 Solve linear equations involving directed numbers.

David Blackwell is, to mathematicians, the most famous, perhaps the greatest, African American mathematician. A biannual prize has been inaugurated in his honour: the Blackwell–Tapia Prize. The first recipient was Dr Arlie O. Petters. Dr Petters emigrated from Belize to the United States in 1979 and became a US citizen in 1990.

BEFORE YOU START

you need to know:
✓ how to add, subtract, multiply and divide simple numbers mentally
✓ the basic properties of the square, rectangle and parallelogram

KEY WORDS

diagonal, directed number, negative, negative coordinate, number line, parallelogram, positive, positive coordinate, quadrilateral, rectangle, square

Negative coordinates

If A(2, 0), B(4, 2) and C(6, 0) are three corners of a square ABCD, we can see that the fourth corner, D, is two squares below the x-axis.

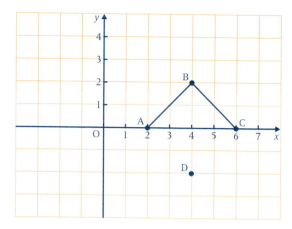

To describe the position of D we need to extend the scale on the y-axis below zero. To do this we use the negative numbers

$$-1, \ -2, \ -3, \ -4, \ \ldots$$

In the same way we can use the negative numbers $-1, -2, -3, \ldots$ to extend the scale on the x-axis to the left of zero.

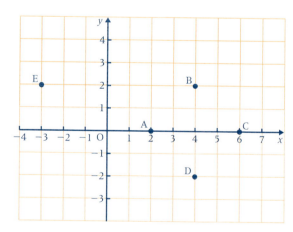

The y-coordinate of the point D is written -2 and is called 'negative 2'.

The x-coordinate of the point E is written -3 and is called 'negative 3'.

The numbers 1, 2, 3, 4, ... are called positive numbers. They could be written as $+1$, $+2$, $+3$, $+4$, ... but we do not usually put the $+$ sign in.

Now D is   4 squares to the right of O so its x-coordinate is 4

and   2 squares below the x-axis so its y-coordinate is -2,

                D is the point $(4, \ -2)$

E is   3 squares to the left of O so its *x*-coordinate is −3

and   2 squares up from O so its *y*-coordinate is 2,

                E is the point (−3, 2)

Exercise 1a

Use this diagram for questions 1 and 2.

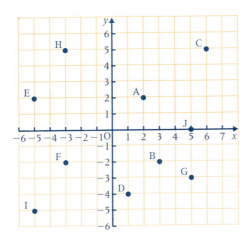

1   Write down the *x*-coordinate of each of the points A, B, C, D, E, F, G, H, I, J and O (the origin).

2   Write down the *y*-coordinate of each of the points A, B, C, D, E, H, I and J.

How many squares above or below the *x*-axis is each of the following points?

3   P: the *y*-coordinate is −5

4   L: the *y*-coordinate is +3

5   M: the *y*-coordinate is −1

6   B: the *y*-coordinate is 10

7   A: the *y*-coordinate is 0

8   D: the *y*-coordinate is −4

Tip A negative *y*-coordinate means go down.

Tip A positive *y*-coordinate means go up.

How many squares to the left or to the right of the *y*-axis is each of the following points?

9   Q: the *x*-coordinate is 3

10   R: the *x*-coordinate is −5

11   T: the *x*-coordinate is +2

12   S: the *x*-coordinate is −7

13   V: the *x*-coordinate is 0

14   G: the *x*-coordinate is −9

15 Write down the coordinates of the points A, B, C, D, E, F, G, H, I and J.

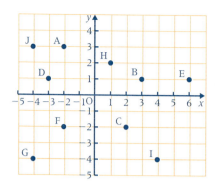

> **Tip** The *x*-coordinate comes first, followed by the *y*-coordinate. Each coordinate may be positive or negative.

In questions **16** to **21** draw your own set of axes and scale each one from −5 to 5:

16 Mark the points A(−3, 4), B(−1, 4), C(1, 3), D(1, 2), E(−1, 1), F(1, 0), G(1, −1), H(−1, −2), I(−3, −2).
Join the points in alphabetical order and join I to A.

17 Mark the points A(4, −1), B(4, 2), C(3, 3), D(2, 3), E(2, 4), F(1, 4), G(1, 3), H(−2, 3), I(−3, 2), J(−3, −1).
Join the points in alphabetical order and join J to A.

18 Mark the points A(2, 1), B(−1, 3), C(−3, 0), D(0, −2).
Join the points to make the figure ABCD. What is the name of the figure?

19 Mark the points A(1, 3), B(−1, −1), C(3, −1).
Join the points to make the figure ABC and describe ABC.

20 Mark the points A(−2, −1), B(5, −1), C(5, 2), D(−2, 2).
Join the points to make the figure ABCD and describe ABCD.

21 Mark the points A(−3, 0), B(1, 3), C(0, −4).
What kind of triangle is ABC?

Exercise *1b*

Draw your own set of axes for each question in this exercise. Mark a scale on each axis from −10 to +10.

In questions **1** to **10** mark the points A and B and then find the length of the line AB:

1 A(2, 2) B(−4, 2)

2 A(−2, −1) B(6, −1)

3 A(−4, −4) B(−4, 2)

4 A(1, −6) B(1, −8)

5 A(3, 2) B(5, 2)

6 A(5, −1) B(5, 6)

7 A(−2, 4) B(−7, 4)

8 A(−1, −2) B(−8, −2)

9 A(−3, 5) B(−3, −6)

10 A(−2, −4) B(−2, 7)

In questions **11** to **20**, the points A, B and C are three corners of a square ABCD. Mark the points and find the point D. Give the coordinates of D:

11 A(1, 1) B(1, −1) C(−1, −1)

12 A(1, 3) B(6, 3) C(6, −2)

13 A(3, 3) B(3, −1) C(−1, −1)

14 A(−2, −1) B(−2, 3) C(−6, 3)

15 A(−5, −3) B(−1, −3) C(−1, 1)

16 A(−3, −1) B(−3, 2) C(0, 2)

17 A(0, 4) B(−2, 1) C(1, −1)

18 A(1, 0) B(3, 2) C(1, 4)

19 A(−2, −1) B(2, −2) C(3, 2)

20 A(−3, −2) B(−5, 2) C(−1, 4)

In questions **21** to **30**, mark the points A and B and the point C, the midpoint of the line AB. Give the coordinates of C:

21 A(2, 2) B(6, 2)

22 A(2, 3) B(2, −5)

23 A(−1, 3) B(−6, 3)

24 A(−3, 5) B(−3, −7)

25 A(−1, −2) B(−9, −2)

26 A(2, 1) B(6, 2)

27 A(2, 1) B(−4, 5)

28 A(−7, −3) B(5, 3)

29 A(−3, 3) B(3, −3)

30 A(−7, −3) B(5, 3)

Straight lines

1

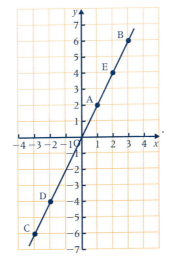

The points A, B, C, D and E are all on the same straight line.

a Write down the coordinates of the points A, B, C, D and E.

b F is another point on the same line. The *x*-coordinate of F is 5. Write down the *y*-coordinate of F.

c G, H, I, J, K, L and M are also points on this line. Fill in the missing coordinates:

G(8, □) H(10, □) I(−4, □) J(□, 12) K(□, 18) L(□, −10) M(a, □).

2

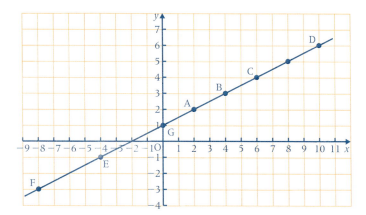

The points A, B, C, D, E, F and G are all on the same straight line.

a Write down the coordinates of the points A, B, C, D, E, F and G.

b How is the y-coordinate of each point related to its x-coordinate?

c H is another point on this line. Its x-coordinate is 8; what is its y-coordinate?

d I, J, K, L, M, N are further points on this line. Fill in the missing coordinates:
   I(12, □),   J(20, □),   K(30, □),   L(−12, □),   M(□, 9),   N(a, □).

3

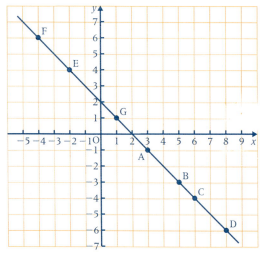

The points A, B, C, D, E, F and G are all on the same straight line.

a Write down the coordinates of the points A, B, C, D, E, F and G.

b H, I, J, K, L, M, N, P and Q are further points on the same line. Fill in the missing coordinates:
   H(7, □), I(10, □), J(12, □), K(20, □), L(−7, □), M(−9, □), N(□, 10), P(□, −8), Q(□, 12).

Exercise 1d

In the following questions we are going to investigate the properties of the diagonals of the special quadrilaterals. You will need your own set of axes for each question. Mark a scale on each axis from −5 to +5. Mark the points A, B, C and D and join them to form the quadrilateral ABCD.

1 A(5, −2), B(2, 4), C(−3, 4), D(0, −2)

  a What type of quadrilateral is ABCD?

  b Join A to C and B to D. These are the diagonals of the quadrilateral. Mark with an E the point where the diagonals cross.

  c Measure the diagonals. Are they the same length?

  d Is E the midpoint of either, or both, of the diagonals?

  e Measure the four angles at E. Do the diagonals cross at right angles?

Now repeat question **1** for the following points:

2 A(2, −2), B(2, 4), C(−4, 4), D(−4, −2)

3 A(2, −2), B(5, 4), C(−3, 4), D(−1, −2)

4 A(2, 0), B(0, 4), C(−2, 0), D(0, −4)

5 A(1, −4), B(1, −1), C(−5, −1), D(−5, −4)

6 Name the quadrilaterals in which the two diagonals are of equal length.

7 Name the quadrilaterals in which the diagonals cut at right angles.

8 Name the quadrilaterals in which the diagonals cut each other in half.

 Investigation

Draw your own set of x- and y-axes and scale each of them from −6 to +8.

Plot the points A(−1, 3), B(3, −1) and C(−1, −5).

  a Can you write down
    i the coordinates of a point D such that ABCD is a square
    ii the coordinates of a point E such that ACBE is a parallelogram
    iii the coordinates of a point F such that CDEF is a rectangle?

  b Can you give the name of the special quadrilateral EDBF?

Use of positive and negative numbers

Positive and negative numbers are collectively known as directed numbers.

Directed numbers can be used to describe any quantity that can be measured above or below a natural zero. For example, a distance of 50 m above sea level and a distance of 50 m below sea level could be written as +50 m and −50 m respectively.

They can also be used to describe time before and after a particular event. For example, 5 seconds before the start of a race and 5 seconds after the start of a race could be written as −5 s and +5 s respectively.

Directed numbers can also be used to describe quantities that involve one of two possible directions. For example, if a car is travelling north at 70 km/h and another car is travelling south at 70 km/h they can be described as going at +70 km/h and −70 km/h respectively.

A familiar use of negative numbers is to describe temperatures. The freezing point of water is 0° centigrade (or Celsius) and a temperature of 5 °C below freezing point is written −5 °C.

Most people would call −5 °C 'minus 5 °C' but we will call it 'negative 5 °C' and there are good reasons for doing so because in mathematics 'minus' means 'take away'.

A temperature of 5 °C above freezing point is called 'positive 5 °C' and can be written as +5 °C. Most people would just call it 5 °C and write it without the positive symbol.

A number without any symbol in front of it is a positive number,

i.e.   2 means +2
and   +3 can be written as 3

Exercise *1e*

Draw a celsius thermometer and mark a scale on it from −10° to +10°. Use your drawing to write the following temperatures as positive or negative numbers:

1 10° above freezing point

2 7° below freezing point

3 3° below zero

4 5° above zero

5 8° below zero

6 freezing point

Write down, in words, the meaning of the following temperatures:

7 −2 °C

8 +3 °C

9 4 °C

10 −10 °C

11 +8 °C

12 0 °C

Which temperature is higher ?

13 +8° or +10°

14 12° or 3°

15 −2° or +4°

16 −3° or −5°

17 −8° or 2°

18 −2° or −5°

19 1° or −1°

20 +3° or −5°

21 −7° or −10°

22 −2° or −9°

23 The contour lines on the map below show distances above sea level as positive numbers and distances below sea level as negative numbers.

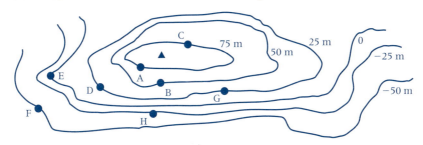

Write down in words the position relative to sea level of the points A, B, C, D, E, F, G and H.

In questions **24** to **34** use positive or negative numbers to describe the quantities.

A ball thrown up a distance of 5 m.

Up is above the start point so a positive number describes this.

+5 m

24 5 seconds before blast off of a rocket.

25 5 seconds after blast off of a rocket.

26 50 c in your purse.

27 50 c owed.

28 1 minute before the train leaves the station.

29 A win of $50 on a lottery.

30 A debt of $5.

31 Walking forwards five paces.

32 Walking backwards five paces.

33 The top of a hill which is 200 m above sea level.

34 A ball thrown down a distance of 5 m.

35 At midnight the temperature was −2 °C. One hour later it was 1° colder. What was the temperature then?

36 At midday the temperature was 18 °C. Two hours later it was 3° warmer. What was the temperature then?

37 A rockclimber started at +200 m and came a distance of 50 m down the rock face. How far above sea level was he then?

38 At midnight the temperature was −5 °C. One hour later it was 2° warmer. What was the temperature then?

39 At the end of the week my financial state could be described as −25 c. I was later given 50 c. How could I then describe my financial state?

40 Positive numbers are used to describe a number of paces forwards and negative numbers are used to describe a number of paces backwards. Describe where you are in relation to your starting point if you walk +10 paces followed by −4 paces.

Extending the number line

If a number line is extended beyond zero, negative numbers can be used to describe points to the left of zero and positive numbers are used to describe points to the right of zero.

On this number line,   5 is to the *right* of 3
and we say that   5 is *greater* than 3
                    or   5 > 3 (> means 'is greater than')

Also   −2 is to the *right* of −4
and we say that   −2 is *greater* than −4
                    or   −2 > −4

So 'greater' means 'higher up the scale'.
(A temperature of −2 °C is higher than a temperature of −4 °C.)

Now   2 is to the *left* of 6
and we say that   2 is *less* than 6
                    or   2 < 6 (< means 'is smaller than')

Also   −3 is to the *left* of −1
and we say that   −3 is *less* than −1
                    or   −3 < −1

So 'less than' means 'lower down the scale'.

Exercise 1f

Draw a number line.

In questions **1** to **12** write either > or < between the two numbers:

1 3   2   **4** −3   −1   **7** 3   −2   **10** −7   3

2 5   1   **5** 1   −2   **8** 5   −10   **11** −1   0

3 −1   −4   **6** −4   1   **9** −3   −9   **12** 1   −1

In questions **13** to **24** write down the next two numbers in the sequence:

13 4, 6, 8	**16** −4, −2, 0	**19** 5, 1, −3	**22** −10, −8, −6
14 −4, −6, −8	**17** 9, 6, 3	**20** 2, 4, 8	**23** −1, −2, −4
15 4, 2, 0	**18** −4, −1, 2	**21** 36, 6, 1	**24** 1, 0, −1

Addition and subtraction of positive numbers

If you were asked to work out $5 - 7$ you would probably say that it cannot be done. But if you were asked to work out where you would be if you walked 5 steps forwards and then 7 steps backwards, you would say that you were two steps behind your starting point.

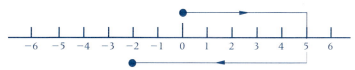

On the number line, $5 - 7$ means

start at 0 and go 5 places to the right

and then go 7 places to the left

So $$5 - 7 = -2$$

i.e. 'minus' a positive number means move to the left
and 'plus' a positive number means move to the right.

In this way $3 + 2 - 8 + 1$ can be shown on the number line as follows:

Therefore $3 + 2 - 8 + 1 = -2$.

Exercise 1g

Find, using a number line if it helps:

1 $3 - 6$	**3** $4 - 6$	**5** $4 - 2$	**7** $-2 + 3$	**9** $-5 - 7$
2 $5 - 2$	**4** $5 - 7$	**6** $5 + 2$	**8** $-3 + 5$	**10** $-3 + 2$

$(+4) - (+3)$

$$(+4) - (+3) = 4 - 3 \quad (+4 = 4 \text{ and } +3 = 3)$$

$$= 1$$

11 $(+3) + (+2)$

12 $(+2) - (+4)$

13 $(+5) - (+7)$

14 $(-3) + (+2)$

15 $(-1) + (+5)$

16 $5 - 2 + 3$

17 $7 - 9 + 4$

18 $5 - 11 + 3$

19 $10 - 4 - 9$

20 $3 + 6 - 10$

21 $-4 + 2 + 5$

22 $-3 + 1 - 4$

23 $5 - 6 - 9$

24 $-3 - 4 + 2$

25 $-2 - 3 + 9$

26 $(+3) + (+4) - (+1)$

27 $(+2) - (+5) + (+6)$

28 $(+9) - (+7) - (+2)$

29 $(-3) + (+5) - (+5)$

30 $(-8) - (+4) + (+7)$

> **Tip** Remember that +3 is the same as 3.

What number does x represent if $x + 2 = 1$?

You can think of this as starting at 2 on the number line. What do you have to do to get to 1?

$$-1 + 2 \text{ is } 1$$

so $$x = -1$$

Find the value of x:

31 $x + 4 = 5$

32 $x - 2 = 0$

33 $x + 2 = 4$

34 $x + 2 = 0$

35 $2 + x = 1$

36 $3 + x = 1$

37 $5 - x = 4$

38 $5 + x = 7$

39 $9 - x = 4$

40 $x - 6 = 10$

Addition and subtraction of negative numbers

Most of you will have some money of your own, from pocket money and other sources. Many of you will have borrowed money at some time.

At any one time you have a *balance* of money, i.e. the total sum that you own or owe!

If you own $2 and you borrow $4, your balance is a debt of $2. We can write this as

$$(+2) + (-4) = (-2)$$

or as $$2 + (-4) = -2$$

But $$2 - 4 = -2$$

∴ $$+(-4) \text{ means } -4$$

If you owe \$2 and then take away that debt, your balance is zero. We can write this as

$$(-2) - (-2) = 0$$

You can pay off a debt on your balance only if someone gives you \$2. So subtracting a negative number is equivalent to adding a positive number, i.e. $-(-2)$ is equivalent to $+2$.

$$-(-2) \quad \text{means} \quad +2$$

Exercise 1h

Find:

1	$3 + (-1)$	9	$-4 + (-10)$
2	$5 + (-8)$	10	$2 - (-8)$
3	$4 - (-3)$	11	$-7 + (-7)$
4	$-1 - (-4)$	12	$-3 - (-3)$
5	$-2 + (-7)$	13	$+4 + (-4)$
6	$-2 - (-5)$	14	$+2 - (-4)$
7	$4 + (-7)$	15	$-3 + (-3)$
8	$-3 - (-9)$		

Tip $+(-1) = -1$

Tip $-(-3) = +3$

$$2 + (-1) - (-4)$$

$$+(-1) = -1 \text{ and } -(-4) = +4$$

$$2 + (-1) - (-4) = 2 - 1 + 4$$

$$= 5$$

16	$5 + (-1) - (-3)$	21	$9 + (-5) - (-9)$
17	$(-1) + (-1) + (-1)$	22	$8 - (-7) + (-2)$
18	$4 - (-2) + (-4)$	23	$10 + (-9) + (-7)$
19	$-2 - (-2) + (-4)$	24	$12 + (-8) - (-4)$
20	$6 - (-7) + (-8)$	25	$9 + (-12) - (-4)$

Addition and subtraction of directed numbers

We can now use the following rules:

$$+(+a) = +a \quad \text{and} \quad -(+a) = -a$$

$$+(-a) = -a \quad \text{and} \quad -(-a) = +a$$

Exercise 1i

Find:

1 $3 + (-2)$	**11** $12 + (-7)$	**21** $7 + (-4) - (-2)$
2 $-3 - (+2)$	**12** $-4 - (+8)$	**22** $3 - (+2) + (-5)$
3 $6 - (-3)$	**13** $3 - (-2)$	**23** $-9 + (-2) - (-3)$
4 $4 + (+4)$	**14** $-5 + (-4)$	**24** $8 + (+9) - (-2)$
5 $-5 + (-7)$	**15** $8 + (-7)$	**25** $7 + (-9) - (+2)$
6 $9 - (+2)$	**16** $4 - (-5)$	**26** $4 + (-1) - (+7)$
7 $7 + (-3)$	**17** $7 + (-3) - (+5)$	**27** $-3 + (+5) - (-2)$
8 $8 + (+2)$	**18** $2 - (-4) + (-6)$	**28** $-4 + (+8) + (-7)$
9 $10 - (-5)$	**19** $5 + (-2) - (+1)$	**29** $-9 - (+4) - (-10)$
10 $-2 - (-4)$	**20** $8 - (-3) + (+5)$	**30** $-2 - (+8) + (-9)$

$$-8 - (4 - 7)$$

$$-8 - (4 - 7) = -8 - (-3) \qquad (\text{brackets first})$$
$$= -8 + 3$$
$$= -5$$

31 $3 - (4 - 3)$	**44** $(4 - 8) - (10 - 15)$
32 $5 + (7 - 9)$	**45** Add $(+7)$ to (-5).
33 $4 + (8 - 12)$	**46** Subtract 7 from -5.
34 $-3 - (7 - 10)$	**47** Subtract (-2) from 1.
35 $6 + (8 - 15)$	**48** Find the value of '8 take away -10'.
36 $(3 - 5) + 2$	**49** Add -5 to $+3$.
37 $5 - (6 - 10)$	**50** Find the sum of -3 and $+4$.
38 $(4 - 9) - 2$	**51** Find the sum of -8 and $+10$.
39 $(7 + 4) - 15$	**52** Subtract positive 8 from negative 7.
40 $8 + (3 - 8)$	**53** Find the sum of -3 and -3 and -3.
41 $(3 - 8) - (9 - 4)$	**54** Find the value of twice negative 3.
42 $(3 - 1) + (5 - 10)$	**55** Find the value of four times -2.
3 $(7 - 12) - (6 - 9)$	

Puzzle

An early Greek mathematician set up a secret society in the sixth century BC. Members swore never to give away any mathematical secrets. One member was killed because he told one of the secrets to a friend who was not a member. The first letters to complete each of the statements below taken in order will spell his name.

The distance around a plane figure is called its _____ .

There are 365 days in a _____ .

A plane figure with three sides is a _____ .

−1 is _____ of −2.

The _____ of a triangle add up to 180 degrees.

> means 'is _____ than'.

The point where the x and y axes intersect is the _____ .

$\frac{4}{5}$ is called the _____ of $\frac{5}{4}$.

To find the _____ of a rectangle we multiply the length by the width.

−8 is the _____ of −6 and −2.

Division of negative numbers by positive numbers

Because $2 \times 3 = 6$, $6 \div 3 = 2$.

In the same way, $(-3) \times 4 = -12$, so $(-12) \div 4 = -3$.

Notice that the order *does* matter in division, e.g.

$$(-12) \div 4 = -3$$

but we shall see that $\quad 4 \div (-12) = \frac{-1}{3}$

Exercise 1j   Find   **a** $(-9) \div 3$ **b** $(-14) \div (+2)$ **c** $\frac{-10}{2}$

a $(-9) \div 3 = -3$

b $(-14) \div (+2) = -7$ $((-14) \div (+2) = -14 \div 2)$

c $\frac{-10}{2} = -5$ $(\frac{-10}{2}$ means $-10 \div 2)$

Find:

1 $(-6) \div 2$

2 $(-10) \div 5$

3 $(-15) \div 3$

4 $(-24) \div 6$

5 $(-12) \div (+3)$

6 $(-18) \div (+9)$

7 $(-30) \div (+3)$

8 $(-36) \div 12$

9 $(-20) \div 4$

10 $(-28) \div (+7)$

11 $(-3) \div 3$

12 $(-10) \div (+5)$

13 $\frac{-8}{4}$

14 $\frac{-12}{6}$

15 $\frac{-16}{4}$

16 $\frac{-27}{3}$

17 $\frac{-36}{9}$

18 $\frac{-30}{15}$

Multiplication of directed numbers

Consider the expression $6x - (x - 3)$.

From $6x$ we have to subtract the number in the bracket which is 3 less than x.

If we start by writing $6x - x$ we have subtracted 3 too many.

To put it right we must add on 3.

Therefore $\qquad 6x - (x - 3) = 6x - x + 3$

Similarly $8x - 3(x - 2)$ means

> 'from $8x$ subtract three times the number that is 2 less than x', which is
>
> $8x - (3x - 6)$

If we write $8x - 3x$ we have subtracted 6 too many. So we must add 6 on again giving

$$8x - 3(x - 2) = 8x - 3x + 6$$

From this, and from the previous exercise, we have

(a) $\qquad (+3) \times (+2) = +6$

> This is just the multiplication of positive numbers,
>
> i.e. $(+3) \times (+2) = 3 \times 2 = 6$

(b) $\qquad (-3) \times (+2) = -6$

> Here we could write $(-3) \times (+2) = -3(2)$.
>
> This is equivalent to subtracting 3 twos, i.e. subtracting 6.

(c) $\qquad (+4) \times (-3) = -12$

> This means four lots of -3,
>
> i.e. $(-3) + (-3) + (-3) + (-3) = -12$

(d) $\qquad (-2) \times (-3) = +6$

> This can be thought of as taking away two lots of -3,
>
> i.e. $-2(-3) = -(-6)$
>
> We have already seen that taking away a negative number is equivalent to adding a positive number, so $(-2) \times (-3) = +6$.

Exercise 1k   Calculate   **a** $(+2) \times (+4)$ **b** 2×4.

a $(+2) \times (+4) = 8$

b $2 \times 4 = 8$

This shows that $(+2) \times (+4)$ means the same as 2×4.

Calculate   **a** $(-3) \times (+4)$ **b** -3×4.

a $(-3) \times (+4) = -12$

b $-3 \times 4 = -12$

This shows that $(-3) \times (+4)$ means the same as -3×4.

Because order does not matter when two quantities are multiplied together, $(+4) \times (-3)$ gives the same answer of -12.

So -3 and $+4$ can be multiplied together in four different ways, but they all mean the same thing.

Calculate   **a** $(-5) \times (-2)$ **b** $-5(-2)$.

a $(-5) \times (-2) = 10$

b $-5(-2) = 10$ $-5(-2)$ means $-5 \times (-2)$

Calculate:

1 $(-3) \times (+5)$

2 $(+4) \times (-2)$

3 $(-7) \times (-2)$

4 $(+4) \times (+1)$

5 $(+6) \times (-7)$

6 $(-4) \times (-3)$

7 $(-6) \times (+3)$

8 $(-8) \times (-2)$

9 $(+5) \times (-1)$

10 $(-6) \times (-3)$

11 $(-3) \times (-9)$

12 $(-2) \times (+8)$

13 $7 \times (-5)$

14 $-6(-4)$

15 -3×5

16 $5 \times (-9)$

17 $-6(4)$

18 $-2(-4)$

19 $-(-3)$

20 $4 \times (-2)$

21 $3(-2)$

22 5×3

23 $6 \times (-3)$

24 $-5(-4)$

25 $6 \times (-4)$

26 $-3(+8)$

27 $(+5) \times (+9)$

28 -4×5

29 $7(-4)$

30 $(-4) \times (-9)$

Exercise 1l

Multiply out $-(x + 2)$

$$-(x + 2) \quad \text{means} \quad -1(x + 2)$$
$$-(x + 2) = -x - 2$$

Multiply out the following brackets:

1 $-6(x - 5)$

2 $-5(3c + 3)$

3 $-2(5e - 3)$

4 $-(3x - 4)$

5 $-8(2 - 5x)$

6 $-7(x + 4)$

7 $-3(2d - 2)$

8 $-2(4 + 2x)$

9 $-7(2 - 3x)$

10 $-(4 - 5x)$

11 $4(3x + 9)$

12 $5(2 + 3x)$

13 $3(2x - 6)$

14 $-7(2 + x)$

15 $-2(3x - 1)$

16 $-(3x + 2)$

17 $8(2 - 3x)$

18 $-3(2y - 4x)$

19 $5(4x - 1)$

20 $-5(1 - 4x)$

21 $6(4 + 5x)$

22 $-6(4 + 5x)$

23 $6(4 - 5x)$

24 $-6(4 - 5x)$

Tip Multiply each term inside the bracket by -6.

25 $-(5a + 5b)$

26 $2(3x + 2y + 1)$

27 $-5(5 + 2x)$

28 $4(x - y)$

29 $-(4c - 5)$

30 $9(2x - 1)$

Tip Multiply each term inside the bracket by 4.

Exercise 1m

Simplify the following expressions:

1 $5x + 4(5x + 3)$

2 $42 - 3(2c + 5)$

3 $2m + 4(3m - 5)$

4 $7 - 2(3x + 2)$

5 $x + (5x - 4)$

6 $9 - 2(4g - 2)$

7 $4 - (6 - x)$

8 $10f + 3(4 - 2f)$

9 $7 - 2(5 - 2s)$

10 $7x + 3(4x - 1)$

11 $7(3x + 1) - 2(2x + 4)$

12 $5(2x - 3) - (x + 3)$

13 $2(4x + 3) + (x - 5)$

14 $7(3 - x) - (6 - 2x)$

15 $5 + 3(4x + 1)$

16 $6x + 2(3x - 7)$

17 $20x - 4(3 + 4x)$

18 $4(x + 1) + 5(x + 3)$

19 $3(2x + 3) - 5(x + 6)$

20 $5(6x - 3) + (x + 4)$

21 $4(x - 1) + 5(2x + 3)$

22 $4(x - 1) - 5(2x + 3)$

23 $4(x - 1) + 5(2x - 3)$

24 $4(x - 1) - 5(2x - 3)$

25 $8(2x - 1) - (x + 1)$

26 $3x + 2(4x + 2) + 3$

27 $5 - 4(2x + 3) - 7x$

28 $3(x + 6) - (x - 3)$

29 $3(x + 6) - (x + 3)$

30 $7x + 8x - 2(5x + 1)$

Solving equations

Exercise 1n

Solve the following equations.

Some equations may have negative answers.

$x + 8 = 6$

$$x + 8 = 6$$

Take 8 from both sides $\qquad x = -2$

1 $x + 4 = 2$ **3** $3 + a = 2$ **5** $4 + w = 2$

2 $x + 6 = 1$ **4** $s + 3 = 2$ **6** $c + 6 = 2$

$x - 6 = 2$

$$x - 6 = 2$$

Add 6 to both sides $\qquad x = 8$

When there are several letter and number terms, deal with the letters first, then the numbers.

$5x + 2 = 2x + 9$

$$5x + 2 = 2x + 9$$

Take $2x$ from both sides $\qquad 3x + 2 = 9$

Take 2 from both sides $\qquad 3x = 7$

Divide both sides by 3 $\qquad x = \frac{7}{3} = 2\frac{1}{3}$

7 $3x + 4 = 2x + 8$ **11** $7x + 3 = 3x + 31$

8 $x + 7 = 4x + 4$ **12** $6z + 4 = 2z + 1$

9 $2x + 5 = 5x - 4$ **13** $7x - 25 = 3x - 1$

10 $3x - 1 = 5x - 11$ **14** $11x - 6 = 8x + 9$

> **Tip** Choose to take away the lower number of x's.

$9 + x = 4 - 4x$

$$9 + x = 4 - 4x$$

Add $4x$ to both sides $\qquad 9 + 5x = 4$

Take 9 from both sides $\qquad 5x = -5$

Divide both sides by 5 $\qquad x = -1$

Check: If $x = -1$, left-hand side $= 9 + (-1)$

$\qquad\qquad\qquad\qquad\qquad = 8$

right-hand side $= 4 - (-4)$

$\qquad\qquad\qquad\qquad\qquad = 8$

So $x = -1$ is the solution.

15 $4x - 3 = 39 - 2x$

16 $5 + x = 17 - 5x$

17 $7 - 2x = 4 + x$

18 $24 - 2x = 5x + 3$

19 $5x - 6 = 3 - 4x$

20 $12 + 2x = 24 - 4x$

21 $32 - 6x = 8 + 2x$

22 $9 - 3x = -5 + 4x$

> **Tip** Remember that negative numbers are lower down the number line than positive numbers.

$9 - 3x = 15 - 4x$

$-4x < -3x$ so add $4x$ to both sides

Take 9 from both sides

$9 - 3x = 15 - 4x$

$9 + x = 15$

$x = 6$

23 $5 - 3x = 1 - x$

24 $16 - 2x = 19 - 5x$

25 $6 - x = 12 - 2x$

26 $-2 - 4x = 6 - 2x$

27 $16 - 6x = 1 - x$

28 $4 - 3x = 1 - 4x$

29 $4 - 2x = 8 - 5x$

30 $3 - x = 5 - 3x$

31 $6 - 3x = 4x - 1$

32 $4z + 1 = 6z - 3$

33 $3 - 6x = 6x - 3$

34 $8 - 4x = 14 - 7x$

35 $13 - 4x = 4x - 3$

36 $7x + 6 = x - 6$

37 $6 - 2x = 9 - 5x$

38 $3 - 2x = 3 + x$

> **Tip** Think of the number line to decide which is the lower number of x's.

$3 - 2x = 5$

There are zero x's on the right-hand side, and $-2x < 0x$

Add $2x$ to both sides

Take 5 from both sides

Divide both sides by 2

$3 - 2x = 5$

$3 = 5 + 2x$

$-2 = 2x$

$x = -1$

39 $13 - 4x = 5$

40 $6 = 2 - 2x$

41 $6 = 8 - 3x$

42 $0 = 6 - 2x$

43 $9x + 4 = 3x + 1$

44 $2x + 3 = 12x$

45 $7 - 2x = 3 - 6x$

46 $3x - 6 = 6 - x$

47 $-4x - 5 = -2x - 10$

48 $5 - 3x = 2$

49 $6 + 3x = 7 - x$

50 $5 - 2x = 4x - 7$

> **Tip** 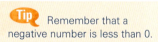 Remember that a negative number is less than 0.

Did you know that

If the sum of two numbers is 24 and the difference is 6, then one of the numbers is $\dfrac{(24+6)}{2} = 15$?

or

If the sum of two numbers is 28 and the difference is 12, then one of them is $\dfrac{(28+12)}{2} = 20$?

Use your algebra to prove that this is always so.

Investigation

Try this on a group of pupils or friends.

Think of a number between 1 and 10.
Add 4.
Multiply the result by 5.
Double your answer.
Divide the result by 10.
Take away the number you first thought of.
Write down your answer.

However many times you try this the answer is always 4.
Investigate what happens when you use numbers other than numbers between 1 and 10. Try, for example, larger whole numbers, decimals, negative whole numbers, fractions.

Is the answer always 4?

Mixed exercises

Exercise 1p

1 Which is the higher temperature, $-5°$ or $-8°$?

2 Write $<$ or $>$ between   **a** -3 2 **b** -2 -4.

Find:

3 $-4 + 6$

4 $3 + 2 - 10$

5 $2 + (-4)$

6 $3 - (-1)$

7 $-2 + (-3) - (-5)$

8 $4 - (2 - 3)$

9 $6 \times (-4)$

10 $-36 \div 3$

11 $(-5) \times (-2)$

12 $-2(7)$

13 $(+4) \times (-6)$

14 $-2(-3) + 1$

15 Simplify   **a** $-3(2 - 5x)$ **b** $3(5 - 2x) - (x - 7)$.

Mixed exercise

Exercise 1q

1 A is the point $(2, -1)$. What are the coordinates of the point immediately

   a 2 above A   **b** 3 to the left of A

   c 4 to the right of A and 2 below it?

> **Tip** Draw your own grid and point A on it.

2 Plot the points $A(-2, -1)$, $B(2, 2)$, $C(4, 0)$, $D(0, -3)$ on axes scaled from -4 to 4 for both x and y.
Join the points to make the figure ABCD.

   a What name is given to this shape?

   b Write down the coordinates of the midpoint of   **i** BC   **ii** AC

3 Simplify   **a** $-3(3b - 4) - 3$ **b** $3(2x - 1) - 5(2 - x)$

4 Solve the equation   **a** $s + 4 = -1$ **b** $27 - 5x = 6 + 2x$ **c** $-3 - 4x = 3 - 2x$

Did you know that Astragalia is the name given to the large number of bones claimed to have been discovered by archaeologists at prehistoric sites?

It is said that these bones were used as dice in ancient games.

IN THIS CHAPTER...

you have seen that:

- directed numbers is the collective name for positive and negative numbers

- directed numbers can be used to describe quantities that can be measured above or below a natural zero

- the rules for addition and subtraction are

  $+(+a)$ and $-(-a)$ both give $+a$ and
  $+(-a)$ and $-(+a)$ both give $-a$

  You can remember these as

  SAME SIGNS GIVE POSITIVE,
  DIFFERENT SIGNS GIVE NEGATIVE

- $5 \times (-3)$, $(+5) \times (-3)$, $(-3) \times 5$, $(-3) \times (+5)$
  ALL MEAN THE SAME

- when a negative number is divided by a positive number, the answer is negative

- negative coordinates allow us to plot points anywhere in the xy plane

- a negative number multiplied by a negative number gives a positive number

- you can solve equations involving directed numbers

2 WORKING WITH NUMBERS

AT THE END OF THIS CHAPTER...

you should be able to:

1. Multiply numbers written in index form.

2. Divide numbers written in index form.

3. Write numbers using negative indices.

4. Find the value of numbers written in index form.

5. Write numbers in standard form.

6. Approximate numbers to a given degree of accuracy.

7. Write numbers correct to a given number of significant figures.

8. Write numbers in standard form.

Did you know that the Googol (10^{100}) is said to have been so named by a teenager?

BEFORE YOU START

you need to know:
- ✓ the meaning of place value in numbers
- ✓ how to work with fractions, decimals and percentages

KEY WORDS

approximation, associative, commutative, decimal place, distributive, identity element, index (plural indices), integer, inverse element, negative indices, positive indices, reciprocal, rough estimate, significant figures, standard form, zero index.

Laws of numbers

From previous work you know that

$$8 + 17 = 17 + 8 \quad \text{and} \quad \text{that } 8 \times 17 = 17 \times 8$$

i.e. for addition and multiplication the order of the numbers does not matter. We say that numbers are *commutative* under addition and multiplication.

On the other hand $8 - 17$ and $17 - 8$ do not give the same answer

neither do $8 \div 17$ and $17 \div 8$

In this case numbers are *non-commutative* under subtraction and division.

You have also seen that

$$5 + (6 + 7) = (5 + 6) + 7 = 5 + 6 + 7$$

and that
$$4 \times (6 \times 8) = (4 \times 6) \times 8 = 4 \times 6 \times 8$$

i.e. the brackets can be removed without changing the answer, so there is no ambiguity in writing $5 + 6 + 7$ or $4 \times 6 \times 8$.
This illustrates the *associative* law for the addition and multiplication of numbers.

On the other hand, for $7 - (5 + 6)$ and $(6 + 9) \div 3$ the answers will be different if the brackets are removed. Subtraction and division of numbers do not satisfy the associative law.

In other calculations you know that $5(10 + 9) = 5 \times 10 + 5 \times 9$

Here multiplication is distributing itself over addition. This is the *distributive* law and is the only law expressing a relation between two basic operations.

The identity element

In a set of numbers when 0 (zero) is added to any number it preserves the identity of that number, i.e. it does not change its value, e.g. $5 + 0 = 5$ and $0 + 7 = 7$ so 0 is the *identity* element for addition.

The identity element for multiplication is 1. Multiplying a number by 1 does not change its value, e.g. $6 \times 1 = 6$ and $1 \times 10 = 10$

The inverse element

For every number, another number can be found so that the result of adding it to the original number is the identity element. The second number is called the *inverse* of the first number.

For addition the identity element is 0 so the inverse of 8 under addition is -8 and the inverse of -4 is 4.

For every number, except 0, another number can be found such that when it multiplies the given number, it produces the identity for multiplication, namely 1. For example, for multiplication, the inverse of 3 is $\frac{1}{3}$, because $3 \times \frac{1}{3} = 1$.

You will also have noticed that if you start with a number from the set $\{\dots -2, -1, 0, 1, 2, \dots\}$ and perform any of the operations $+$, $-$ or $\times$ the result is still a member of the set. However if you divide one member of the set by another you may get a number that is not in the set, e.g. $17 \div 8$.

We say that the numbers in the set $\{\dots -2, -1, 0, 1, 2, \dots\}$ are closed under $+$, $-$ and $\times$ but not under $\div$.

Exercise 2a

Which law, if any (associative, commutative or distributive), does each of the following statements illustrate?

1 $3 \times (4 \times 3) = (3 \times 4) \times 3$

2 $10 + (5 + 7) = (10 + 5) + 7$

3 $9 + 12 = 12 + 9$

4 $7 \times (3 + 5) = 7 \times 3 + 7 \times 5$

5 $7 + 3 + 2 = 2 + 7 + 3$

6 $(4 \times 6) \times 2 = 4 \times (6 \times 2)$

7 $8 \times (4 - 1) = 8 \times 4 + 8 \times (-1)$

8 $3 \times (3 + 2) = 3 \times 3 + 3 \times 2$

9 $5 \times 4 = 4 \times 5$

10 $6 \times (6 - 2) = 6 \times 6 - 6 \times 2$

11 In the statement $10 + 0 = 0 + 10$ how would you describe the zero (0)?

12 In the statement $1 \times 8 = 8 \times 1$ how would you describe the 1?

13 In the statement $10 + (-10) = 0$ how would you describe the -10?

14 What is the inverse of 9 under addition?

15 What is the inverse of 9 under multiplication?

16 What is the inverse of -5 under addition?

In questions **17** to **24**, write down the letter that goes with the correct answer.

17 The inverse of 4 under addition is

    A 4   **B** -4 **C** $\frac{1}{4}$ **D** 0

18 The inverse of 4 under multiplication is

    A 4   **B** -4 **C** $\frac{1}{4}$ **D** 0

19 $6 + 0 = 6$ so 0 is

    A the inverse of 6

    B the identity element for addition

    C not a member of the set of integers

20 a and b are integers and $a + b = c$. c is

   A an integer   **B** the identity element

   C the inverse under addition

> **Tip** An integer is a number of the set $\{\ldots -2, -1, 0, 1, 2, \ldots\}$

21 x and y are even numbers and $x + y = z$. z is

   A an even number   **B** an odd number

   C a fraction less than 1   **D** the identity element under addition

22 p and q are odd numbers and $p + q = r$. r is

   A a fraction smaller than 1   **B** the inverse of p under addition

   C an odd number   **D** an even number

23 a is an odd number, b is an even number, and $a + b = c$. c is

   A the identity element under addition   **B** the inverse of b under addition

   C an odd number   **D** an even number

24 a is an odd number, b is twice the value of a, and $a + b = c$. c is

   A the identity element under addition   **B** $\frac{1}{2}$

   C an odd number   **D** an even number

25 a, b and c are integers. Which of the following statements are always true?

   a $a + b = b + a$ **b** $a - b = b - a$

   c $a + b - c = a - c + b$ **d** $a \times b$ is an integer

   e $a \div b$ is an integer   **f** $a + b \times c = a \times c + b$

   g $a \div b \times c = a \times c \div b$

Changing between fractions, decimals and percentages

To change a fraction to a decimal, divide the numerator by the denominator.
The fraction $\frac{3}{8}$ means $3 \div 8$

You can calculate $3 \div 8$:

$$3 \overline{)8.000} \quad \frac{0.375}{}$$

So $\frac{3}{8} = 0.375$

Any fraction can be treated like this.

To change a decimal to a fraction express it as a number of tenths or hundredths, etc. and, if possible, simplify.
The decimal 0.6 can be written $\frac{6}{10}$, which simplifies to $\frac{3}{5}$

and the decimal 1.85 can be written $1\frac{85}{100}$, which simplifies to $1\frac{17}{20}$

To change a percentage to a decimal divide the percentage by 100.
To express a percentage as a decimal, start by expressing it as a fraction, but *do not simplify*, because dividing by 100, or by a multiple of 100, is easy.

Changing between fractions, decimals and percentages

For example $44\% = \frac{44}{100} = 44 \div 100 = 0.44$ and $12.5\% = \frac{12.5}{100} = 12.5 \div 100 = 0.125$

To change a decimal to a percentage simply multiply by 100
For example $0.34 = 34\%$ and $1.55 = 155\%$

To change a percentage to a fraction divide by 100 and simplify.
We know that 20% of the cars in a car park means $\frac{20}{100}$ of the cars there.

Now $\frac{20}{100}$ can be simplified to the equivalent fraction $\frac{1}{5}$, i.e. $20\% = \frac{1}{5}$.

Similarly 45% of the sweets in a bag means the same as $\frac{45}{100}$ of them and $\frac{45}{100} = \frac{9}{20}$

i.e. $45\% = \frac{9}{20}$

To change a fraction to a percentage change it to a decimal, then multiply by 100.
You can write a fraction as a percentage in two steps.

First write the fraction as a decimal; you do this by dividing the top by the bottom.

For example, $\frac{4}{5} = 4 \div 5 = 0.8$

then change the decimal to a percentage: $0.8 = 80\%$.

Exercise 2b

1 Work out each fraction as a decimal.

   a $\frac{3}{4}$ **b** $\frac{3}{5}$ **c** $\frac{3}{10}$ **d** $\frac{3}{20}$ **e** $\frac{7}{8}$ **f** $\frac{6}{25}$

2 Work out $\frac{9}{20}$ as a decimal.

   Now decide which is larger, $\frac{9}{20}$ or 0.47?

3 Write each decimal as a fraction in its lowest terms, using mixed numbers where necessary.

   a 0.06   **b** 0.004   **c** 15.5   **d** 2.01   **e** 3.25

In questions **4** and **5** write each decimal as a fraction in its lowest terms.

4 It is estimated that 0.86 of the families in Northgate Street own a car.

5 There were 360 seats on the aircraft and only 0.05% of them were vacant.

6 Write these decimals as percentages.

   a 0.3   **b** 0.2   **c** 0.7   **d** 0.035   **e** 0.925

7 Write these decimals as percentages.

   a 1.32   **b** 1.5   **c** 2.4   **d** 1.05   **e** 2.555

   (Remember that 1 is 100%, so $1.66 = 100\% + 66\% = 166\%$)

8 Write these percentages as decimals.

   a 45%   **b** 60%   **c** 95%   **d** 5.5%   **e** 12.5%

9 Express each percentage as a fraction in its lowest terms.

   a 40%   **b** 65%   **c** 54%   **d** 25%

10 Express each fraction as a percentage.

   a $\frac{2}{5}$ **b** $\frac{3}{20}$ **c** $\frac{21}{50}$ **d** $\frac{15}{40}$

In questions **11** to **14** express the given percentage as a fraction in its lowest terms.

11 Last summer 60% of the pupils in my class went on holiday.

12 At my youth club only 35% of the members are boys.

13 The Post Office claims that 95% of the letters posted arrive the following day.

14 A survey showed that 32% of the pupils in a year group needed to wear glasses.

In each question from **15** to **18** express the fraction as a percentage.

15 At a youth club $\frac{17}{20}$ of those present took part in at least one sporting activity.

16 About $\frac{17}{50}$ of first-year pupils watch more than 20 hours of television a week.

17 Approximately $\frac{3}{5}$ of sixteen-year-olds have a Saturday job.

18 Recently, at the local garage, $\frac{1}{8}$ of the cars tested failed to get a certificate.

19 Copy and complete the following table

Fraction	Percentage	Decimal
$\frac{3}{5}$	60%	0.6
$\frac{4}{5}$		
	75%	
		0.7
$\frac{11}{20}$		
	44%	

20 The registers showed that only 0.05 of the pupils in the first year had 100% attendance last term.

   a What fraction is this?

   b What percentage of the first-year pupils had a 100% attendance last term?

21 Marion spends $\frac{21}{50}$ of her income on food and lodgings.

   a What percentage is this?

   b As a decimal, what part of her total income does she spend on food and lodging?

22 Marmalade consists of 28% fruit, $\frac{3}{5}$ sugar and the remainder water.

   a What fraction of the marmalade is fruit?

   b What percentage of the marmalade is sugar?

   c What percentage is water?

23 An alloy is 60% copper, $\frac{7}{20}$ nickel and the remainder is tin.

    a What fraction is copper?

    b What percentage is **(i)** nickel **(ii)** either nickel or copper?

    c Express the part that is tin as a decimal.

    d What is the ratio of the amount of copper to the amount of tin?

Positive indices

We have seen that 3^2 means 3×3

      and that $2 \times 2 \times 2$ can be written as 2^3.

The small number at the top is called the *index* or *power*. (The plural of index is indices.)

It follows that 2 can be written as 2^1, although we would not normally do so.

$$5^1 \text{ means } 5$$

Exercise 2c

Find:

1 3^2

2 4^1

3 10^2

4 5^3

5 10^3

6 3^4

7 2^7

8 10^1

9 4^3

10 10^4

11 10^6

12 3^3

> **Tip** $3^2 = 3 \times 3$

> **Tip** $2^7 = 2 \times 2 \times 2 \times 2 \times 2 \times 2 \times 2$

Find the value of:

13 7.2×10^3

14 8.93×10^2

15 6.5×10^4

16 3.82×10^3

17 2.75×10^1

18 5.37×10^5

19 4.63×10^1

20 5.032×10^2

21 7.09×10^2

22 6.978×10^1

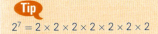

> **Tip** $10^3 = 10 \times 10 \times 10$

Multiplying numbers written in index form

We can write $2^2 \times 2^3$ as a single number in index form because

$$2^2 \times 2^3 = (2 \times 2) \times (2 \times 2 \times 2)$$
$$= 2 \times 2 \times 2 \times 2 \times 2$$
$$= 2^5$$
$$\therefore \qquad 2^2 \times 2^3 = 2^{2+3} = 2^5$$

But we cannot do the same with $2^2 \times 5^3$ because the numbers multiplied together are not all 2s (nor are they all 5s).

We can multiply together different powers of the *same* number by adding the indices but we cannot multiply together powers of different numbers in this way.

Exercise 2d

Write as a single expression in index form:

1 $3^5 \times 3^2$

2 $7^5 \times 7^3$

3 $9^2 \times 9^8$

4 $2^4 \times 2^7$

5 $b^3 \times b^2$

6 $5^4 \times 5^4$

7 $12^4 \times 12^5$

8 $p^6 \times p^8$

9 $4^7 \times 4^9$

10 $r^5 \times r^3$

Tip $3^5 \times 3^2 = 3^{5+2}$

Dividing numbers written in index form

If we want to write $2^5 \div 2^2$ as a single number in index form then

$$2^5 \div 2^2 = \frac{2^5}{2^2} = \frac{\cancel{2} \times \cancel{2} \times 2 \times 2 \times 2}{\cancel{2} \times \cancel{2}} = 2^3$$

i.e. $\qquad \dfrac{2^5}{2^2} = 2^{5-2} = 2^3$

We can divide different powers of the *same* number by subtracting the indices.

Exercise 2e

Write as a single expression in index form:

1 $4^4 \div 4^2$

2 $7^9 \div 7^3$

3 $5^6 \div 5^5$

4 $10^8 \div 10^3$

5 $q^9 \div q^5$

6 $15^8 \div 15^4$

7 $6^{12} \div 6^7$

8 $b^7 \div b^5$

9 $9^{15} \div 9^{14}$

10 $p^4 \div p^3$

11 $6^4 \times 6^7$

12 $3^9 \div 3^6$

13 $2^8 \div 2^7$

14 $a^9 \times a^3$

15 $c^6 \div c^3$

16 $2^2 \times 2^4 \times 2^3$

17 $4^2 \times 4^3 \div 4^4$

18 $a^2 \times a^2 \div a^3$

19 $3^6 \div 3^2 \times 3^4$

20 $b^2 \times b^3 \times b^4$

Investigation

a Everyone has two biological parents.

Going back one generation, each of your parents has two biological parents.

Copy and complete the tree – fill in the number of ancestors for five generations back.

Do not fill in names!

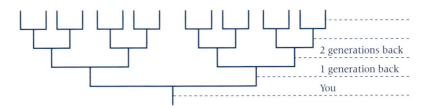

2 generations back

1 generation back

You

b Giving your answers as a power of 2, how many ancestors does this table suggest you have

i five generations back
ii six generations back
iii ten generations back?

c If we assume that each generation spans 25 years, how many generations are needed to go back 1000 years?

d Find the number of ancestors the table suggests that you would expect to have 6000 years back. Give your answer as a power of 2. What other assumptions are made to get this answer?

e About 6000 years ago, according to the bible, Adam and Eve were the only people on Earth. This contradicts the answer from part **d**. Suggest some reasons for this contradiction.

Negative indices

Consider $2^3 \div 2^5$

Subtracting the indices gives $2^3 \div 2^5 = 2^{3-5} = 2^{-2}$

But, as a fraction,
$$2^3 \div 2^5 = \frac{2^3}{2^5} = \frac{2 \times 2 \times 2}{2 \times 2 \times 2 \times 2 \times 2} = \frac{1}{2^2}$$

Therefore
$$2^{-2} \text{ means } \frac{1}{2^2}$$

In the same way, 5^{-3} means $\dfrac{1}{5^3} = \dfrac{1}{5 \times 5 \times 5} = \dfrac{1}{125}$

$\dfrac{1}{5^3}$ is called the *reciprocal* of 5^3, so also 5^{-3} is the reciprocal of 5^3

In general, a^{-b} is the *reciprocal* of $a^b \left(\text{i.e. } a^{-b} = \dfrac{1}{a^b} \right)$

Exercise 2f

Find the value of:

1	2^{-2}	**5**	7^{-1}	**9**	3^{-2}	**13**	15^{-1}	**17**	10^{-2}
2	3^{-3}	**6**	4^{-2}	**10**	4^{-1}	**14**	6^{-1}	**18**	2^{-3}
3	2^{-4}	**7**	3^{-4}	**11**	4^{-3}	**15**	7^{-2}	**19**	10^{-1}
4	3^{-1}	**8**	5^{-1}	**12**	6^{-2}	**16**	5^{-3}	**20**	8^{-2}

Find the value of 1.7×10^{-2}

$$1.7 \times 10^{-2} = 1.7 \times \frac{1}{10^2}$$

$$= \frac{1.7}{100} = 0.017$$

Find the value of:

21	3.4×10^{-3}	**25**	5.38×10^{-4}	**28**	2.805×10^{-2}
22	2.6×10^{-1}	**26**	4.67×10^{-5}	**29**	51.73×10^{-4}
23	6.2×10^{-2}	**27**	3.063×10^{-1}	**30**	30.04×10^{-1}
24	8.21×10^{-3}				

Write as a single number in index form:

31	$5^2 \div 5^4$	**36**	$10^3 \div 10^6$
32	$3 \div 3^4$	**37**	$b^5 \div b^9$
33	$6^4 \div 6^7$	**38**	$4^8 \div 4^3$
34	$2^5 \div 2^3$	**39**	$c^5 \div c^4$
35	$a^5 \div a^7$	**40**	$2^a \div 2^b$

> **Tip** Remember to subtract the indices. Some answers give a negative index.

The meaning of a^0

Consider $2^3 \div 2^3$

Subtracting indices gives $\qquad 2^3 \div 2^3 = 2^0$

Simplifying $\dfrac{2^3}{2^3}$ gives $\qquad \dfrac{\cancel{2} \times \cancel{2} \times \cancel{2}}{\cancel{2} \times \cancel{2} \times \cancel{2}} = 1$

So 2^0 means 1

In the same way $a^3 \div a^3 = a^0$ (subtracting indices)

But $a^3 \div a^3 = \dfrac{a \times a \times a}{a \times a \times a} = 1$ (simplifying the fraction)

> Any non-zero number with an index of zero is equal to 1
> i.e. $a^0 = 1$

Mixed questions on indices

Exercise 2g

Find the value of:

1 2^2	**3** 4^3	**5** 7^0	**7** 3^4	**9** 4^1	**11** 10^{-3}
2 5^{-2}	**4** 3^{-1}	**6** 5^3	**8** 2^0	**10** 6^{-2}	**12** 7^{-2}

13 2.41×10^3	**18** 1.074×10^{-1}
14 7.032×10^{-1}	**19** 7.834×10^2
15 4.971×10^2	**20** 3.05×10^3
16 7.805×10^{-3}	**21** 5.99×10^0
17 5.92×10^4	**22** 3.8601×10^{-4}

Write as a single number in index form:

23 $2^3 \times 2^4$	**28** $5^4 \times 5^{-2}$	**33** $2^2 \times 2^4 \times 2^3$	**38** $a^3 \times a^2 \times a^5$
24 $4^6 \div 4^3$	**29** $3^5 \div 3^5$	**34** $a^2 \times a^4 \times a^6$	**39** $3^2 \div 3^6 \times 3^2$
25 $3^{-2} \times 3^4$	**30** $b^3 \div b^3$	**35** $3^5 \times 3^2 \div 3^3$	**40** $b^3 \times b^{-3}$
26 $a^4 \times a^3$	**31** $4^{-2} \times 4^6$	**36** $7^3 \times 7^3 \div 7^6$	**41** $5^{-2} \times 5^{-3}$
27 $a^7 \div a^3$	**32** $5^3 \div 5^9$	**37** $\dfrac{4^2 \times 4^6}{4^3}$	**42** $\dfrac{a^3 \times a^4}{a^7}$

Standard form

The nearest star to us (Alpha Centauri) is about 25 million million miles away. Written in figures this very large number is 25 000 000 000 000.

The diameter of an atom is roughly 2 ten-thousand-millionths of a metre, or 0.000 000 000 2 metres and this is very small.

These numbers are cumbersome to write down and, until we have counted the zeros, we cannot tell their size. We need a way of writing such numbers in a shorter form from which it is easier to judge their size: the form that we use is called *standard form* (sometimes called *scientific notation*).

Written in standard form the first number is 2.5×10^{13}

and the second number is 2×10^{-10}

Standard form is a number between 1 and 10
multiplied by a power of 10.

So 1.3×10^2, 2.86×10^4 and 3.72×10^{-2} are in standard form,

but 13×10^3 and 0.36×10^{-2} are not in standard form because the first number is not between 1 and 10.

Exercise 2h

Each of the following numbers is written in standard form. Write them as ordinary numbers.

1 3.78×10^3

2 1.26×10^{-3}

3 5.3×10^6

4 7.4×10^{14}

5 1.3×10^{-4}

6 3.67×10^{-6}

7 3.04×10^4

8 8.503×10^{-4}

9 4.25×10^{12}

10 6.43×10^{-8}

Changing numbers into standard form

To change 6800 into standard form, the decimal point has to be placed between the 6 and the 8 to give a number between 1 and 10.

Counting then tells us that, to change 6.8 to 6800, we need to move the decimal point three places to the right (i.e. to multiply by 10^3)

i.e. $6800 = 6.8 \times 1000 = 6.8 \times 10^3$

To change 0.019 34 into standard form, the point has to go between the 1 and the 9 to give a number between 1 and 10.

This time counting tells us that, to change 1.934 to 0.019 34, we need to move the point two places to the left (i.e. to divide by 10^2)

so $0.019\,34 = 1.934 \div 100 = 1.934 \times \frac{1}{100} = 1.934 \times 10^{-2}$

Exercise *2i*

Change the following numbers into standard form:

1	2500	**6**	39 070	**11**	26 030
2	630	**7**	4 500 000	**12**	547 000
3	15 300	**8**	530 000 000	**13**	30 600
4	260 000	**9**	40 000	**14**	4 060 000
5	9900	**10**	80 000 000 000	**15**	704

Write 0.006 043 in standard form.

$$0.006\,043 = 6.043 \times 10^{-3}$$

Write the following numbers in standard form:

16	0.026	**31**	79.3	**46**	88.92
17	0.0048	**32**	0.005 27	**47**	0.000 050 6
18	0.053	**33**	80 600	**48**	0.000 000 057
19	0.000 018	**34**	0.9906	**49**	503 000 000
20	0.52	**35**	0.0705	**50**	99 000 000
21	0.79	**36**	60.5	**51**	84
22	0.0069	**37**	0.003 005	**52**	351
23	0.000 007 5	**38**	0.600 05	**53**	0.09
24	0.000 000 000 4	**39**	7 080 000	**54**	0.007 05
25	0.684	**40**	560 800	**55**	36
26	0.907	**41**	5 300 000 000 000	**56**	5090
27	0.0805	**42**	0.000 000 050 2	**57**	268 000
28	0.088 08	**43**	0.007 008 09	**58**	30.7
29	0.000 704 4	**44**	708 000	**59**	0.005 05
30	0.000 000 000 073	**45**	40.5	**60**	0.000 008 8

Investigation

If you read about computers, you will notice specifications such as '700 Mb capacity writable CDs' or '40 Gb hard disk'. Mb stands for megabytes and Gb stand for gigabytes.

Mega and Giga are prefixes used to describe very large numbers.

There are other prefixes used to describe very small numbers.

Find out what Mega and Giga mean.

Find out what other prefixes are used to describe very large and very small numbers and what they mean.

Approximations: whole numbers

We saw in Book 1 that it is sometimes necessary to approximate given numbers by rounding them off to the nearest 10, 100, ... For example, if you measured your height in millimetres as 1678 mm, it would be reasonable to say that you were 1680 mm tall to the nearest 10 mm.

The rule is that if you are rounding off to the nearest 10 you look at the units. If there are 5 or more units you add one on to the tens. If there are less than 5 units you leave the tens alone.

Similar rules apply to rounding off to the nearest 100 (look at the tens); to the nearest 1000 (look at the hundreds); and so on.

Exercise **2j**   Round off 1853 to

a the nearest ten

b the nearest hundred

c the nearest thousand

a 185|3 = 1850 to the nearest 10 (Put a cut-off line (|) after the 10s)

b 18|53 = 1900 to the nearest 100 (Put a cut-off line (|) after the 100s)

c 1|853 = 2000 to the nearest 1000 (Put a cut-off line (|) after the 1000s)

Round off each of the following numbers to

a   the nearest ten   b   the nearest hundred   c   the nearest thousand:

| 1 | 1547 | 3 | 2750 | 5 | 68 414 | 7 | 4066 | 9 | 53 804 | 11 | 4981 |
| 2 | 8739 | 4 | 36 835 | 6 | 5729 | 8 | 7507 | 10 | 6007 | 12 | 8699 |

A building firm stated that, to the nearest 100, it built 2600 homes last year. What is the greatest number of homes that it could have built and what is the least number of homes that it could have built?

Look at this number line.

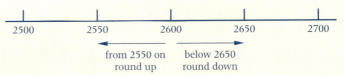

The smallest whole number that can be rounded up to 2600 is 2550.

The biggest whole number that can be rounded down to 2600 is 2649.

So the firm built at most 2649 homes and at least 2550 homes.

13 A bag of marbles is said to contain 50 marbles to the nearest 10. What is the greatest number of marbles that could be in the bag and what is the least number of marbles that could be in the bag?

14 To the nearest thousand, the attendance at a particular First Division football match was 45 000. What is the largest number that could have been there and what is the smallest number that could have attended?

15 1500 people came to the school fête. If this number is correct to the nearest hundred, give the maximum and the minimum number of people that could have come.

16 The annual accounts of Scrub plc (soap manufacturers) gave the company's profit as $3 000 000 to the nearest million. What is the least amount of profit that the company could have made?

17 The chairman of A. Brick (Builders) plc said that they employ 2000 people. If this number is correct to the nearest 100, what is the least number of employees that the company can have?

Approximations: decimals

If you measure your height in centimetres as 167.8 cm, it would be reasonable to say that, to the nearest centimetre, you are 168 cm tall.

We write $167.8 = 168$ correct to the nearest unit.

If you measure your height in metres as 1.678 m, it would be reasonable to say that, to the nearest $\frac{1}{100}$ m, you are 1.68 m tall.

Hundredths are represented in the second decimal place so we say that $1.678 = 1.68$ correct to 2 decimal places.

Exercise 2k

Give each of the following numbers correct to

a 2 decimal places   **b** 1 decimal place   **c** the nearest unit:

1 2.758	**6** 3.896		
2 7.371	**7** 8.936		
3 16.987	**8** 73.649		
4 23.758	**9** 6.896		
5 9.858	**10** 55.575		

> **Tip** You may find it helpful to draw a cut-off line.

Give the following numbers correct to the number of decimal places given in brackets:

11 5.07	(1)	**16** 0.9752	(3)
12 0.0087	(3)	**17** 5.5508	(3)
13 7.897	(2)	**18** 285.59	(1)
14 34.82	(1)	**19** 6.749	(1)
15 0.007 831	(4)	**20** 9.999	(2)

 ## Puzzle

What is the largest number you can make with two nines?

Significant figures

In the previous two sections we used a height of 1678 mm as an example. This height was measured in three different units and then rounded off:

in the first case to   1680 mm   correct to the nearest 10 mm,

in the second case to   168 cm   correct to the nearest centimetre,

in the third case to   1.68 m   correct to 2 d.p.

We could also give this measurement in kilometres, to the same degree of accuracy, as 0.001 68 km correct to 5 d.p.
Notice that the three figures 1, 6 and 8 occur in all four numbers and that it is the 8 that is the corrected figure.

The figures 1, 6 and 8 are called the *significant figures* and in all four cases the numbers are given correct to 3 significant figures.

Using significant figures rather than place values (i.e. tens, units, first d.p., second d.p., ...) has advantages. For example, if you are asked to measure your height and give the answer correct to 3 significant figures, then you can choose any convenient unit. You do not need to be told which unit to use and which place value in that unit to correct your answer to.

Writing a number in standard form is an easy way of finding the first significant figure: it is the number to the left of the decimal point.

For example $170.6 = \underline{1}.706 \times 10^2$

So 1 is the first significant figure in 170.6.
The second significant figure is the next figure to the right (7 in this case).
The third significant figure is the next figure to the right again (0 in this case), and so on.

Exercise 2l

Write down

a   the first significant figure

b   the third significant figure, in 0.001 503

First write the number in standard form

$$0.001\,503 = 1.503 \times 10^{-3}$$

a   The first s.f. is 1

b   The third s.f. is 0

In each of the following numbers write down the significant figure specified in the bracket:

1	36.2	(1st)	**6**	5.083	(3rd)
2	378.5	(3rd)	**7**	34.807	(4th)
3	0.0867	(2nd)	**8**	0.076 03	(3rd)
4	3.786	(3rd)	**9**	54.06	(3rd)
5	47 632	(2nd)	**10**	5.7087	(4th)

Exercise 2m

Give 32 685 correct to 1 s.f.

(First write 32 685 in standard form.)

$$32\,685 = 3.\!2685 \times 10^4$$

(As before, to correct to 1 s.f. we look at the second s.f.: if it is 5 or more we add one to the first s.f.; if it is less than 5 we leave the first s.f. alone.)

So $3\!2\,685 = 30\,000$ to 1 s.f.

Give the following numbers correct to 1 s.f.:

1 59 727 | **4** 586 359 | **7** 51 488 | **10** 908

2 4164 | **5** 80 755 | **8** 4099 | **11** 26

3 4 396 185 | **6** 476 | **9** 667 505 | **12** 980

Give the following numbers correct to 2 s.f.:

13 4673 | **16** 892 759 | **19** 72 601 | **22** 53 908

14 57 341 | **17** 6992 | **20** 444 | **23** 476

15 59 700 | **18** 9973 | **21** 50 047 | **24** 597

Give 0.021 94 correct to 3 s.f.

$$0.021\,94 = 2.19|4 \times 10^{-2}$$

(The fourth s.f. is 4 so we leave the third s.f. alone.)

So $\qquad 0.021\,9|4 = 0.0219$ to 3 s.f.

Give the following numbers correct to 3 s.f.:

25 0.008 463 | **27** 5.8374 | **29** 46.8451 | **31** 7.5078 | **33** 0.989 624

26 0.825 716 | **28** 78.49 | **30** 0.007 854 7 | **32** 369.649 | **34** 53.978

Give each of the following numbers correct to the number of significant figures indicated in the brackets. Try to do this without converting to standard form.

35 46.931 06 (2) | **40** 4537 (1)

36 0.006 845 03 (4) | **41** 37.856 72 (3)

37 576 335 (1) | **42** 6973 (2)

38 497 (2) | **43** 0.070 865 (3)

39 7.824 38 (3) | **44** 0.067 34 (1)

Find $50 \div 8$ correct to 2 s.f.

(To give an answer correct to 2 s.f. we first work to 3 s.f.)

$$8\overline{\smash{)}50.00} \quad 6.2|5$$

So $\qquad 50 \div 8 = 6.3$ to 2 s.f.

Give, correct to 2 s.f.

45 $20 \div 6$ | **47** $25 \div 2$ | **49** $125 \div 9$ | **51** $73 \div 3$ | **53** $0.23 \div 9$

46 $10 \div 6$ | **48** $53 \div 4$ | **50** $143 \div 5$ | **52** $0.7 \div 3$ | **54** $0.0013 \div 3$

 **Puzzle**

Everton Giles stands on the middle rung of a ladder. He climbs 3 rungs higher but has forgotten something so descends 7 rungs to get it. He now goes up 16 rungs and reaches the top of the ladder. How many rungs are there to the ladder?

Rough estimates

If you were asked to find 1.397×62.54 you could do it by long multiplication or you could use a calculator. Whichever method you choose, it is essential first to make a rough estimate of the answer. You will then know whether the actual answer you get is reasonable or not.

One way of estimating the answer to a calculation is to write each number correct to 1 significant figure.

So $$1.397 \times 62.57 \approx 1 \times 60 = 60$$

Exercise 2n   Correct each number to 1 s.f. and hence give a rough answer to

   a 9.524×0.0837 **b** $54.72 \div 0.761$

   a $9.524 \times 0.0837 \approx 10 \times 0.08 = 0.8$

   b $\dfrac{54.72}{0.761} \approx \dfrac{50}{0.8} = \dfrac{500}{8}$

   $= 60$ (giving $500 \div 8$ to 1 s.f.)

Correct each number to 1 s.f. and hence give a rough answer to each of the following calculations:

1 4.78×23.7	**6** $82.8 \div 146$	**11** 34.7×21	**16** $0.0326 \div 12.4$
2 56.3×0.573	**7** 0.632×0.845	**12** 8.63×0.523	**17** $0.007\,24 \times 0.783$
3 $0.0674 \div 5.24$	**8** 0.0062×574	**13** $34.9 \div 15.8$	**18** $3581 \div 45$
4 354.6×0.0475	**9** $7.835 \div 6.493$	**14** $0.47 \div 0.714$	**19** 1097×94
5 576×256	**10** 4736×729	**15** $985 \div 57.2$	**20** 45.07×0.0327

Correct each number to 1 s.f. and hence estimate $\dfrac{0.048 \times 3.275}{0.367}$ to 1 s.f.

$$\dfrac{0.048 \times 3.275}{0.367} \approx \dfrac{0.05 \times 3}{0.4} = \dfrac{0.15}{0.4} = \dfrac{1.5}{4}$$

$$= 0.4 \text{ (to 1 s.f.)}$$

21 $\dfrac{3.87 \times 5.24}{2.13}$ **25** $\dfrac{43.8 \times 3.62}{4.72}$ **28** $\dfrac{8.735}{5.72 \times 5.94}$

22 $\dfrac{0.636 \times 2.63}{5.47}$ **26** $\dfrac{89.03 \times 0.079\,37}{5.92}$ **29** $\dfrac{0.527}{6.41 \times 0.738}$

23 $\dfrac{21.78 \times 4.278}{7.96}$ **27** $\dfrac{975 \times 0.636}{40.78}$ **30** $\dfrac{57.8}{0.057 \times 6.93}$

24 $\dfrac{6.38 \times 0.185}{0.628}$

Calculations: multiplication and division

Display

When you key in a number on your calculator it appears on the display. Check that the number on display is the number that you intended to enter.

Also check that you press the correct operator, i.e. press $\times$ to multiply and $\div$ to divide.

Exercise 2p

First make a rough estimate of the answer. Then use your calculator to give the answer correct to 3 s.f.

1 2.16×3.28 **11** $9.571 \div 2.518$ **21** 63.8×2.701

2 2.63×2.87 **12** $5.393 \div 3.593$ **22** $40.3 \div 2.74$

3 1.48×4.74 **13** $7.384 \div 2.51$ **23** $400 \div 35.7$

4 4.035×2.116 **14** $4.931 \div 3.204$ **24** $(34.2)^2$

5 3.142×2.925 **15** $8.362 \div 5.823$ **25** 5007×2.51

6 6.053×1.274 **16** 23.4×56.7 **26** $5703 \div 154.8$

7 2.304×3.251 **17** 384×21.8 **27** 39.03×49.94

8 8.426×1.086 **18** 45.8×143.7 **28** $2000 \div 52.66$

9 $5.839 \div 3.618$ **19** $537.8 \div 34.6$ **29** $(36.8)^2$

10 $6.834 \div 4.382$ **20** $45.35 \div 6.82$ **30** $29\,006 \div 2.015$

31 0.366×7.37

32 0.0526×0.372

33 6.924×0.00793

34 0.638×825

35 52×0.0895

36 0.0826×0.582

37 24.78×0.0724

38 0.00835×0.617

39 0.5824×6.813

40 $(0.74)^2$

41 $0.583 \div 4.82$

42 $0.628 \div 7.61$

43 $0.493 \div 1.253$

44 $0.518 \div 5.047$

45 $82.7 \div 593$

46 $89.5 \div 0.724$

47 $38.07 \div 0.682$

48 $5.71 \div 0.0623$

49 $7.045 \div 0.0378$

50 $6.888 \div 0.0072$

51 $45.37 \div 0.925$

52 $8.41 \div 0.000748$

53 $6.934 \div 0.0829$

54 $0.824 \div 0.362$

55 $0.572 \div 0.851$

56 $0.528 \div 0.0537$

57 $0.571 \div 0.824$

58 $0.0455 \div 0.0613$

59 $0.006 \div 0.04703$

60 $0.824 \div 0.00008$

61 $5000 \div 0.789$

62 $(0.078)^2$

63 0.0608×573

64 $(78.5)^3$

65 $\dfrac{3.782 \times 0.467}{4.89}$

66 4.88×0.00417

67 $0.9467 \div 7683$

68 $0.0467 \div 0.000074$

69 $(0.00031)^2$

70 $\dfrac{54.9 \times 36.6}{0.406}$

71 $68.41 \div 392.9$

72 $0.0482 \div 0.00289$

73 $(0.0527)^3$

74 $\dfrac{0.857 \times 8.109}{0.5188}$

Mixed exercises

Exercise 2q

1 Find the value of 4^{-2}.

2 Simplify $b^2 \div b^5$.

3 Find the value of $\dfrac{3^2 \times 3^3}{3^5}$.

4 Write $36\,400$ in standard form.

5 Write $0.005\,07$ in standard form.

6 Give $57\,934$ correct to 1 s.f.

7 Give $0.061\,374$ correct to 3 s.f.

8 Find 0.582×6.382, giving your answer correct to 3 s.f.

9 Find $45.823 \div 15.89$, giving your answer correct to 3 s.f.

Exercise 2r

1 Find the value of 6^3.

2 Write $\dfrac{2^4 \times 2^2}{2^8}$ as a single number in index form.

3 Find the value of $5^6 \div 5^7$.

4 Simplify $a^2 \times a^4 \times a$.

5 Write 650 000 000 in standard form.

6 Give 45 823 correct to 2 s.f.

7 The organisers of a calypso show hope that, to the nearest thousand, 8000 people will buy tickets. What is the minimum number of tickets that they hope to sell?

8 Find the value of $12.07 \div 0.008\,97$ giving your answer correct to 3 s.f.

9 Find the value of $(0.836)^2$ giving your answer correct to 3 s.f.

10 Change 35% into

    a a fraction in its lowest terms   **b** a decimal.

Exercise 2s

1 Find the value of $5^{-2} \times 5^3$.

2 Simplify $\dfrac{a^4}{a^3 \times a^2}$.

3 Find the value of $3^2 \times 3^4 \div 3^6$.

4 Write 0.005 708 in standard form.

5 Give 9764 correct to 1 s.f.

6 Give 0.050 806 correct to 3 s.f.

7 Correct to 1 significant figure, there are 70 matches in a box. What is the difference between the maximum and the minimum number of matches that could be in the box?

8 Find $0.0468 \div 0.004\,73$ giving your answer correct to 3 s.f.

9 Find $\dfrac{56.82 \times 0.714}{8.625}$ giving your answer correct to 3 s.f.

10 Change $\frac{5}{8}$ into

    a a percentage   **b** a decimal.

Bet you didn't know that

1. 1729 is the smallest positive integer that can be represented in two ways as the sum of two cubes

$$9^3 + 10^3 \quad \text{or} \quad 1^3 + 12^3$$

2. The total number of gifts given in the song 'The Twelve Days of Christmas' is 364. One for each day of the year except Christmas Day. Check it.

3. $2592 = 2^5 \times 9^2$

IN THIS CHAPTER...

you have seen that:

- the associative and commutative laws are true for the addition and multiplication of whole numbers but not for subtraction and division,

  e.g. $7 + 8 = 8 + 7$ and $7 \times 8 = 8 \times 7$

  but $7 - 8$ is not equal to $8 - 7$ and $7 \div 8$ is not equal to $8 \div 7$

- the identity element is 0 for addition and 1 for multiplication

- the inverse element for any whole number under addition is minus it, and for multiplication it is the reciprocal,

  e.g. the inverse of 4 under multiplication is $1 \div 4$ (or $\times \frac{1}{4}$)

- you can interchange fractions, percentages and decimals using these rules

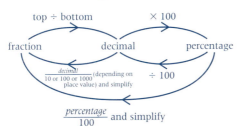

- you can multiply different powers of the same number by adding the indices,

  e.g. $3^4 \times 3^3 = 3^{4+3} = 3^7$

- you can divide different powers of the same number by subtracting the indices,

  e.g. $5^7 \div 5^2 = 5^{7-2} = 5^5$

- a negative index means the reciprocal of the quantity, e.g. $4^{-3} = \dfrac{1}{4^3}$

- any number, except 0, raised to the power zero is 1, e.g $4^0 = 1$

- a number in standard form is a number between 1 and 10 multiplied by a power of 10, e.g. 1.2×10^5 and 1.6×10^{-4} are in standard form

- the first significant figure is the first non-zero figure in a number. The next figure (zero or otherwise) is the second significant figure, and so on

- to correct a number to a given degree of accuracy, place a cut-off line after the place value required and look at the next figure – if it is 5 or more, round up, otherwise round down

- you can make a rough estimate of a calculation by correcting each number to one significant figure

- you need to be careful when you use a calculator to work out accurate answers.

3 PARALLEL LINES AND ANGLES

What did the 60° say to the 30° angle?

We compliment each other beautifully.

Parallel lines

Two straight lines that are always the same distance apart, however far they are drawn, are called parallel lines.

The lines in your exercise books are parallel. You can probably find many other examples of parallel lines.

Exercise 3a

1   Using the lines in your exercise book, draw three lines that are parallel. Do not make them all the same distance apart. For example

(We use arrows to mark lines that are parallel.)

2   Using the lines in your exercise book, draw two parallel lines. Make them fairly far apart. Now draw a slanting line across them. For example

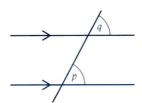

Mark the angles in your drawing that are in the same position as those in the diagram. Are they acute or obtuse angles? Measure your angles marked p and q.

3   Draw a grid of parallel lines like the diagram below. Use the lines in your book for one set of parallels and use the two sides of your ruler to draw the slanting parallels.

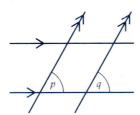

Mark your drawing like the diagram. Are your angles p and q acute or obtuse? Measure your angles p and q.

4   Repeat question 3 but change the direction of your slanting lines.

5 Draw three slanting parallel lines like the diagram below, with a horizontal line cutting them. Use the two sides of your ruler and move it along to draw the third parallel line.

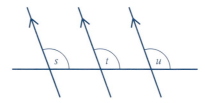

Mark your drawing like the diagram. Decide whether angles *s*, *t* and *u* are acute or obtuse and then measure them.

6 Repeat question 5 but change the slope of your slanting lines.

Corresponding angles

In the exercise above, lines were drawn that crossed a set of parallel lines.

A line that crosses a set of parallel lines is called a *transversal*.

When you have drawn several parallel lines you should notice that

two parallel lines on the same flat surface will never meet however far they are drawn.

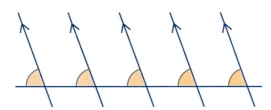

If you draw the diagram above by moving your ruler along you can see that all the shaded angles are equal. These angles are all in corresponding positions: they are all above the transversal and to the left of the parallel lines. Angles like these are called *corresponding angles*.

When two or more parallel lines are cut by a transversal, the corresponding angles are equal.

Exercise **3b**

In the diagrams below write down the letter that corresponds to the shaded angle:

1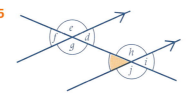

5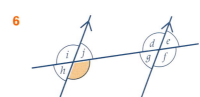

2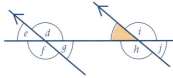

6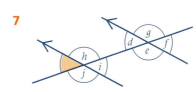

3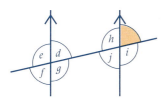

7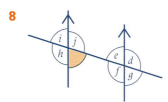

4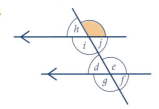

8

Drawing parallel lines (using a protractor)

The fact that the corresponding angles are equal gives us a method for drawing parallel lines.

If you need to draw a line through the point C that is parallel to the line AB, first draw a line through C to cut AB.

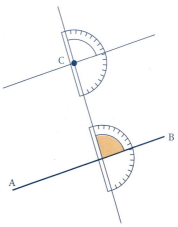

Use your protractor to measure the shaded angle. Place your protractor at C as shown in the diagram. Make an angle at C the same size as the shaded angle and in the corresponding position.

You can now extend the arm of your angle both ways, to give the parallel line.

Exercise 3c

1   Using your protractor draw a grid of parallel lines like the one in the diagram. (It does not have to be an exact copy.)

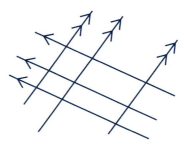

2   Trace the diagram below.

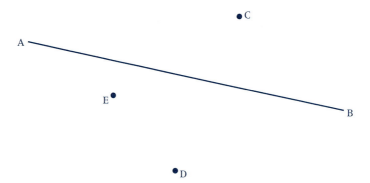

Now draw lines through the points C, D and E so that each line is parallel to AB.

3   Draw a sloping line on your exercise book. Mark a point C above the line. Use your protractor to draw a line through C parallel to your first line.

4   Trace the diagram below.

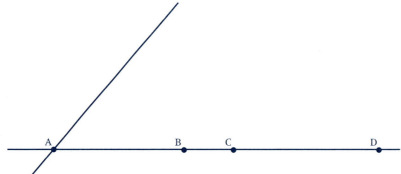

Measure the acute angle at A. Draw the corresponding angles at B, C and D. Extend the arms of your angles so that you have a set of four parallel lines.

In questions **5** to **8** remember to draw a rough sketch before doing the accurate drawing.

5 Draw an equilateral triangle with the equal sides each 8 cm long. Label the corners A, B and C. Draw a line through C that is parallel to the side AB.

6 Draw an isosceles triangle ABC with base AB which is 10 cm long and base angles at A and B which are each 30°. Draw a line through C which is parallel to AB.

7 Draw the triangle as given in question **5** again and this time draw a line through A which is parallel to the side BC.

8 Make an accurate drawing of the figure below where the side AB is 7 cm, the side AD is 4 cm and $\widehat{A} = 60°$.

(A figure like this is called a *parallelogram*.)

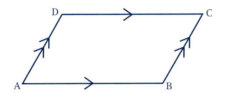

Puzzle

Copy this grid.

How many different-shaped parallelograms can you draw on this grid?

Each vertex must be on a dot. One has been drawn for you.

Do not include squares and rectangles.

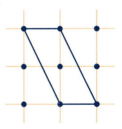

Problems involving corresponding angles

The simplest diagram for a pair of corresponding angles is an F shape.

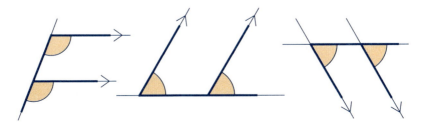

Looking for an F shape may help you to recognise the corresponding angles.

Exercise **3d**

Write down the size of the angle marked *d* in each of the following diagrams:

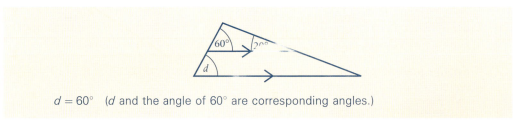

$d = 60°$ (*d* and the angle of 60° are corresponding angles.)

1

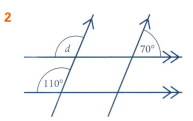

> **Tip** Look for an F shape round the angle you need to find.

2

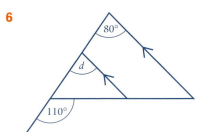

6

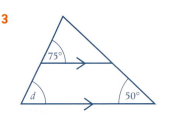

3

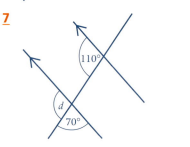

7

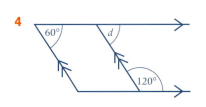

4

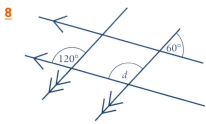

8

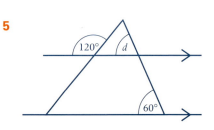

5

9

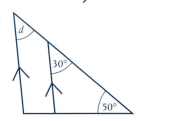

10

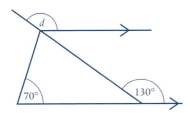

11

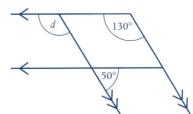

Reminder:

Vertically opposite angles are equal.　　Angles on a straight line add up to 180°.

Angles at a point add up to 360°.　　The angles of a triangle add up to 180°.

 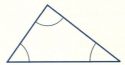

You will need these facts in the next exercise. If you cannot see immediately the angle you want, copy the diagram. On your diagram, write down the size of any angles you can, including those that are not marked. This should help you to find the size of other angles in the diagram, including those that you need. Remember you can use any facts you know about angles.

Exercise 3e

Find the size of each marked angle:

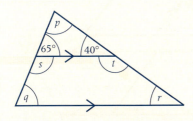

$p = 75°$ (angles of △ add up to 180°)

$q = 65°$ (corresponding angles)

$s = 115°$ (s and 65° add up to 180°)

$r = 40°$ (corresponding angles)

$t = 140°$ (t and 40° add up to 180°)

1

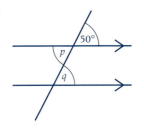

2

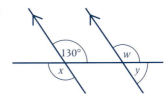

3

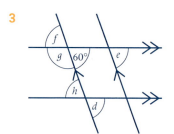

10

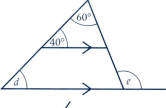

4

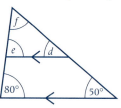

11

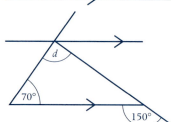

5

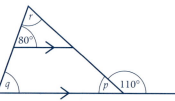

12

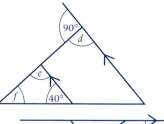

6

13

7

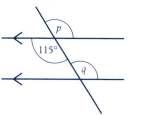

14

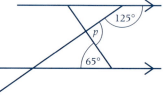

8

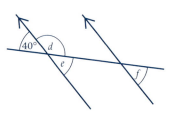

15

9

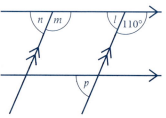

16

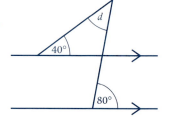

Find the size of angle *d* in questions **17** to **24**:

17

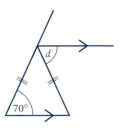

21

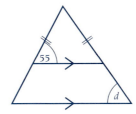

18

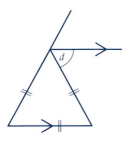

22

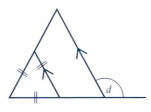

19

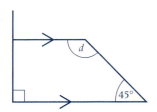

23

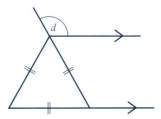

20

24

Alternate angles

Draw a large letter Z. Use the lines of your exercise book to make sure that the outer arms of the Z are parallel.

This letter has rotational symmetry about the point marked with a cross. This means that the two shaded angles are equal. Measure them to make sure.

Draw a large N, making sure that the outer arms are parallel.

This letter also has rotational symmetry about the point marked with a cross, so once again the shaded angles are equal. Measure them to make sure.

The pairs of shaded angles like those in the Z and N are between the parallel lines and on alternate sides of the transversal.

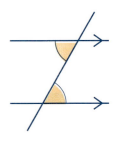

Angles like these are called *alternate angles*.

> When two parallel lines are cut by a transversal, the alternate angles are equal.

The simplest diagram for a pair of alternate angles is a Z shape.

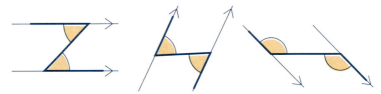

Looking for a Z shape may help you to recognise the alternate angles.

Exercise 3f

Tip Look for a Z shape around the angle you want to find.

Write down the angle that is alternate to the shaded angle in the following diagrams:

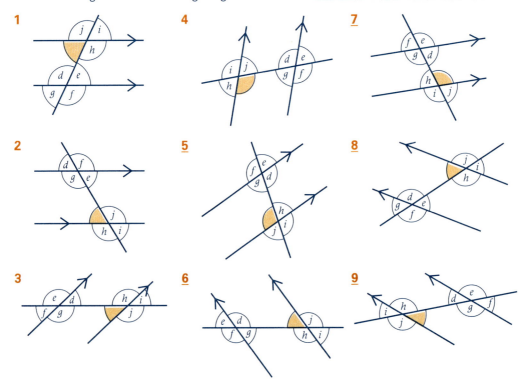

Problems involving alternate angles

Without doing any measuring we can show that alternate angles are equal by using the facts that we already know:

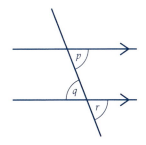

$p = r$ because they are corresponding angles

$q = r$ because they are vertically opposite angles

$\therefore$ $p = q$ and these are alternate angles

Exercise 3g

Find the size of each marked angle:

1

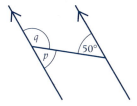

> **Tip** Remember that you can use any angle you know.

2

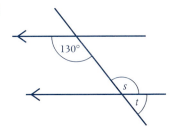

4

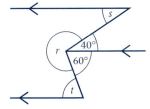

3

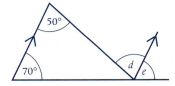

5

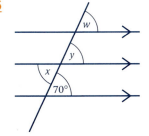

6

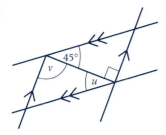

9

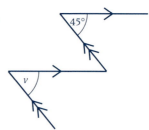

7

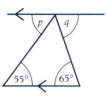

10

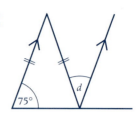

8

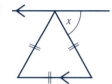

11

Investigation

This diagram represents a child's billiards table.

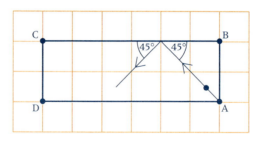

There is a pocket at each corner.

The ball is projected from the corner A at 45° to the sides of the table. It carries on bouncing off the sides at 45° until it goes down a pocket. (This is a very superior toy – the ball does not lose speed however many times it bounces!)

a How many bounces are there before the ball goes down a pocket?

b Which pocket does it go down?

c What happens if the table is 2 squares by 8 squares?

d Can you predict what happens for a 2 by 20 table?

e Now try a 2 by 3 table.

f Investigate for other sizes of tables. Start by keeping the width at 2 squares, then try other widths. Copy this table and fill in the results.

Size of table	Number of bounces	Pocket
2 × 6		
2 × 8		
2 × 3		
2 × 5		

g Can you predict what happens with a 3 × 12 table?

Interior angles

In the diagram on the right, *f* and *g* are on the same side of the transversal and 'inside' the parallel lines.

Pairs of angles like *f* and *g* are called *interior angles*.

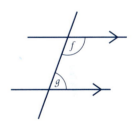

Exercise **3h**

In the following diagrams, two of the marked angles are a pair of interior angles.

Name them:

Tip You may find it helpful to look for a U shape.

1

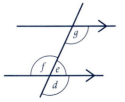

2

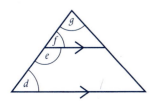

3

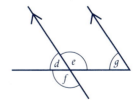

4

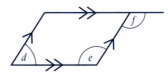

5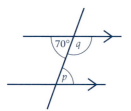

6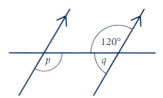

In the following diagrams, use the information given to find the size of p and of q. Then find the sum of p and q:

7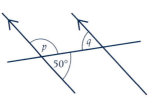

10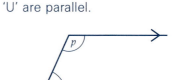

8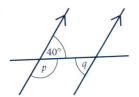

11 Make a large copy of the diagram below. Use the lines of your book to make sure that the outer arms of the 'U' are parallel.

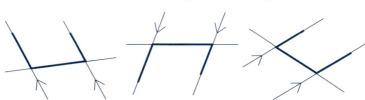

Measure each of the interior angles p and q. Add them together.

9

> The sum of a pair of interior angles is 180°.

You will probably have realised this fact by now.
We can show that it is true from the following diagram.

$$d + f = 180°$$ because they are angles on a straight line

$$d = e$$ because they are alternate angles

So $$e + f = 180°$$

The simplest diagram for a pair of interior angles is a U shape.

Looking for a U shape may help you to recognise a pair of interior angles.

Exercise 3i

Find the size of each marked angle:

1

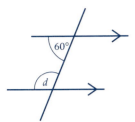

6

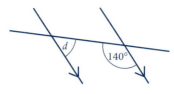

2

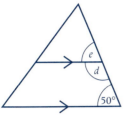

7

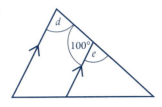

3

8

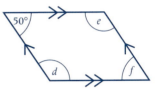

4

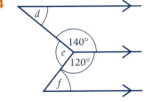

9

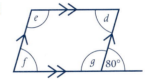

5

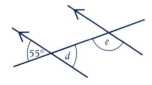

10

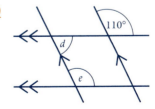

Mixed exercises

You now know that when a transversal cuts a pair of parallel lines

> the corresponding (F) angles are equal
>
> the alternate (Z) angles are equal
>
> the interior (U) angles add up to 180°

You can use any of these facts, together with the other angle facts you know, to answer the questions in the following exercises.

Exercise 3j

Find the size of each marked angle:

1

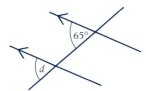

2

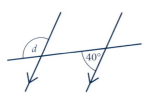

3

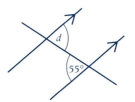

4

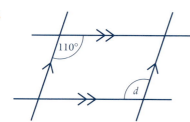

5

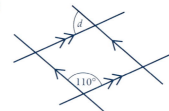

6

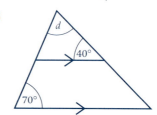

7

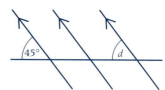

8 Construct a triangle ABC in which AB = 12 cm, BC = 8 cm and AC = 10 cm. Find the midpoint of AB and mark it D. Find the midpoint of AC and mark it E. Join ED. Measure AD̂E and AB̂C. What can you say about the lines DE and BC?

Exercise 3k

Find the size of each marked angle:

1

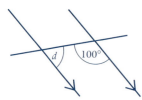

2

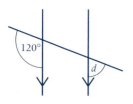

3

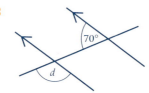

4

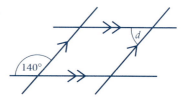

5

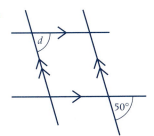

6

7

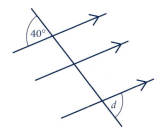

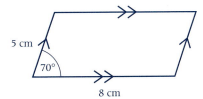

8 Construct the parallelogram below, making it full size.

5 cm

70°

8 cm

MATHS IS OUT THERE

What did one of two parallel lines say to the transversal?

You double-crosser!

IN THIS CHAPTER...

you have seen that:

- you can draw parallel lines

- parallel lines cut by a transversal give different types of angles – some are called corresponding angles, some alternate angles and others interior angles

- corresponding angles are equal; they can be recognised by an F shape

- alternate angles are equal; they can be recognised by a Z shape

- interior angles add up to 180°; they can be recognised by a U shape

- geometry problems can often be solved by starting with a copy of the diagram and filling in the sizes of the angles you know

MATHS IS OUT THERE

Isaac Newton (1642–1727) wrote:

The description of right lines and circles, upon which geometry is founded, belongs to mechanics. Geometry does not teach us to draw these lines, but requires them to be drawn.

KEY WORDS

alternate angles, angles at a point, angles on a straight line, angle sum of a quadrilateral, angle sum of a triangle, arc, bisect, chord, circle, circumcircle, compasses, construct, corresponding angles, cross-section, diagonal, equilateral, incircle, interior angles, isosceles, midpoint, net, parallel, perpendicular, perpendicular bisector, plane face, polyhedron, prism, protractor, pyramid, radius, regular, rhombus, right angle, square, supplementary, symmetry, tetrahedron (plural tetrahedra), transversal, vertically opposite angles.

Angles and triangles

Reminder:

Vertically opposite angles are equal.

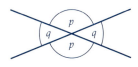

Angles at a point add up to 360°.

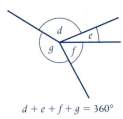

$$d + e + f + g = 360°$$

Angles on a straight line add up to 180°.

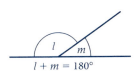

$$l + m = 180°$$

The sum of the three angles in any triangle is 180°.

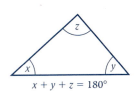

$$x + y + z = 180°$$

The sum of the four angles in any
quadrilateral is 360°.

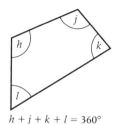

$$h + j + k + l = 360°$$

An equilateral triangle has all three sides the same
length and each of the three angles is 60°.

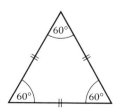

An isosceles triangle has two equal sides and the two
angles at the base of the equal sides are equal.

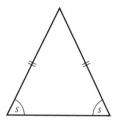

When a transversal cuts a pair of parallel lines:

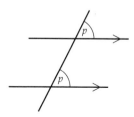

the corresponding angles, or F angles, are equal

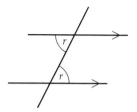

the alternate angles, or Z angles, are equal

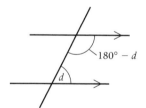

the interior angles are supplementary
(add up to 180°)

Exercise **4a**

Find the sizes of the marked angles. If two angles are marked with the same letter they are the same size.

1

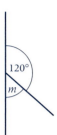

120°

m

6

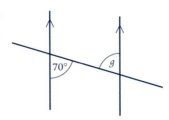

70°   g

2

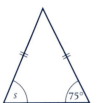

s   75°

7

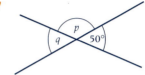

p

q   50°

3

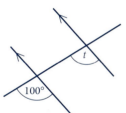

t

100°

8

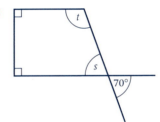

t

s

70°

4

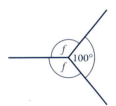

f

f   100°

9

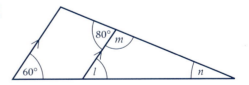

80° m

60°   l   n

5

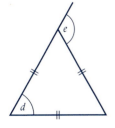

e

d

10

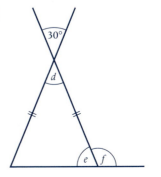

30°

d

e   f

In Book 1 you learnt how to construct triangles. This is revised on page 74. Before you start a construction, remember to make a rough sketch and to put all the information that you are given on to that sketch. Then decide which method to use.

Construct

11 △ABC in which AB = 5 cm, BC = 7 cm and AC = 6 cm

12 △PQR in which $\hat{P}$ = 60°, $\hat{Q}$ = 40° and PQ = 8 cm

13 △LMN in which $\hat{M}$ = 45°, LM = 7 cm and MN = 8 cm

14 △XYZ in which $\hat{X}$ = 100°, $\hat{Y}$ = 20° and XY = 5 cm

15 △RST in which RS = 10 cm, ST = 6 cm and RT = 7 cm

 ## Puzzle

In ten years' time the combined age of four brothers will be 100. What will it be in five years' time?

Constructing angles without using a protractor

Some angles can be made without using a protractor: one such angle is 60°.

Every equilateral triangle, whatever its size, has three angles of 60°. To make an angle of 60° we construct an equilateral triangle but do not draw the third side.

To construct an angle of 60°

Start by drawing a straight line and marking a point, A, near one end.

Next open your compasses to a radius of about 4 cm (this will be the length of the sides of your equilateral triangle).

With the point of your compasses on A, draw an arc to cut the line at B, continuing the arc above the line.

Move the point to B and draw an arc above the line to cut the first arc at C.

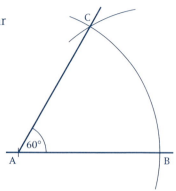

Draw a line through A and C. Then $\hat{A}$ is 60°.

△ABC is the equilateral triangle so *be careful not to alter the radius on your compasses during this construction*. Why is △ABC always equilateral?

Bisecting angles

Bisect means 'cut exactly in half'.

The construction for bisecting an angle makes use of the fact that, in an isosceles triangle, the line of symmetry cuts $\hat{A}$ in half.

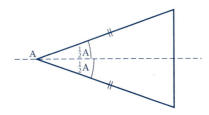

To bisect $\hat{A}$, open your compasses to a radius of about 6 cm.

With the point on A, draw an arc to cut both arms of $\hat{A}$ at B and C. (If we joined BC, △ABC would be isosceles.)

With the point on B, draw an arc between the arms of $\hat{A}$.

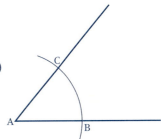

Move the point to C (being careful not to change the radius) and draw an arc to cut the other arc at D.

Join AD.

The line AD then bisects $\hat{A}$.

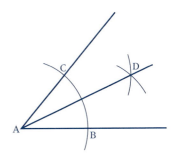

Exercise 4b

1 Construct an angle of 60°.

2 Draw an angle of about 50°. Bisect this angle. Measure both halves of your angle.

3 Construct an angle of 60°. Now bisect this angle. What size should each new angle be? Measure both of them.

4 Use what you learned from the last question to construct an angle of 30°.

5 Draw a straight line and mark a point A near the middle.

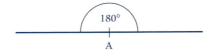

You now have an angle of 180° at A.

6 Draw an angle of 180° and then bisect it. What is the size of each new angle? Measure each of them.

7 Use what you learned from the last question to construct an angle of 90°.

8 Construct an angle of 45°. (Begin by constructing an angle of 90° and then bisect it.)

9 Construct an angle of 15°. (Start by constructing an angle of 60° and bisect as often as necessary.)

10 Construct an angle of 22.5°. (Start with 90° and bisect as often as necessary.)

Construction of angles of 60°, 30°, 90°, 45°

You constructed these angles in the last exercise. Here is a summary of these constructions.

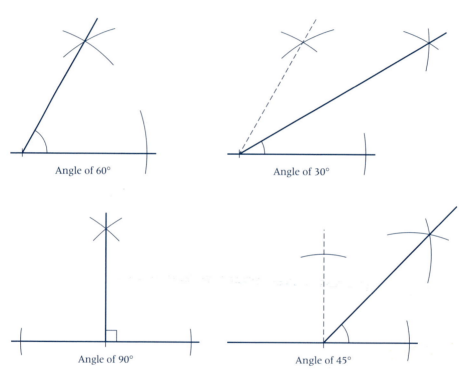

Angle of 60°   Angle of 30°

Angle of 90°   Angle of 45°

Exercise 4c

Tip Make sure that your pencil is sharp.

Construct the following figures using only a ruler and a pair of compasses:

1
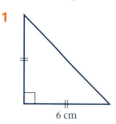
6 cm

2
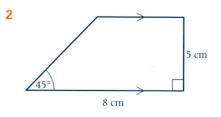
45°
8 cm
5 cm

3

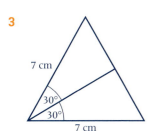

7

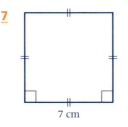

4

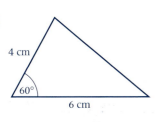

8

5

9

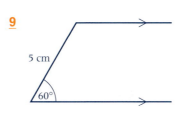

6

10

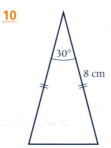

For questions **11** to **15**, draw a rough sketch before starting the construction.

11 Draw a line, AB, 12 cm long. Construct an angle of 60° at A. Construct an angle of 30° at B. Label with C the point where the arms of Â and B̂ cross. What size should Ĉ be? Measure Ĉ as a check on your construction.

12 Construct a triangle, ABC, in which AB is 10 cm long, Â is 90° and AC is 10 cm long. What size should Ĉ and B̂ be? Measure Ĉ and B̂ as a check.

13 Construct a square, ABCD, with a side of 6 cm.

14 Construct a quadrilateral, ABCD, in which AB is 12 cm, Â is 60°, AD is 6 cm, B̂ is 60° and BC is 6 cm. What can you say about the lines AB and DC?

15 Construct an angle of 120°. Label it BAC (so that A is the vertex and B and C are at the ends of the arms). At C, construct an angle of 60° so that Ĉ and Â are on the same side of AC. You have constructed a pair of parallel lines; mark them and devise your own check.

The rhombus

1 Draw a line 12 cm long across your page. Label the ends A and C. Open your compasses to a radius of 9 cm. With the point on A, draw an arc above AC and another arc below AC. Keeping the same radius, move the point of your compasses to C. Draw arcs above and below AC to cut the first pair of arcs. Where the arcs intersect (i.e. cross), label the points B and D.

   Join A to B, B to C, C to D and D to A. ABCD is called a rhombus.

Questions **2** to **9** refer to the figure that you have constructed in question **1**.

2 Without measuring them, what can you say about the lengths of AB, BC, CD and DA?

3 ABCD has two lines of symmetry. Name them.

4 If ABCD is folded along BD, where is A in relation to C?

5 If ABCD is folded along AC, where is D in relation to B?

6 Where AC and BD cut, label the point E. With ABCD unfolded, where is E in relation to A and C?

7 Where is E in relation to B and D?

8 If ABCD is folded first along BD and then folded again along AE, what is the size of the angle at E?

9 With ABCD unfolded, what are the sizes of the four angles at E?

Properties of the diagonals of a rhombus

From the last exercise you should be convinced that

the diagonals of a rhombus bisect each other at right angles.

These properties form the basis of the next two constructions.

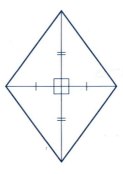

Construction to bisect a line

To bisect a line we have to find the midpoint of that line. To do this we construct a rhombus with the given line as one diagonal, but we do not join the sides of the rhombus.

To bisect XY, open your compasses to a radius that is about $\frac{3}{4}$ of the length of XY.

With the point on X, draw arcs above and below XY.

Move the point to Y (being careful not to change the radius) and draw arcs to cut the first pair at P and Q.

Join PQ.

The point where PQ cuts XY is the midpoint of XY.

(XPYQ is a rhombus because the same radius is used to draw all the arcs, i.e. XP = YP = YQ = XQ. PQ and XY are the diagonals of the rhombus so PQ bisects XY.)

Note. When you are going to bisect a line, draw it so that there is plenty of space for the arcs above *and* below the line.

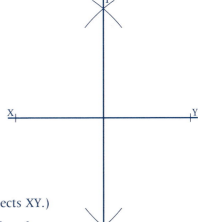

Dropping a perpendicular from a point to a line

If you are told to drop a perpendicular from a point, C, to a line, AB, this means that you have to draw a line through C which is at right angles to the line AB.

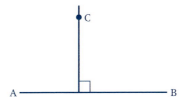

To drop a perpendicular from C to AB, open your compasses to a radius that is about $1\frac{1}{2}$ times the distance of C from AB.

With the point on C, draw arcs to cut the line AB at P and Q.

Move the point to P and draw an arc on the other side of AB. Move the point to Q and draw an arc to cut the last arc at D.

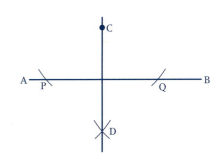

Join CD.

CD is then perpendicular to AB.

Remember to keep the radius unchanged throughout this construction: you then have a rhombus, PCQD, of which CD and PQ are the diagonals.

Constructing triangles

To construct a triangle given the lengths of the three sides, start by drawing one side, AB.

With the point of the compasses at A, and the radius equal to AC, draw an arc.

Then with the point of the compasses at B and the radius equal to BC, draw an arc to cut the first arc.

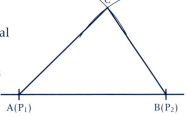

To construct a triangle given two sides and the angle between them, start by drawing one of the sides, say PQ.

Next use your protractor to measure the angle at P.

Then with the point of your compasses at P, radius PR, draw an arc to cut the arm of your angle.

This gives R. Join R and Q.

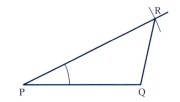

To construct a triangle given one side and two angles, start by drawing the side, say BC.

Now measure the angle you want at B and the angle you want at C.

(If either ∠B or ∠C is not known you can find it using $\angle A + \angle B + \angle C = 180°$)

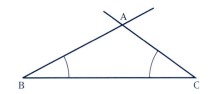

The point of intersection of the two lines drawn to make the angles gives A.

Exercise 4e

1 Construct a triangle ABC, in which AB = 6 cm, BC = 8 cm and CA = 10 cm. Using a ruler and compasses only, drop a perpendicular from B to AC.

> **Tip** Label your sketch and mark the measurements given. Remember also to use a sharp pencil.

2 Construct a triangle ABC, in which AB = 8 cm, AC = 10 cm and CB = 9 cm.
Drop a perpendicular from C to AB.

3 Construct a triangle XYZ, in which XY = 12 cm, XZ = 5 cm and YZ = 9 cm.
Drop a perpendicular from Z to XY.

4 Construct the isosceles triangle LMN in which LM = 6 cm, LN = MN = 8 cm.
Construct the perpendicular bisector of the side LM. Explain why this line is a line of symmetry of △LMN.

5 Construct the isosceles triangle PQR, in which PQ = 5 cm, PR = RQ = 7 cm.
Construct the perpendicular bisector of the side PR. This line is not a line of symmetry of △PQR; why not?

6 The figure below is a circle whose centre is C, with a line, AB, drawn across the circle.

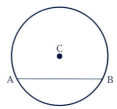

(AB is called a *chord*.)

This figure has one line of symmetry that is not shown. Make a rough sketch of the figure and mark the line of symmetry. Explain what the line of symmetry is in relation to AB.

7 Draw a circle of radius 6 cm and mark the centre, C. Draw a chord, AB, about 9 cm long. (Your drawing will look like the one in question 6.) Construct the line of symmetry.

8 Construct a triangle ABC, in which AB = 8 cm, BC = 10 cm and AC = 9 cm. Construct the perpendicular bisector of AB. Construct the perpendicular bisector of BC. Where these two perpendicular bisectors intersect (i.e. cross), mark G. With the point of your compasses on G and with a radius equal to the length of GA, draw a circle.

This circle should pass through B and C, and it is called the *circumcircle* of △ABC.

9 Repeat question 8 with a triangle of your own.

10 Construct a square ABCD, such that its sides are 5 cm long. Construct the perpendicular bisector of AB and the perpendicular bisector of BC. Label with E the point where the perpendicular bisectors cross. With the point of your compasses on E and the radius equal to the distance from E to A, draw a circle.

This circle should pass through all four corners of the square. It is called the circumcircle of ABCD.

11 Construct a triangle ABC, in which AB = 10 cm, AC = 8 cm and BC = 12 cm. Construct the bisector of Â and the bisector of B̂. Where these two angle bisectors cross, mark E. Drop the perpendicular from E to AB. Label G, the point where this perpendicular meets AB. With the point of your compasses on E and the radius equal to EG, draw a circle.

This circle should touch all three sides of △ABC and it is called the *incircle* of △ABC.

12 Repeat question 11 with the equilateral triangle ABC, with sides that are 10 cm long.

13 Repeat question 11 with a triangle of your own.

14 Construct a square ABCD, of side 8 cm. Construct the incircle (i.e. the circle that *touches* all four sides of the square) of ABCD. First decide how you are going to find the centre of the circle.

Puzzle

How many cubes can you see?

Making solids from nets

To make a solid object from a sheet of flat paper you need to construct a *net*: this is the shape that has to be cut out, folded and stuck together to make the solid. A net should be drawn as accurately as possible, otherwise you will find that the edges will not fit together properly.

Exercise 4f

Each solid in this exercise has flat faces (called *plane* faces) and is called a polyhedron. 'Poly' is a prefix used quite often; it means 'many'.

1 Cube
   This net will make a cube
   a Which edge meets AB?
   b Which other corners meet at H?

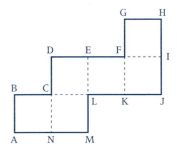

2 Tetrahedron
   This net consists of four equilateral triangles. Construct the net accurately, making the sides of each triangle 6 cm long. Start by drawing one triangle of side 12 cm; mark the midpoints of the sides and join them up. Draw flaps on the edges shown. (These are not part of the net.)

Cut out the net. Score the solid lines (use a ruler and ballpoint pen – an empty one is best) and fold the outer triangles up so that their vertices meet. Use the flaps to stick the edges together.

This solid is called a *regular* tetrahedron. A regular solid is one in which all the faces are identical. These make good Christmas tree decorations if painted or if made out of foil-covered paper.

3 Octahedron

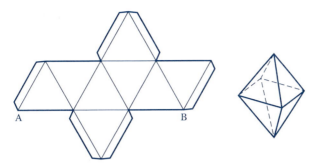

This net consists of equilateral triangles: make the sides of each triangle 4 cm long, and start by making AB 12 cm long. Is this octahedron a regular solid?

4 Square-based pyramid

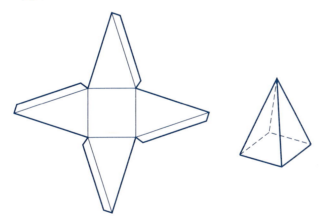

This net consists of a square with an isosceles triangle on each side of the square. Make the sides of the square 6 cm and the equal sides of the triangles 10 cm long. Is this a regular solid?

5 Prism with triangular cross-section

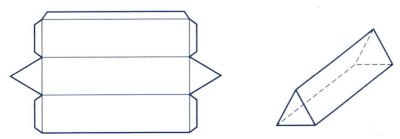

This net consists of three rectangles, each 8 cm long and 4 cm wide, and two equilateral triangles (sides 4 cm).

6 Eight-pointed star (stella octangula)

This model needs time and patience. If you have both it is worth the effort!

It consists of a regular octahedron (see question 2) with a regular tetrahedron (see question 1) stuck on each face.
You will need 8 tetrahedra. In all the nets make the triangles have sides of length 4 cm.

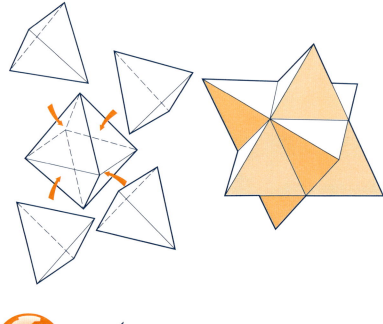

A circle by any other name is just as round.

IN THIS CHAPTER...

you have seen that:

- you can construct an angle of 60° by constructing an equilateral triangle

- you can construct an angle of 90° by constructing a perpendicular to a line

- you can construct an angle of 30° by bisecting an angle of 60° and an angle of 45° by bisecting an angle of 90°

- the diagonals of a rhombus bisect each other at right angles

- a net is a flat shape that can be folded to make a solid.

5 SCALE DRAWING

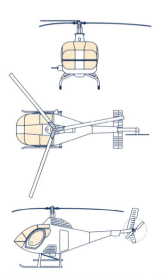

This is a scale drawing of a helicopter

Accurate drawing with scaled down measurements

If you are asked to draw a car park that is a rectangle measuring 50 m by 25 m, you obviously cannot draw it full size. To fit it on to your page you will have to scale down the measurements. In this case you could use 1 cm to represent 5 m on the car park. This is called the *scale*; it is usually written as 1 cm ≡ 5 m, and must *always* be stated on any scale drawing.

Exercise **5a**

Start by making a rough drawing of the object you are asked to draw to scale. Mark all the full-size measurements on your sketch. Next draw another sketch and put the scaled measurements on this one. Then do the accurate scale drawing. Always give the scale on your drawing.

The end wall of a bungalow is a rectangle with a triangular top. The rectangle measures 6 m wide by 3 m high. The base of the triangle is 6 m and the sloping sides are 4 m long. Using a scale of 1 cm to 1 m, make a scale drawing of this wall. Use your drawing to find, to the nearest tenth of a metre, the distance from the ground to the ridge of the roof.

Rough sketch of wall to give measurements

Rough sketch of scale drawing with scale measurements

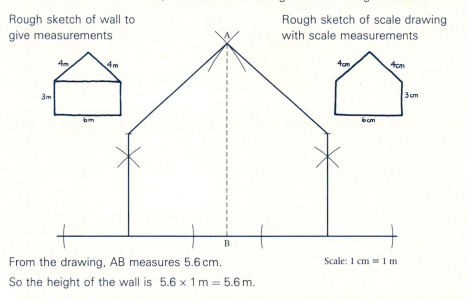

From the drawing, AB measures 5.6 cm.

Scale: 1 cm ≡ 1 m

So the height of the wall is 5.6 × 1 m = 5.6 m.

In questions **1** to **5**, use a scale of 1 cm to 1 m to make a scale drawing.

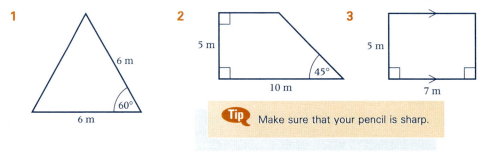

1

6 m
6 m
60°

2

5 m
10 m
45°

3

5 m
7 m

Tip Make sure that your pencil is sharp.

4   **5**

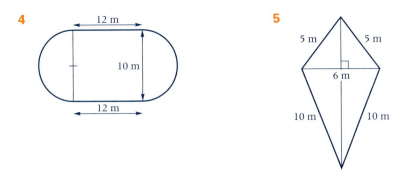

In questions **6** to **10**, choose your own scale.

Choose a scale that gives lines that are long enough to draw easily; in general, the lines on your drawing should be at least 5 cm long. Avoid scales that give lengths involving awkward fractions of a centimetre, such as thirds; $\frac{1}{3}$ cm cannot be read from your ruler.

6   **8**

7   **9**

A casement window with equally spaced glazing bars

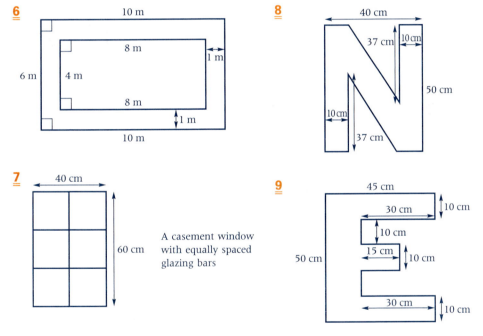

10   A rectangular door with four rectangular panels, each 35 cm by 70 cm, and 10 cm from the edges of the door.

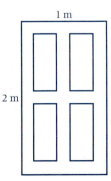

11 A field is rectangular in shape. It measures 300 m by 400 m. A land drain goes in a straight line from one corner of the field to the opposite corner. Using a scale of 1 cm to 50 m, make a scale drawing of the field and use it to find the length of the land drain.

12 The end wall of a ridge tent is a triangle. The base is 2 m and the sloping edges are each 2.5 m. Using a scale of 1 cm to 0.5 m, make a scale drawing of the triangular end of the tent and use it to find the height of the tent.

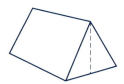

13 The surface of a swimming pool is a rectangle measuring 25 m by 10 m. Choose your own scale and make a scale drawing of the pool.
Now compare and discuss your drawing with other pupils.

14 The whole class working together can collect the information for this question. Measure your classroom and make a rough sketch of the floor plan. Mark the position and width of doors and windows. Choosing a suitable scale, make an accurate scale drawing of the floor plan of your classroom.

Practical work

The sketch shows the measurements of Ken's bathroom. There is only one outside wall and the bottom of the window is 120 cm above the level of the floor. Draw an accurate diagram of the floor, using a scale of 1 cm to 10 cm.

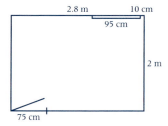

Ken wants a new bathroom suite. The units he would like to install, together with their measurements, are:

   bath: 170 cm × 75 cm
   handbasin: 60 cm × 42 cm, the longest edge against a wall
   shower tray: 80 cm square
   toilet: 70 cm × 50 cm, the shorter measurement against a wall
   bidet: 55 cm × 35 cm, the shorter measurement against a wall.

Using the same scale make accurate drawings of the plans of these units, then cut them out and see if you can place them on your plan in acceptable positions.

If they will not all fit into the room, which unit(s) would you be prepared to do without?

Give reasons for your answer.

Is it possible to arrange your chosen units so that all the plumbing is against

a   the outside wall

b   not more than two walls at right angles, one of which is the outside wall?

Illustrate your answer with a diagram.

Angles of elevation

If you are standing on level ground and can see a tall building, you will have to look up to see the top of that building.

If you start by looking straight ahead and then look up to the top of the building, the angle through which you raise your eyes is called the *angle of elevation* of the top of the building.

Angle of elevation

You

There are instruments for measuring angles of elevation. A simple one can be made from a large card protractor and a piece of string with a weight on the end.

You can read the size of $\widehat{A}$.

Then the angle of elevation, $\widehat{B}$, is given by $\widehat{B} = 90° - \widehat{A}$.

(Note that this method is not very accurate.)

If your distance from the foot of the building and the angle of elevation of the top are both known, you can make a scale drawing of $\triangle PQR$.

This drawing can then be used to work out the height of the building.

From a point A on the ground, 50 m from the base of a tree, the angle of elevation of the top of the tree is 22°. Using a scale of 1 cm ≡ 5 m, make a scale drawing and use it to find the height of the tree.

Rough sketch to give actual measurements

Rough sketch of scale drawing with scale measurements

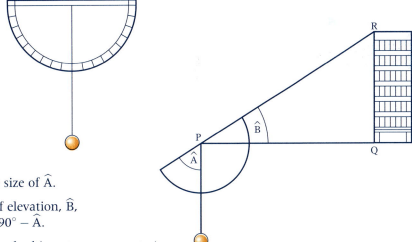

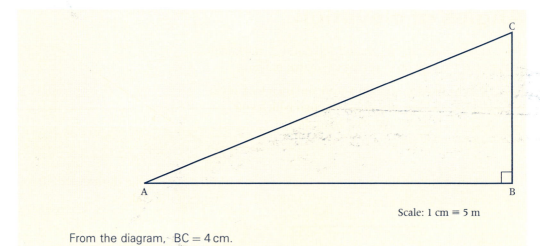

Scale: 1 cm ≡ 5 m

From the diagram, BC = 4 cm.

∴ the tree is 4 × 5 m = 20 m high.

In questions **1** to **4**, A is a place on the ground, Â is the angle of elevation of C, the top of BC. Using a scale of 1 cm ≡ 5 m, make a scale drawing to find the height of BC.

1

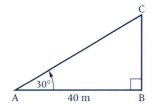

2

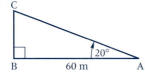

3

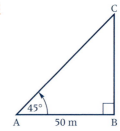

4

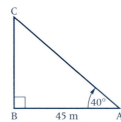

In questions **5** to **7**, use a scale of 1 cm ≡ 10 m.

5 From A, the angle of elevation of C is 35°. Find BC.

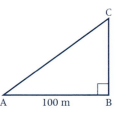

6 From P, the angle of elevation of R is 15°. Find QR.

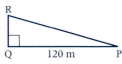

7 From N, the angle of elevation of L is 30°. Find ML.

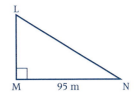

8 From a point D on the ground, 100 m from the foot of a church tower, the angle of elevation of the top of the tower is 30°. Use a scale of 1 cm to 10 m to make a scale drawing. Use your drawing to find the height of the tower.

9 From the opposite side of the road, the angle of elevation of the top of the roof of my house is 37°. The horizontal distance from the point where I measured the angle to the house is 12 m. Make a scale drawing, using a scale of 1 cm to 1 m, and use it to find the height of the top of the roof.

10 From a point P on the ground, 150 m from the base of the Barbados Central Bank, the angle of elevation of the top is 17°. Use a scale of 1 cm to 20 m to make a scale diagram and find the height of the Barbados Central Bank.

11 The top of a radio mast is 76 m from the ground. From a point, P, on the ground, the angle of elevation of the top of the mast is 40°. Use a scale of 1 cm to 10 m to make a scale drawing to find how far away P is from the mast. (You will need to do some calculations before you can do the scale drawing.)

Angles of depression

An *angle of depression* is the angle between the line looking straight ahead and the line looking *down* at an object below you.

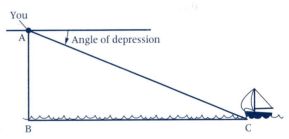

If, for example, you are standing on a cliff looking out to sea, the diagram shows the angle of depression of a boat.

If the angle of depression and the height of the cliff are both known, you can make a scale drawing of △ABC. Then you can work out the distance of the boat from the foot of the cliff.

Exercise 5c

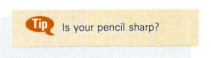
Tip Is your pencil sharp?

In questions **1** to **4**, use a scale of 1 cm ≡ 10 m to make a scale drawing.

1 From A, the angle of depression of C is 25°. Find BC.

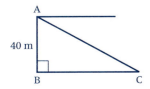

2 From L, the angle of depression of N is 40°. Find MN.

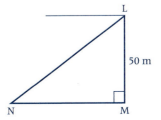

3 From P, the angle of depression of R is 35°. Find RQ.

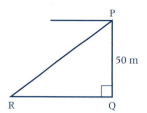

4 From Z, the angle of depression of X is 42°. Find XY.

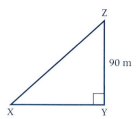

5 From the top of Blackpool Tower, which is 158 m high, the angle of depression of a ship at sea is 25°. Use a scale of 1 cm to 50 m to make a scale drawing to find the distance of the ship from the base of the tower.

6 From the top of Hackelton's Cliff in Barbados, which is 300 m high, the angle of depression of a house is 20°. Use a scale of 1 cm to 50 m to make a scale drawing and find the distance of the house from the base of the cliff.

7 From the top of a vertical cliff, which is 30 m high, the angle of depression of a yacht is 15°. Using a scale of 1 cm to 5 m, make a scale drawing to find the distance of the yacht from the foot of the cliff.

8 An aircraft flying at a height of 300 m measures the angle of depression of the end of the runway as 18°. Using a scale of 1 cm to 100 m, make a scale diagram to find the horizontal distance of the aircraft from the runway.

9 The Sears Tower in Chicago is an office building and it is 443 m high. From the top of this tower, the angle of depression of a ship on a lake is 40°. How far away from the base of the building is the ship? Use a scale of 1 cm to 50 m to make your scale drawing.

For the remaining questions in this exercise, make a scale drawing choosing your own scale.

10 From a point on the ground 60 m away, the angle of elevation of the top of a factory chimney is 42°. Find the height of the chimney.

11 From the top of a hill, which is 400 m above sea level, the angle of depression of a boathouse is 20°. The boathouse is at sea level. Find the distance of the boathouse from the top of the hill.

12 An aircraft flying at 5000 m measures the angle of depression of a point on the coast as 30°. At the moment that it measures the angle, how much further has the plane to fly before passing over the coastline?

13 A vertical radio mast is 250 m high. From a point A on the ground, the angle of elevation of the top of the mast is 30°. How far is the point A from the foot of the mast?

14 An automatic lighthouse is stationed 500 m from a point, A, on the coast. There are high cliffs at A and from the top of these cliffs, the angle of depression of the lighthouse is 15°. How high are the cliffs?

15 An airport controller measures the angle of elevation of an approaching aircraft as 20°. If the aircraft is then 1.6 km from the control building, at what height is it flying?

16 A surveyor standing 400 m from the foot of a church tower, on level ground, measures the angle of elevation of the top of the tower. If this angle is 35° how high is the tower?

 **Puzzle**

What number should go in the centre of the last cross?

Three-figure bearings

A bearing is a compass direction.

If you are standing at a point, A, and looking at a tree, B, in the distance, as shown in the diagram below, then using compass directions you could say that

from A, the bearing of B is SE

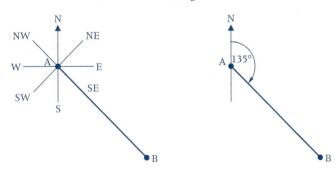

Using the modern method of a three-figure bearing we first look north and then turn clockwise until we are looking at B. The angle turned through is the three-figure bearing. In this case

from A, the bearing of B is 135°

A three-figure bearing is a clockwise angle measured from the north.

If the angle is less than 100°, it is made into a three-figure angle by putting zero in front, e.g. 20° becomes 020°.

Exercise 5d   From a ship, C, the bearing of a ship, D, is 290°. Make a rough sketch and mark the angle.

Start by marking C. Next draw a north line from C. Then mark the angle turning clockwise around C.

Draw a rough sketch to illustrate each of the following bearings. Mark the angle in your sketch.

1   From a ship, P, the bearing of a yacht, Q, is 045°

2   From a control tower, F, the bearing of an aeroplane, A, is 090°

3   From a point, A, the bearing of a radio mast, M, is 120°

4   From a town, T, the bearing of another town, S, is 180°

5   From a point, H, the bearing of a church, C, is 210°

6   From a ship, R, the bearing of a port, P, is 300°

7   From an aircraft, A, the bearing of an airport, L, is 320°

8   From a town, D, the bearing of another town, E, is 260°

9   From a helicopter, G, the bearing of a landing pad, P, is 060°

10   From a point, L, the bearing of a tree, T, is 270°

11   The bearing of a ship, A, from the pier, P, is 225°

12   The bearing of a radio mast, S, from a point, P, is 140°

13   The bearing of a yacht, Y, from a tanker, T, is 075°

14   The bearing of a town, Q, from a town, R, is 250°

15   The bearing of a tree, X, from a hill top, Y, is 025°

16   From a point, A, the bearing of a house, H, is 190°

17   From a town, T, the bearing of another town, S, is 290°

18   From a barn, B, the bearing of a tree, T, is 020°

19   The bearing of a boat, B, from a jetty, J, is 030°

20   The bearing of a flagpole, F, from a tent, T, is 300°

Using bearings to find distances

If we measure the bearing of a distant object from two different positions and then make a scale diagram, we can use this diagram to find the distance of that object from one or other of the positions.

Exercise 5e   From one end, A, of a road the bearing of a building, L, is 015°. The other end of the road, B, is 300 m due east of A. From B the bearing of the building is 320°. Using a scale of 1 cm to 50 m, make a scale diagram to find the distance of the building from A.

Rough sketch to give   Rough sketch of scale drawing
actual measurements   with scale measurements

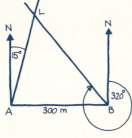

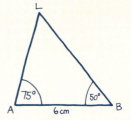

Scale: 1 cm ≡ 50 m

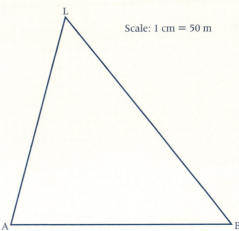

From the diagram, LA = 5.6 cm

∴ the distance of the building from A is 5.6 × 50 m = 280 m

1 From a point, A, the bearing of a tree, C, is 060°. From a second point, B, which is 100 m due east of A, the bearing of the tree is 330°. Use a scale of 1 cm to 10 m to make a scale diagram and find the distance of the tree from A.

2 From a point, A, the bearing of a ship, C, is 140°. From a second point, B, which is 200 m due east of A, the bearing of the ship is 210°. Using a scale of 1 cm to 20 m make a scale diagram and use it to find the distance of the ship from B.

3 From a point, A, the bearing of a tower, T, is 030°. From a second point, B, which is 400 m due north of A, the bearing of the tower is 140°. Using a scale of 1 cm to 50 m, make a scale drawing and use it to find the distance of the tower from A.

4 From a point, A, the bearing of a radar mast, M, is 060°. From a second point, B, which is 40 m due east of A, the bearing of the radar mast is 010°. Use a scale of 1 cm to 10 m and make a scale drawing to find the distance of the radar mast from A.

5 From a ship, P, the bearing of a fishing boat, S, is 020°. From a second ship, Q, which is 1000 m due north of P, the bearing of the fishing boat is 070°. Using a scale of 1 cm to 200 m, make a scale drawing to find the distance of the fishing boat from P.

Mixed exercises

Exercise 5f

1 Using a scale of 1 cm to 100 cm, make a scale drawing of the figure on the right. Use your drawing to find the length of the diagonal AC.

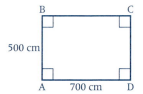

For each of the following questions, make a rough sketch to show all the given information.

2 From the top of a tower which is 150 m tall, the angle of depression of a house is 17°.

3 From a point, A, the bearing of a point, B, is 270°.

4 An aircraft is flying at a height of 2000 m. From a point on the ground its angle of elevation is 40°.

5 An aircraft is flying at a height of 500 m when it measures the angle of depression of the end of the runway as 30°.

Exercise 5g

1 Use a scale of 1 cm to 10 m to make a scale drawing of the figure on the right. Use your scale drawing to find the length of AC.

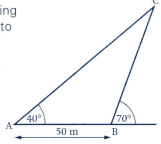

For each of the following questions, make a rough sketch showing all the given information.

2 The bearing of a ship, R, from a ship, S, is 075°.

3 From a position, A, on the ground, the angle of elevation of the top of an office block is 25°. The office block is 75 m tall.

4 From the top of a cliff which is 50 m high, the angle of depression of a boat is 34°.

5 From a ship, A, the bearing of an oil tanker, T, is 300°. From a second ship, B, which is 1000 m due west of A, the bearing of the oil tanker is 060°. Explain why the oil tanker is the same distance from A as it is from B.

Exercise 5h

1 Use a scale of 1 cm to 50 m to make a scale drawing of the figure on the right. Use your drawing to find the distance CD.

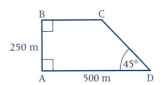

For the following questions, make rough sketches showing all the given information.

2 From the top window of a house the angle of depression of the end of the garden is 32°. The garden is 10 m long.

3 The bearing of an aircraft, X, from the control tower, T, is 157°.

4 The angle of elevation of a helicopter, H, from the landing pad, L, is 45°. The helicopter is at a height of 45 m.

In the following questions you are also asked to find bearings.

5 The diagram shows the positions of three places in Barbados: St Patricks, Shorey and Holetown.

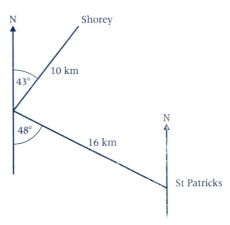

  a What is the bearing of

   i Shorey from Holetown

   ii St Patricks from Holetown

   iii Holetown from Shorey?

  b Using 1 cm to represent 2 km and the distances given on the diagram, make a scale diagram to show the positions of the three cities.

  c Join St Patricks to Shorey and measure the length of your line to the nearest millimetre. Hence find the distance from St Patricks to Shorey.

  d **i** By measuring a suitable angle find the bearing of Shorey from St Patricks.

   ii What is the bearing of St Patricks from Shorey?

6 a Using the diagram, find the bearing of

    i C from B   **ii** A from B

    iii B from A   **iv** B from C.

b Find the angles in triangle ABC.

c Given that the distance represented by AB is 5 km, construct △ABC using 1 cm to represent 1 km and hence find the actual distance between A and C.

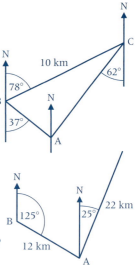

7 A helicopter flies 12 km on a bearing of 125°. It then changes course and flies 22 km on a bearing of 025°.

Draw a scale diagram using 1 cm to represent 2 km.

a How far is the helicopter due east of its starting point?

b How far is the helicopter due north of its starting point?

 Puzzle

What is the smallest whole number that will divide exactly by every whole number from 2 to 12?

Did you know that up to the middle of the 18th century, because of the difficulty of measuring longitude, it was impossible to fix your exact position at sea? Because of this, thousands of sailors had perished. The problem was solved by John Harrison (1693–1776), a self-taught Yorkshire clockmaker, who spent 40 years designing and building a clock that would keep perfect time at sea. John Harrison's clock, tested on a journey from the UK to Jamaica, gave an error of 5 seconds or less than 1 nautical mile.

IN THIS CHAPTER...

you have seen that:

- you can use scale diagrams to find heights and distances

- an angle of elevation is the angle you turn your eyes through from the horizontal to look *up* at an object

- an angle of depression is the angle you turn your eyes through from the horizontal to look *down* at an object

- a three-figure bearing is a clockwise angle measured from the north direction.

6

EQUATIONS AND FORMULAE

AT THE END OF THIS CHAPTER...

you should be able to:

1   Solve simple equations.

2   Form simple equations and use them to solve problems.

3   Construct formulae from given information.

4   Substitute numerical values in a formula.

5   Change the subject of a given formula.

Did you know that the German mathematician Carl Friedrich Gauss (1777–1855) was the leading algebraist and theoretical astronomer of the day. He was born of poor parents but, because of his prodigious talent, his education was paid for by the Duke of Brunswick. Nearly all his fundamental mathematical discoveries were made before he was 22.

BEFORE YOU START

you need to know:
- ✓   how to simplify expressions containing brackets
- ✓   how to collect like terms
- ✓   how to multiply and divide one fraction by another
- ✓   the order in which to do multiplication, division, addition and subtraction
- ✓   how to multiply directed numbers
- ✓   the meaning of a formula
- ✓   what the lowest common multiple means

KEY WORDS

area, breadth, denominator, directed number, eliminate, equation, equilateral triangle, equivalent, formula (plural formulae), fraction, like terms, lowest common multiple, negative number, numerator, perimeter, rectangle, subject of a formula, substitute, x and y coordinates

Equations

Imagine a balance.

On the left-hand side there are two bags each containing the same (but unknown) number of apples and three loose apples.

On the right-hand side there are thirteen apples.

Using the letter x to stand for the unknown number of apples in each bag we can write this as an equation:

$$2x + 3 = 13$$

We can solve this equation (i.e. find the number that x represents) as follows:

take three apples off each pan $2x = 10$

halve the contents of each pan $x = 5$

Remember that we want to isolate x and that we can do anything as long as we do it to both sides of the equation.

Exercise 6a

Solve the equation $5x - 4 = 6$

$$5x - 4 = 6$$

Add 4 to each side $5x = 10$

Divide each side by 5 $x = 2$

Check: $LHS = 5 \times 2 - 4 = 6$ $RHS = 6$

Solve the following equations:

1 $2x = 8$

2 $x - 3 = 1$

3 $x + 4 = 16$

4 $2x + 3 = 7$

5 $3x + 5 = 14$

6 $3x - 2 = 10$

7 $5 + 2x = 7$

8 $5x - 4 = 11$

9 $3 + 6x = 15$

10 $7x - 6 = 15$

Solve the equation $3x + 4 = 12 - x$

(We need to start by getting the terms containing x on one side of the equation and the terms without x on the other side. The left-hand side has the greater number of x s, so we will collect them on this side.)

$$3x + 4 = 12 - x$$

Add x to each side	$4x + 4 = 12$
Take 4 from each side	$4x = 8$
Divide each side by 4	$x = \frac{8}{4} = 2$

Solve the equation $4 - x = 6 - 3x$

(There are fewer x s missing from the LHS so we will collect them on this side.)

$$4 - x = 6 - 3x$$

Add $3x$ to each side	$4 - x + 3x = 6$
	$4 + 2x = 6$
Take 4 from each side	$2x = 2$
Divide each side by 2	$x = 1$

Solve the following equations:

11 $2x + 5 = x + 9$

12 $3x + 2 = 2x + 7$

13 $x - 4 = 2 - x$

14 $3 - 2x = 7 - 3x$

15 $2x + 1 = 4 - x$

16 $x + 4 = 4x + 1$

17 $3x - 2 = 2x + 1$

18 $1 - 3x = 9 - 4x$

19 $2 - 5x = 6 - 3x$

20 $5 - 3x = 1 + x$

Solve the equation $4x + 2 - x = 7 + x - 3$

$$4x + 2 - x = 7 + x - 3$$

Collect like terms	$3x + 2 = 4 + x$
Take x from each side	$2x + 2 = 4$
Take 2 from each side	$2x = 2$
Divide each side by 2	$x = 1$

Solve the following equations:

21 $x + 2 + 2x = 8$

22 $x - 4 = 3 - x + 1$

23 $3x + 1 - x = 5$

24 $4 + 3x - 1 = 6$

25 $7 + 4x = 2 - x + 10$

26 $3 + x - 1 = 3x$

27 $x - 4 + 2x = 5 + x - 1$

28 $x + 5 - 2x = 3 + x$

29 $x + 17 - 4x = 2 - x + 6$

30 $8 - 3x - 3 = x - 4 + 2x$

31 $5x - 8 = 2$

32 $4 - x = 3x$

33 $5 - x = 7 + 2x - 4$

34 $4 - 2x = 8 - 4x$

35 $15 = 21 - 2x$

36 $x + 4 - 3x = 2 - x$

37 $3x - 7 = 9 - x + 6$

38 $x + 4 = 6x$

39 $8 - 3x = 5x$

40 $5 - 4x + 7 = 2x$

Brackets

Reminder: If we want to multiply both x and 3 by 4, we group x and 3 together in a bracket and write $4(x + 3)$.

So $4(x + 3)$ means that *both x and* 3 are to be multiplied by 4. (Note that the multiplication sign is invisible, as it is in 5a.)

i.e. $\qquad\qquad 4(x + 3) = 4x + 12$

($4x$ and 12 are unlike terms so $4x + 12$ cannot be simplified)

Exercise 6b

Multiply out the following brackets:

1 $6(x + 4)$

2 $3(2x + 1)$

3 $4(x - 3)$

4 $2(3x - 5)$

5 $4(3 - 2x)$

6 $5(4x + 2)$

7 $3(2 - 3x)$

8 $7(5 - 4x)$

9 $2(5x - 7)$

10 $6(7 + 2x)$

Simplify:

11 $2(3 + x) + 3(2x + 4)$

12 $7(2x + 3) + 4(3x - 2)$

13 $4(6x + 3) + 5(2x - 5)$

14 $2(2x - 4) + 4(x + 3)$

> **Tip** First multiply the brackets, then collect like terms.

15 $5(3x - 2) + 3(2x + 5)$

16 $3(3x + 1) + 4(x + 4)$

17 $5(2x + 3) + 6(3x + 2)$

18 $6(2x - 5) + 2(3x - 7)$

19 $8(2 - x) + 3(3 + 4x)$

20 $5(7 - 2x) + 4(3 - 5x)$

21 $3(2x - 1) + 4(x + 2)$

22 $5(2 - x) + 2(2x + 1)$

23 $3(x - 4) + 7(2x - 3)$

24 $2(2x + 1) + 4(3 - 2x)$

25 $6(2 - x) + 2(1 - 2x)$

26 $5(4 + 3x) + 3(2 + 7x)$

27 $4(3 + 2x) + 5(4 - 3x)$

28 $8(x + 1) + 7(2 - x)$

29 $3(2x + 7) + 5(3x - 8)$

30 $9(x - 2) + 5(4 - 3x)$

Solve the following equations:

31 $2(x+2)=8$

32 $4(2-x)=2$

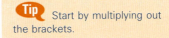 **Tip** Start by multiplying out the brackets.

33 $5(3x+1)=20$

34 $2(2x-1)=6$

35 $3(2x+5)=18$ **41** $8(x-1)=4$

36 $3(3-2x)=3$ **42** $3(1-4x)=11$

37 $2(x+4)=3(2x+1)$ **43** $5(3-2x)=3(4-3x)$

38 $4(2x-3)=2(3x-5)$ **44** $7(1+2x)=21$

39 $6(3x+5)=12$ **45** $7(2x-1)=5(3x-2)$

40 $6(x+3)=2(2x+5)$ **46** $4(3x+2)=14$

 ## Puzzle

James bought seventeen pens, some black and some red, for $7.20. He paid 10c more for each red pen than for each black pen.

How many of each colour did he buy?

Multiplication and division of fractions

Remember that, to multiply fractions, the numerators are multiplied together and the denominators are multiplied together:

i.e. $\dfrac{3}{4} \times \dfrac{5}{7} = \dfrac{3 \times 5}{4 \times 7} = \dfrac{15}{28}$

Also $\dfrac{1}{6}$ of x means $\dfrac{1}{6} \times x = \dfrac{1}{6} \times \dfrac{x}{1} = \dfrac{x}{6}$ (1)

Remember that, to divide by a fraction, that fraction is turned upside down and multiplied:

i.e. $\dfrac{2}{3} \div \dfrac{5}{7} = \dfrac{2}{3} \times \dfrac{7}{5} = \dfrac{14}{15}$

and $x \div 6 = \dfrac{x}{1} \div \dfrac{6}{1} = \dfrac{x}{1} \times \dfrac{1}{6} = \dfrac{x}{6}$ (2)

Comparing (1) and (2) we see that

$\dfrac{1}{6}$ of x, $\dfrac{1}{6}x$, $x \div 6$ and $\dfrac{x}{6}$ are all equivalent

97

Exercise 6c Simplify $12 \times \dfrac{x}{3}$

$$12 \times \frac{x}{3} = \frac{\cancel{12}^{\,4}}{1} \times \frac{x}{\cancel{3}_{\,1}}$$

$$= 4x$$

Simplify $\dfrac{2x}{3} \div 8$

$$\frac{2x}{3} \div 8 = \frac{2x}{3} \div \frac{8}{1}$$

$$= \frac{\cancel{2x}^{\,1}}{3} \times \frac{1}{\cancel{8}_{\,4}} \qquad \text{(Now cancel common factors: remember that } 2x = 2 \times x)$$

$$= \frac{x}{12}$$

Simplify:

1 $4 \times \dfrac{x}{8}$

2 $\dfrac{1}{2} \times \dfrac{x}{3}$

3 $9 \times \dfrac{x}{6}$

4 $\dfrac{1}{3}$ of $2x$

5 $\dfrac{2x}{3} \times \dfrac{6}{5}$

6 $\dfrac{1}{5}$ of $10x$

7 $\dfrac{2}{5} \times \dfrac{3x}{4}$

8 $\dfrac{3}{4} \times 2x$

9 $\dfrac{2}{3}$ of $9x$

10 $\dfrac{x}{2} \times \dfrac{x}{3}$

11 $\dfrac{5x}{2} \div 4$

12 $\dfrac{4x}{9} \div 8$

13 $\dfrac{x}{3} \div \dfrac{1}{6}$

14 $\dfrac{x}{4} \div \dfrac{1}{2}$

15 $\dfrac{2x}{3} \div \dfrac{5}{6}$

16 $\dfrac{3}{4} \times \dfrac{2x}{5}$

Tip $4 = \dfrac{4}{1}$

17 $\dfrac{4x}{9} \div \dfrac{2}{3}$

18 $\dfrac{3}{5}$ of $15x$

19 $\dfrac{3x}{2} \div \dfrac{1}{6}$

20 $\dfrac{5x}{3} \times \dfrac{6x}{15}$

Fractional equations

Exercise 6d Solve the equation $\dfrac{x}{3} = 2$

$$\left(\text{As } \frac{x}{3} \text{ means } \frac{1}{3} \text{ of } x, \text{ to find } x \text{ we need to make } \frac{x}{3} \text{ three times larger.} \right)$$

$$\frac{x}{3} = 2$$

Multiply each side by 3

$$\frac{x}{\cancel{3}_{\,1}} \times \frac{\cancel{3}^{\,1}}{1} = 2 \times 3$$

$$x = 6$$

Solve the following equations:

1 $\dfrac{x}{5} = 3$ **3** $\dfrac{x}{6} = 8$ **5** $16 = \dfrac{9x}{2}$ **7** $\dfrac{4x}{7} = 8$

2 $\dfrac{x}{2} = 4$ **4** $\dfrac{2x}{3} = 8$ **6** $\dfrac{2x}{5} = 9$ **8** $\dfrac{6x}{5} = 10$

Solve the equation $\dfrac{2x}{5} = \dfrac{1}{3}$

$$\dfrac{2x}{5} = \dfrac{1}{3}$$

Multiply each side by 5 to get rid of the fraction on the left-hand side.

$$\dfrac{2x}{\cancel{5}_1} \times \dfrac{\cancel{5}^1}{1} = \dfrac{1}{3} \times \dfrac{5}{1}$$

$$2x = \dfrac{5}{3}$$

Divide each side by 2

$$x = \dfrac{5}{3} \div 2$$

$$x = \dfrac{5}{3} \times \dfrac{1}{2}$$

$$x = \dfrac{5}{6}$$

Solve the following equations:

9 $\dfrac{3x}{2} = \dfrac{1}{4}$ **11** $\dfrac{2x}{9} = \dfrac{1}{3}$ **13** $\dfrac{3x}{8} = \dfrac{1}{2}$ **15** $\dfrac{3x}{5} = \dfrac{1}{4}$

10 $\dfrac{4x}{3} = \dfrac{1}{5}$ **12** $\dfrac{6x}{5} = \dfrac{2}{3}$ **14** $\dfrac{5x}{7} = \dfrac{3}{4}$ **16** $\dfrac{4x}{7} = \dfrac{2}{5}$

Solve the equation $\dfrac{x}{5} + \dfrac{1}{2} = 1$

(Both 5 and 2 divide into 10, so by multiplying each side by 10 we can eliminate all fractions from this equation before we start to solve for *x*.)

$$\dfrac{x}{5} + \dfrac{1}{2} = 1$$

Multiply both sides by 10 $10\left(\dfrac{x}{5} + \dfrac{1}{2}\right) = 10 \times 1$

$$\dfrac{\cancel{10}^2}{1} \times \dfrac{x}{\cancel{5}_1} + \dfrac{\cancel{10}^5}{1} \times \dfrac{1}{\cancel{2}_1} = 10$$

$$2x + 5 = 10$$

Take 5 from each side $2x = 5$

Divide each side by 2 $x = 2\dfrac{1}{2}$

Solve the following equations:

17 $\dfrac{x}{3} + \dfrac{1}{4} = 1$ **19** $\dfrac{x}{5} + \dfrac{2x}{3} = 3$ **21** $\dfrac{2x}{3} - \dfrac{1}{2} = 4$ **23** $\dfrac{x}{3} - \dfrac{2}{9} = 4$ **25** $\dfrac{3}{4} - \dfrac{x}{5} = 1$

18 $\dfrac{x}{5} - \dfrac{3}{4} = 2$ **20** $\dfrac{5x}{7} + \dfrac{x}{2} = 2$ **22** $\dfrac{x}{3} + \dfrac{5}{6} = 2$ **24** $\dfrac{3x}{4} - \dfrac{x}{2} = 5$ **26** $\dfrac{5}{7} + \dfrac{3x}{4} = 2$

Solve the following equations:

27 $\dfrac{x}{3} + \dfrac{1}{4} = \dfrac{1}{2}$

28 $\dfrac{x}{5} + \dfrac{2}{3} = \dfrac{14}{15}$

Tip Multiply both sides by the lowest common multiple of the denominators.

29 $\dfrac{x}{4} - \dfrac{1}{2} = \dfrac{9}{4}$ **35** $\dfrac{5x}{12} - \dfrac{1}{3} = \dfrac{x}{8}$ **41** $\dfrac{3}{11} - \dfrac{x}{2} = \dfrac{2x}{11} + \dfrac{1}{4}$

30 $\dfrac{2x}{3} + \dfrac{2}{7} = \dfrac{1}{3}$ **36** $\dfrac{2x}{5} - \dfrac{x}{15} = \dfrac{5}{9}$ **42** $\dfrac{3}{5} - \dfrac{x}{9} = \dfrac{2}{15} - \dfrac{2x}{45}$

31 $\dfrac{x}{2} - \dfrac{3}{7} = \dfrac{1}{2}$ **37** $\dfrac{3x}{4} + \dfrac{1}{3} = \dfrac{x}{2} + \dfrac{5}{8}$ **43** $\dfrac{4}{7} + \dfrac{2x}{9} = \dfrac{15}{9} - \dfrac{4x}{21}$

32 $\dfrac{3x}{5} + \dfrac{2}{9} = \dfrac{11}{15}$ **38** $\dfrac{2x}{7} - \dfrac{3}{4} = \dfrac{x}{14} + \dfrac{1}{2}$ **44** $\dfrac{x}{3} + \dfrac{1}{4} - \dfrac{x}{6} = \dfrac{7}{12}$

33 $\dfrac{5x}{6} + \dfrac{x}{8} = \dfrac{3}{4}$ **39** $\dfrac{5x}{7} - \dfrac{2}{3} = \dfrac{3}{7} - \dfrac{x}{3}$ **45** $\dfrac{5}{8} - \dfrac{x}{6} + \dfrac{1}{12} = \dfrac{3}{4}$

34 $\dfrac{3x}{4} + \dfrac{1}{8} = \dfrac{1}{2}$ **40** $\dfrac{2x}{9} - \dfrac{3}{4} = \dfrac{7}{18} - \dfrac{5x}{12}$ **46** $\dfrac{5}{9} - \dfrac{7x}{12} = \dfrac{1}{6} - \dfrac{x}{8}$

Problems

Exercise 6e

Form an equation for each of the following problems and then solve the equation.

A bag of sweets was divided into three equal shares. David had one share and he got 8 sweets. How many sweets were there in the bag?

Let x stand for the number of sweets in the bag.

One share is $\frac{1}{3}$ of x ∴ $\frac{1}{3}$ of $x = 8$

$$\dfrac{x}{3} = 8$$

Multiply each side by 3 $x = 24$

Therefore there were 24 sweets in the bag.

1. Tracy Brown came first in the St James, Barbados Golf Tournament and won $100. This was $\frac{2}{3}$ of the total prize money paid out. Find the total prize money.

> **Tip** Start by letting x or some other letter stand for the number you need to find.

2. Peter lost 8 marbles in a game. This number was one-fifth of the number that he started with. Find how many he started with.

3. The width of a rectangle is 12 cm. This is two-fifths of its length. Find the length of the rectangle.

4. I think of a number, halve it and the result is 6. Find the number that I first thought of.

5. The length of a rectangle is 8 cm and this is $\frac{1}{3}$ of its perimeter. Find its perimeter.

6. In an equilateral triangle, the perimeter is 15 cm. Find the length of one side of the triangle.

7. I think of a number, take $\frac{1}{3}$ of it and then add 4. The result is 7. Find the number I first thought of.

8. I think of a number and divide it by 3. The result is 2 less than the number I first thought of. Find the number I first thought of.

9. I think of a number and add $\frac{1}{3}$ of it to $\frac{1}{2}$ of it. The result is 10. Find the number I first thought of.

10. John Smith won the singles competition of a local tennis tournament, for which he got $\frac{1}{5}$ of the total prize money. He also won the doubles competition, for which he got $\frac{1}{20}$ of the prize money. He got $250 altogether. How much was the total prize money?

Investigation

Meg wanted to find out Malcolm's age without asking him directly.

The following conversation took place.

Meg: Think of your age but don't tell me what it is.

Malcolm: Right.

Meg: Multiply it by 5, add 4 and take away your age.

Malcolm: Yes.

Meg: Divide the result by 4 and tell me your answer.

Malcolm: 15

Meg: That means you are 14.

Malcolm: Correct. How did you know that?

Investigate whether this always works.

Use algebra in your investigation.

Directed numbers

Reminder:
$$(+2) \times (+3) = +6 \quad (+2) \times (-3) = -6$$
$$(-2) \times (+3) = -6 \quad (-2) \times (-3) = +6$$

Another way to remember these rules when multiplying directed numbers is like signs give positive, unlike signs give negative.

Division is the reverse of multiplication, e.g. $4 \div 4 = 1$, so $-4 \div (-4) = 1$

This means that dividing a negative number by a negative number gives a positive number,

e.g. $-8 \div (-2) = +4$

In the same way $(+8) \div (-2) = -4$

and $(-8) \div (+2) = -4$

Hence we can use the rules of multiplication of directed numbers for the division of directed numbers.

Exercise 6f

Evaluate:

1 $(+2) \times (-4)$ **4** $(-\frac{1}{2}) \times (+6)$ **7** $(+12) \div (-3)$

2 $(-3) \times (-5)$ **5** $(-16) \div (-8)$ **8** $(+\frac{1}{2}) \times (+\frac{2}{3})$

3 $(-6) \times (+4)$ **6** $(-4) \times (-7)$ **9** $(-6) \div (-2)$

Remember that the positive sign is often omitted, i.e. 6 means +6.

Simplify $4(x-3) - 3(2-3x)$

Multiply out the brackets. Remember that $-3(2-3x) = -3 \times 2 - 3x(-3x)$

$$4(x-3) - 3(2-3x) = 4x - 12 - 6 + 9x$$

Collect like terms
$$= 13x - 18$$

Simplify:

10 $7 - 2(x-5)$ **13** $4 - 7(2x-3)$ **16** $2(3x+5) - 2(4+3x)$

11 $2x + 5(3x-4)$ **14** $3x - 4(5-3x)$ **17** $5(2x-8) - 3(2-5x)$

12 $3x - 6(3x+5)$ **15** $3(x-4) + 6(3-2x)$ **18** $7(x-2) - (2x+3)$

Solve the following equations:

19 $4x - 2(x-3) = 8$

20 $7 - 3(5-2x) = 10$

21 $4x + 2(2x-5) = 6$

22 $3(x-4) - 7 = 2(x-3)$

Tip Start by multiplying out brackets, then collect like terms.

23 $4 - 3x = 3 + 4(2x - 3)$

24 $3x - 2(4 - 5x) = 5 - 3x$

25 $2x - \frac{1}{2}(6 + 2x) = 7$

26 $10 - \frac{1}{4}(4x - 8) = 5$

27 $3 - \frac{2}{3}(6x + 9) = 5 - 2x$

28 $\frac{3}{4}(4 - 8x) = 2x - \frac{2}{3}(6 - 12x)$

 ## Puzzle

My dog weighs nine-tenths of its weight plus nine-tenths of a kilogram. What does it weigh?

Formulae

For all rectangles it is true that the area is equal to the length multiplied by the breadth, provided that the length and breadth are measured in the same units.

If we use letters for the unknown quantities (A for area, l for length, b for breadth) we can write the first sentence more briefly as a formula: $A = l \times b$.

The multiplication sign is usually left out giving

$$A = lb$$

Exercise 6g

The letters in the diagrams all stand for a number of centimetres.

The perimeter of the square below is P cm. Write down a formula for P.

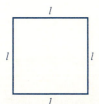 Start by writing the perimeter in terms of the letters in the diagram: this is $l + l + l + l$ (cm). As we are told that P cm is the perimeter we can write

$$P = l + l + l + l$$

Collect like terms $P = 4l$

In each of the following figures the perimeter is P cm. Write down a formula for P starting with $P =$

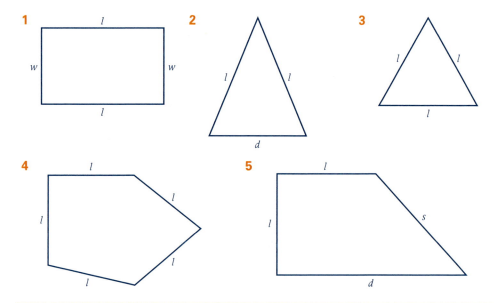

If G is the number of girls in a class and B is the number of boys, write down a formula for the total number, T, of children in the class.

Start by writing the total number in terms of the letters other than T.

$$T = G + B$$

6 I buy x lb of apples and y lb of pears. Write down a formula for W if W lb is the mass of fruit that I have bought.

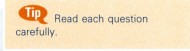 Read each question carefully.

7 If l m is the length of a rectangle and b m is the breadth, write down a formula for P if the perimeter of the rectangle is P m.

8 I start a game with N marbles and win another M marbles. Write down a formula for the number, T, of marbles that I finish the game with.

9 I start a game with N marbles and lose L marbles. Write down a formula for the number, T, of marbles that I finish with.

10 The side of a square is l m long. Write down a formula for A if the area of the square is A m^2.

11 Peaches cost n cents each. Write down a formula for N if the cost of 10 peaches is N cents.

12 Oranges cost x cents each and I buy n of these oranges. Write down a formula for C where C cents is the total cost of the oranges.

13 I have a piece of string which is l cm long. I cut off a piece which is d cm long. Write down a formula for L if the length of string which is left is L cm.

14 A rectangle is $2l$ m long and l m wide. Write down a formula for P where P m is the perimeter of the rectangle.

15 Write down a formula for A where $A\,m^2$ is the area of the rectangle described in question 14.

16 I had a bag of sweets with S sweets in it; I then ate T of them. Write down a formula for the number, N, of sweets left in the bag.

17 A lorry has mass T tonnes when empty. Steel girders with a total mass of S tonnes are then loaded on to the lorry. Write down a formula for W where W tonnes is the mass of the loaded lorry.

18 I started the term with a new packet of N felt-tipped pens. During the term I lost L of them and R of them ran dry. Write down a formula for the number, S, that I had at the end of the term.

19 A truck travels p km in one direction and then it comes back q km in the opposite direction. If it is then r km from its starting point, write down a formula for r.

20 One box of tinned fruit has mass K kg. The mass of n such boxes is W kg. Write down a formula for W.

21 Two points have the same y-coordinate. The x-coordinate of one point is a and the x-coordinate of the other point is b. If d is the distance between the two points, write down a formula for d given that a is less than b. Make a sketch to illustrate this problem.

22 A letter costs x cents to post. The cost of posting 20 such letters is $\$q$. Write down a formula for q.

Tip Be careful – look at the units given.

23 One grapefruit costs y cents. The cost of n such grapefruit is $\$L$. Write down a formula for L.

Tip Look carefully at the units.

24 A rectangle is l m long and b cm wide. The area is A cm^2. Write down a formula for A.

25 On my way to work this morning the bus I was travelling on broke down. I spend t hours on the bus and s minutes walking. Write down a formula for T if the total time that my journey took was T hours.

Substituting numerical values into a formula

The formula for the area of a rectangle is $A = lb$.

If a rectangle is 3 cm long and 2 cm wide, we can substitute the number 3 for l and the number 2 for b to give $A = 3 \times 2 = 6$.

So the area of that rectangle is 6 cm^2.

When you substitute numerical values into a formula you may have a mixture of operations, i.e. ( ), $\times$, $\div$, $+$, $-$, to perform. Remember the order from the capital letters of 'Bless My Dear Aunt Sally'.

Exercise 6h If $v = u + at$, find v when $u = 2$, $a = \frac{1}{2}$ and $t = 4$.

$$v = u + at$$

When $u = 2$, $a = \frac{1}{2}$, $t = 4$, $\quad v = 2 + \frac{1}{2} \times 4 \quad$ (Do multiplication first)

$$= 2 + 2$$

$$= 4$$

1 If $N = T + G$, find N when $T = 4$ and $G = 6$.

2 If $T = np$, find T when $n = 20$ and $p = 5$.

3 If $P = 2(l + b)$, find P when $l = 6$ and $b = 9$.

4 If $L = x - y$, find L when $x = 8$ and $y = 6$.

5 If $N = 4(l - s)$, find N when $l = 7$ and $s = 2$.

6 If $S = n(a + b)$, find S when $n = 20$, $a = 2$ and $b = 8$.

7 If $V = lbw$, find V when $l = 4$, $b = 3$ and $w = 2$.

8 If $A = \dfrac{PRT}{100}$, find A when $P = 100$, $R = 3$ and $T = 5$.

9 If $w = u(v - t)$, find w when $u = 5$, $v = 7$ and $t = 2$.

10 If $s = \frac{1}{2}(a + b + c)$, find s when $a = 5$, $b = 7$ and $c = 3$.

If $v = u - at$, find v when $u = 5$, $a = -2$, $t = -3$

$$v = u - at$$

When $u = 5$, $a = -2$, $t = -3$, $\quad v = 5 - (-2) \times (-3)$

$$= 5 - (+6)$$

$$= 5 - 6$$

$$= -1$$

(Notice that where negative numbers are substituted for letters they have been put in brackets. This makes sure that only one operation at a time is carried out.)

11 If $N = p + q$, find N when $p = 4$ and $q = -5$.

12 If $C = RT$, find C when $R = 4$ and $T = -3$.

Tip Put negative numbers in brackets.

13 If $z = w + x - y$, find z when $w = 4$, $x = -3$ and $y = -4$.

14 If $r = u(v - w)$, find r when $u = -3$, $v = -6$ and $w = 5$.

15 Given that $X = 5(T - R)$, find X when $T = 4$ and $R = -6$.

16 Given that $P = d - rt$, find P when $d = 3$, $r = -8$ and $t = 2$.

17 Given that $v = l(a + n)$, find v when $l = -8$, $a = 4$ and $n = -6$.

18 If $D = \dfrac{a - b}{c}$, find D when $a = -4$, $b = -8$ and $c = 2$.

19 If $Q = abc$, find Q when $a = 3$, $b = -7$ and $c = -5.9$

20 If $l = \frac{2}{3}(x + y - z)$, find l when $x = 4$, $y = -5$ and $z = -6$.

Given that $2S = d(a + l)$, find a when $S = 20$, $d = 2$ and $l = 16$

$$2S = d(a + l)$$

Substituting $S = 20$, $d = 2$, $l = 16$ gives

$$40 = 2(a + 16)$$

(We can now solve this equation for a.)

Multiply out the brackets $\qquad\qquad 40 = 2a + 32$

Take 32 from each side $\qquad\qquad\quad 8 = 2a$

Divide by 2 $\qquad\qquad\qquad\qquad 4 = a$ or $a = 4$

21 Given that $N = G + B$, find B when $N = 40$ and $G = 25$.

22 If $R = t \div c$, find t when $R = 10$ and $c = 20$.

23 Given that $d = st$, find t when $d = 50$ and $s = 15$.

24 If $N = 2(p + q)$, find q when $N = 24$, and $p = 5$.

25 Given that $L = P(2 - a)$, find a when $L = 10$ and $P = 40$.

26 Given that $s = \frac{1}{3}(a - b)$, find b when $s = 15$ and $a = 24$.

27 Given that $v = u + at$, find u when $v = 32$, $a = 8$ and $t = 4$.

28 If $v^2 = u^2 + 2as$, find a when $v = 3$, $u = 2$ and $s = 12$.

29 If $d = \frac{1}{2}(a + b + c)$, find a when $d = 16$, $b = 4$ and $c = -3$.

30 If $H = P(Q - R)$, find Q when $H = 12$, $P = 4$ and $R = -6$.

Problems

1 Given that $v = at$, find the value of

    a v when $a = 4$ and $t = 12$ **b** v when $a = -3$ and $t = 6$

    c t when $v = 18$ and $a = 3$ **d** a when $v = 25$ and $t = 5$

2 Given that $N = 2(n - m)$, find the value of

    a N when $n = 6$ and $m = 4$ **b** N when $n = 7$ and $m = -3$

    c n when $N = 12$ and $m = 2$ **d** m when $N = 16$ and $n = -4$

3 If $A = P + QT$, find the value of

    a A when $P = 50$, $Q = \frac{1}{2}$ and $T = 4$

    b A when $P = 70$, $Q = 5$ and $T = -10$

    c P when $A = 100$, $Q = \frac{1}{4}$ and $T = 16$

    d T when $A = 25$, $P = -15$ and $Q = -10$

4 Given that $s = \frac{1}{2}(a - b)$, find the value of

    a s when $a = 16$ and $b = 6$ **b** s when $a = -4$ and $b = -10$

    c a when $s = 15$ and $b = 8$ **d** b when $s = 10$ and $a = -4$

5 Given that $z = x - 3y$, find the value of

    a z when $x = 3\frac{1}{2}$ and $y = \frac{3}{4}$ **b** z when $x = \frac{3}{8}$ and $y = -1\frac{1}{2}$

    c x when $z = 5\frac{1}{3}$ and $y = 2\frac{1}{2}$ **d** y when $z = \frac{1}{4}$ and $x = \frac{7}{8}$

6 If $P = 100r - t$, find the value of

    a P when $r = 0.25$ and $t = 10$ **b** P when $r = 0.145$ and $t = 15.6$

    c r when $P = 18.5$ and $r = 0.026$ **d** t when $P = 50$ and $t = -12$

A rectangle is $3l$ cm long and l cm wide. If the area of the rectangle is A cm^2, write down a formula for A.

Use your formula to find the area of this rectangle if it is 5 cm wide.

$$\text{Area} = \text{length} \times \text{width}$$

$$\therefore \quad A = 3l \times l$$

$$A = 3l^2$$

When $l = 5$, $A = 3 \times 5 \times 5$

$$= 75$$

$$\therefore \quad \text{Area} = 75 \text{ cm}^2$$

7 Oranges cost n c each. If the cost of a box of 50 of these oranges is C cents, write down a formula for C. Use your formula to find the cost of a box of oranges if each orange costs 12 c.

8 Lemons cost n c each. The cost of a box of 50 lemons is \$$L$. Write down a formula for L. Use your formula to find the cost of a box of these lemons when they cost 10 c each.

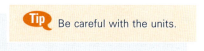
Tip Be careful with the units.

9 A rectangular box is l cm long, b cm wide and d cm deep. The volume of the box is V cm³. Write down a formula for V. Use your formula to find the volume of a box measuring 20 cm by 12 cm by 5 cm.

10 A rectangle is a cm long and b cm wide. Write down a formula for P if P cm is the perimeter of the rectangle. Use your formula to find the perimeter of a rectangle measuring 20 cm by 15 cm.

11 The length of a rectangle is twice its width. If the rectangle is x cm wide, write down a formula for P if its perimeter is P cm. Use your formula to find the width of a rectangle that has a perimeter of 36 cm.

12 A roll of paper is L m long. N pieces each of length r m are cut off the roll. If the length of paper left is P m, write down a formula for P. Use your formula to find the length of paper left from a roll that was 20 m long after 10 pieces, each of length 1.5 m, are cut off.

13 An equilateral triangle has sides each of length a cm. If the perimeter of the triangle is P cm, write down a formula for P. Use your formula to find the lengths of the sides of an equilateral triangle whose perimeter is 72 cm.

14 Tins of baked beans weigh a g each. N of these tins are packed into a box. The empty box weighs p g. Write down a formula for W where W g is the weight of the full box. Use your formula to find the number of tins that are in a full box if the full box weighs 10 kg, the empty box weighs 1 kg and each tin weighs 200 g.

15 The rectangular box in the diagram is l cm long, w cm wide and h cm high. Write down a formula for A if A cm² is the total surface area of the box (i.e. the area of all six faces). Use your formula to find the surface area of a rectangular box measuring 50 cm by 30 cm by 20 cm.

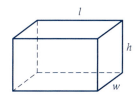

16 A person whose weight on Earth is W finds his weight on certain planets from these formulae.

a Weight on Venus 0.85 W

b Weight on Mars 0.38 W

c Weight on Jupiter 2.64 W.

Calculate your weight on each of the above planets.

Changing the subject of a formula

Suppose that we have to use the formula $A = lb$ to find the value of l when $A = 20$ and $b = 5$. There are two ways of doing this.

Either we can substitute the numbers directly, giving $20 = l \times 5$, and solve this equation for l, which gives $l = 4$

or, by dividing both sides of the formula by b, we can rearrange the formula to $l = \dfrac{A}{b}$,

then substitute in the numbers to give $l = \frac{20}{5} = 4$

When the formula is in the form $A = lb$, A is called the subject of the formula.

When the formula is in the form $l = \dfrac{A}{b}$, l is called the subject of the formula.

Changing from $A = lb$ to $l = \dfrac{A}{b}$ is called changing the subject of the formula.

Exercise 6j

Make r the subject of the formula $p = q + r$

(To make r the subject of $p = q + r$ we have to 'solve' the formula for r. This means we have to isolate r on one side of the '=' sign.)

$$p = q + r$$

Take q from both sides $\qquad p - q = r$

$\qquad\qquad\qquad\qquad\qquad$ or $r = p - q$

Make the letter in brackets the subject of the following formulae:

1	$N = T + G$	(T)		**11**	$P = a + b$	(a)
2	$z = xy$	(x)		**12**	$N = R + T$	(T)
3	$S = \dfrac{d}{t}$	(d)		**13**	$b = a + c + d$	(c)
4	$L = X - Y$	(X)		**14**	$v = rt + u$	(u)
5	$s = a + 2b$	(a)		**15**	$N = rn$	(n)
6	$v = u + t$	(u)		**16**	$x = y - z$	(y)
7	$S = d - t$	(d)		**17**	$P = ab + c$	(c)
8	$P = 2y + z$	(z)		**18**	$L = \dfrac{m}{n}$	(m)
9	$C = RT$	(T)		**19**	$v = u + at$	(u)
10	$L = a + b + c$	(a)		**20**	$s = ax + y$	(y)

Expressions, equations and formulae

In this chapter you have worked with expressions, equations and formulae.

Remember that:

An **expression** is a collection of one or more algebraic terms, for example $2x$, $5x + 2y$, $a^2 - 4b$ and $6(2x - 3)$ are expressions.

An **equation** is an equality between two expressions, for example $2x = 4$ and $y + 2 = 3x + 1$ are equations.

A **formula** is a general rule for finding one quantity in terms of other quantities, for example the formula for finding the area, $A \text{ cm}^2$, of a rectangle measuring l cm by b cm is $A = l \times b$. A is called the subject of the formula.

($A = l \times b$ is also an equation.)

Exercise 6k

For each question write down whether what is given is an expression, an equation or a formula.

1   $P = 2(l + b)$

2   $5x - 1 = 4x + 2$

3   $5x - 9 - \frac{1}{2}x = 0$

4   $4(a - 3) + 2(b + 4)$

5   $y = \frac{1}{2}(3x + 3z)$

6   $3(x - 2) + 8$

7   $5a + 2b - 3(a - b)$

8   $\frac{1}{3}(x + 3) = 4$

9   $3x + 2y = 6$

10   $A = \pi r^2$

11   $r = C/2\pi$

12   $4(7a + 4)$

Mixed exercises

Exercise 6l

1   Solve the equation $8 = 3 + 2x$.

2   Solve the equation $x - 4 = 5 - 2x + 1$.

3   Multiply out $3(2x - 8)$.

4   Find $\frac{3}{5}$ of $10x$.

5   Solve the equation $\dfrac{2x}{3} = 8$.

6   Find the value of x if $\dfrac{x}{2} + \dfrac{1}{6} = \dfrac{1}{3}$.

7   Simplify the expression $3x - 2(4 - x)$.

8   Write down a formula for P if P cm is the perimeter of the figure in the diagram. (Each letter stands for a number of centimetres.)

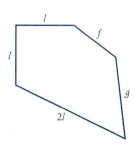

9 If $P = a - b$, find the value of P when $a = 2$ and $b = 5$.

10 Make N the subject of the formula $R = N - D$.

Exercise **6m**

1 Solve the equation $3 - x = 2 + 2x$.

2 Solve the equation $3(2x + 2) = 10$.

3 Simplify $\frac{3}{4} \times 8x$.

4 Simplify $5x \div \frac{2}{3}$.

5 Solve the equation $\dfrac{3x}{5} = \dfrac{9}{10}$.

6 Simplify $6(3 - 2x) - 4(2 - x)$.

7 Solve the equation $\dfrac{x}{4} - \dfrac{3}{5} = \dfrac{7}{8}$.

8 If $z = x - 2y$, find z when $x = 3$ and $y = -6$.

9 There are three classes in the first year of Appletown School. There are a children in one class, b children in another class and c children in the third class. Write down a formula for the number, N, of children in the first year.

10 Make N the subject of the formula $n = N - ab$.

Exercise **6n**

1 Find $\frac{3}{8}$ of $10x$.

2 Solve the equation $5(3 - 4x) = x - 2(3x - 5)$.

3 I think of a number and double it, then I add on 3 and double the result: this gives 14. If x stands for the number I first thought of, form an equation for x and then solve it.

4 Simplify $\dfrac{3x}{4} \div \dfrac{9}{11}$.

5 Find $\frac{3}{8}$ of $\frac{2}{3}x$.

6 Simplify the expression $5x - \frac{2}{3}(6x - 9)$.

7 Solve the equation $\dfrac{3}{8} - \dfrac{5x}{6} = \dfrac{2}{3}$.

8 Given that $r = s - vt$, find the value of r when $s = 4$, $v = 3$ and $t = -2$.

9 A rectangle is twice as long as it is wide. If it is a cm wide, write down a formula for P where P cm is the perimeter of the rectangle.

10 Make p the subject of the formula $L = 3pq$.

MATHS IS OUT THERE

Did you know that Cardan (1501–1576) was one of the greatest 16th century mathematicians? He was also considered to be a very strange man. There is a story that he cut off the ears of his youngest son when the boy was bad. Cardan 'read the stars', and claimed that he could, through them, tell the exact day of his death. When the day he had predicted came in 1576, Cardan made his claim come true by killing himself.

IN THIS CHAPTER...

you have seen that:

- to solve a linear equation where the unknown quantity is on both sides of the equals sign, collect the terms containing the unknown on the side where there are more of them

- when an equation contains brackets, multiply these out first

- fractions can be eliminated from an equation by multiplying both sides by the lowest common multiple of the denominators

- when you substitute numerical values into a formula, place negative numbers in brackets

- the subject of a formula is the letter on its own on one side of the equals sign

- to change the subject of a formula, solve it for the letter that is to be the new subject.

7 PROBABILITY

AT THE END OF THIS CHAPTER...

you should be able to:

1 Draw a possibility space for two events and use it to find probabilities.

2 Construct and use tree diagrams to find probabilities.

3 Estimate the probability of certain events by experiment.

Did you know that the French mathematician Francois Viete (1540–1603) was the first mathematician to use letters of the alphabet to denote both known and unknown quantities? The best known example today of a letter that stands for a number is the Greek letter π.

BEFORE YOU START

you need to know:
- ✓ how to simplify a fraction
- ✓ how to add and subtract fractions
- ✓ what an ordinary pack of playing cards consists of

KEY WORDS

approximation, equally likely, possibility space, probability tree, random, tree diagram, unbiased

Single events

If we toss an unbiased die, each of the six numbers is equally likely to appear. We say that there are six equally likely *outcomes* to this experiment. The probability of scoring 2 is $\frac{1}{6}$, because out of six equally likely outcomes, only one is 'successful', i.e. is a 2

We write
$$P(2) = \frac{1}{6}$$

Similarly
$$P(\text{even number}) = \frac{3}{6} = \frac{1}{2}$$

$$P(10) = 0$$

$$P(1 \text{ or } 2 \text{ or } 3 \text{ or } 4 \text{ or } 5 \text{ or } 6) = \frac{6}{6} = 1$$

$$P(\text{not } 2) = 1 - \frac{1}{6}$$

To summarise, a successful event can be denoted by A and the probability of it happening is then $P(A)$. If A is 'throwing a 2' then $P(A) = \frac{1}{6}$.

$$P(\text{successful event}) = \frac{\text{number of successful outcomes}}{\text{total number of possible outcomes}}$$

$P(\text{certainty}) = 1$

$P(\text{impossibility}) = 0$

$0 \leqslant P(A) \leqslant 1$

$P(\text{not } A) = 1 - P(A)$

Possibility space for two events

Suppose a 5 c coin and a 10 c coin are tossed together. One possibility is that the 5 c coin will land head up and that the 10 c coin will also land head up.

If we use H for a head on the 5 c coin and *H* for a head on the 10 c coin, we can write this possibility more briefly as the ordered pair (H, *H*).

To list all the possibilities, an organised approach is necessary, otherwise we may miss some. We use a table called a *possibility space*. The possibilities for the 10 c coin are written across the top and the possibilities for the 5 c coin are written down the side:

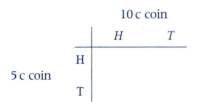

When both coins are tossed we can see all the combinations of heads and tails that are possible and then fill in the table.

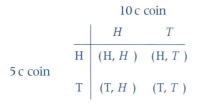

10 c coin

	H	T
H	(H, H)	(H, T)
T	(T, H)	(T, T)

5 c coin

Exercise 7a

1 Two bags each contain 3 white counters and 2 black counters. One counter is removed at random from each bag. Copy and complete the following possibility space for the possible combination of two counters.

1st bag

	○	○	○	●	●
○	(○,○)	(○,○)	(○,○)	(○,●)	
○					
○					
●					(●,●)
●	(●,○)				

2nd bag

2 An ordinary six-sided die is tossed and a 10 c coin is tossed. Copy and complete the following possibility space.

Die

	1	2	3	4	5	6
H		(H, 2)				
T				(T, 4)		

10 c coin

3 One bag contains 2 red counters, 1 yellow counter and 1 blue counter. Another bag contains 2 yellow counters, 1 red counter and 1 blue counter. One counter is taken at random from each bag. Copy and complete the following possibility space.

1st bag

	R	R	Y	B
R		(R, R)		
Y				(Y, B)
Y				
B	(B, R)			

2nd bag

4 A top like the one in the diagram is spun twice. Copy and complete the possibility space.

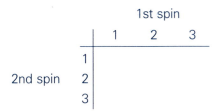

1st spin

	1	2	3
1			
2			
3			

2nd spin

5 A boy goes into a shop to buy a pencil and a rubber. He has a choice of a red, a green or a yellow pencil and a round, a square or a triangular shaped rubber. Make your own possibility space for the possible combinations of one pencil and one rubber that he could buy.

Using a possibility space

When there are several entries in a possibility space it can take a long time to fill in the ordered pairs. To save time we use a cross in place of each ordered pair. We can see which ordered pair a particular cross represents by looking at the edges of the table.

Exercise 7b Two ordinary six-sided dice are tossed. Draw up a possibility space showing all the possible combinations in which the die may land.

Use the possibility space to find the probability that a total score of at least 10 is obtained.

1st die

2nd die		1	2	3	4	5	6
	1	×	×	×	×	×	×
	2	×	×	×	×	×	×
	3	×	×	×	×	×	×
	4	×	×	×	×	×	⊗
	5	×	×	×	×	⊗	⊗
	6	×	×	×	⊗	⊗	⊗

(There are 36 entries in the table and 6 of these give a score of 10 or more.)

$$P(\text{score of at least }10) = \frac{6}{36} = \frac{1}{6}$$

1 Use the possibility space in the example above to find the probability of getting a score of

a 4 or less   **b** 9   **c** a double.

2 Use the possibility space for question 1 of Exercise 7a to find the probability that the two counters removed

   a are both black   **b** contain at least one black.

3 Use the possibility space for question 2 of Exercise 7a to find the probability that the coin lands head up and the die gives a score that is less than 3.

4 Use the possibility space for question 3 of Exercise 7a to find the probability that the two counters removed are

   a both blue   **b** both red

   c one blue and one red   **d** such that at least one is red.

5 A 5 c coin and a 1 c coin are tossed together. Make your own possibility space for the combinations in which they can land. Find the probability of getting two heads.

6 A six-sided die has two of its faces blank and the other faces are numbered 1, 3, 4 and 6. This die is tossed with an ordinary six-sided die (faces numbered 1, 2, 3, 4, 5, 6). Make a possibility space for the ways in which the two dice can land and use it to find the probability of getting a total score of

   a 6   **b** 10   **c** 1   **d** at least 6.

7 One bag of coins contains three 10 c coins and two 25 c coins. Another bag contains one 10 c coin and one 25 c coin. One coin is removed at random from each bag. Make a possibility space and use it to find the probability that a 25 c coin is taken from each bag.

8 One bookshelf contains two storybooks and three textbooks. The next shelf holds three storybooks and one textbook. Draw a possibility space showing the various ways in which you could pick up a pair of books, one from each shelf. Use this to find the probability that

   a both books are storybooks   **b** both are textbooks.

9 The four aces and the four kings are removed from an ordinary pack of playing cards. One card is taken from the set of four aces and one card is taken from the set of four kings. Make a possibility space for the possible combinations of two cards and use it to find the probability that the two cards

   a are both black   **b** are both spades

   c include at least one black card   **d** are both of the same suit.

 ## Puzzle

A man asked a farmer how many animals he had. The farmer replied, "They're all cows but 2, all sheep but 2 and all pigs but 2."

How many animals did the farmer own?

Addition of probabilities

If we select a card at random from a pack of 52, the probability of drawing a jack is $\frac{4}{52}$ and the probability of drawing a red four is $\frac{2}{52}$.

There are 4 jacks and 2 red fours so if we want to find the probability of drawing either a jack or a red four there are 6 cards that we would count as 'successful'.

The probability of drawing a jack or a red four is therefore $\frac{6}{52}$, which is the same as $\frac{4}{52} + \frac{2}{52}$, i.e. the *sum* of the separate probabilities.

> We add probabilities if there are several *separate* events
> we would count as 'successful'.

Notice though that the probability of drawing a jack or a heart is not $\frac{4}{52} + \frac{13}{52}$ because the jack of hearts is in both groups. The two events are not completely separate.

Sometimes it is useful to draw a diagram to represent the information about the probabilities.

If a box contains 3 red, 4 blue and 2 white beads and we draw one bead at random then $P(\text{red}) = \frac{3}{9}$, $P(\text{blue}) = \frac{4}{9}$ and $P(\text{white}) = \frac{2}{9}$; this information can be shown on a diagram.

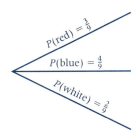

If we want either a red or a blue bead then we *add* the probabilities,

i.e. $$P(\text{red or blue}) = \frac{3}{9} + \frac{4}{9} = \frac{7}{9}$$

Notice that all three probabilities add up to 1

i.e. $$P(\text{red or blue or white}) = \frac{3}{9} + \frac{4}{9} + \frac{2}{9}$$

$$= \frac{9}{9} = 1$$

We cannot draw a diagram to show the probabilities of picking a jack or a heart because they are not completely separate events.

This type of diagram is called a *tree diagram*; later we use it to show more information and it grows more branches.

Finding the number of outcomes

If we know that there are 24 beads in a bag and that the probability of drawing a red bead is $\frac{3}{8}$ then we know that the number of red beads is $\frac{3}{8}$ of 24, i.e. there are 9 red beads.

Exercise 7c

In a prize draw, the prizes are grocery hampers and garden centre vouchers. The possibility that the first ticket drawn is for a hamper is $\frac{1}{200}$ and the probability that it is for a voucher is $\frac{3}{200}$.

What is the probability that the first ticket is for

a either a hamper or a voucher

b neither of these?

a The prizes of a hamper or a voucher are separate events, so we can add the probabilities

$$P(\text{hamper or voucher}) = \frac{1}{200} + \frac{3}{200} = \frac{4}{200} = \frac{1}{50}$$

b Remember that $P(\text{event does not happen}) = 1 - P(\text{event happens})$

$$P(\text{neither a hamper nor a voucher}) = 1 - \frac{1}{50} = \frac{49}{50}$$

---

1 A card is drawn at random from the 12 court cards (jacks, queens and kings).

What is the probability that the card is

a a red queen   **b** a black king

c either a red queen or   black king?

> **Tip** $P(\text{event A or event B}) = P(\text{event A}) + P(\text{event B})$ as long as A and B are separate events.

2 Sophy is looking for her keys. The probability that she has put them in a pocket is $\frac{2}{9}$ and the probability that she has put them in her bag is $\frac{1}{3}$. Find the probability that

a she has put them in a pocket or her bag

b she has put them somewhere else.

3 The probability that Sean wins a race is $\frac{1}{10}$ and that Ewan wins is $\frac{1}{5}$.

There are no dead heats.

What is the probability that

a either Sean or Ewan wins the race

b some other person wins the race?

The probability that the Blackwells' newspaper will be delivered before 8 a.m. is $\frac{3}{16}$, between 8 and 9 a.m. is $\frac{1}{2}$ and after 9 a.m. is $\frac{5}{16}$.

a Draw a tree diagram to show this information.

b Find the probability that the paper will be delivered at or after 8 a.m.

a

$P(\text{before 8 a.m.}) = \frac{3}{16}$

$P(\text{8 a.m.–9 a.m.}) = \frac{1}{2}$

$P(\text{after 9 a.m.}) = \frac{5}{16}$

b The results along two of the branches are acceptable so we add the probabilities.

$$P(\text{at or after 8 a.m.}) = \frac{1}{2} + \frac{5}{16}$$

$$= \frac{13}{16}$$

In questions **4** to **6**, draw a tree diagram to show the given information.

4 Tony is waiting at a bus stop, where only buses numbered 4, 25 or 72 stop.

The probability is $\frac{2}{7}$ that the next bus is a 4, $\frac{3}{7}$ that it is a 25 and $\frac{2}{7}$ that it is a 72.

Find the probability that the next bus is

a either a 4 or a 25 **b** either a 4 or a 72.

5 When I draw the curtains in the morning the probability is $\frac{1}{8}$ that the first bird I see is a blackbird, $\frac{1}{4}$ that it is a blue tit and $\frac{5}{8}$ that it is a sparrow.

Find the probability that the first bird I see is

a either a blackbird or a sparrow **b** either a blue tit or a blackbird.

6 Danny goes to work by bus or cycle or on foot. The probability that he chooses to go by bus is $\frac{5}{12}$, on his cycle is $\frac{1}{4}$ and on foot is $\frac{1}{3}$.

Find the probability that on Monday he chooses to go

a by bus or cycle **b** by bus or on foot.

A board is divided into 16 small squares and some of these are coloured grey.

If a small square is picked at random, the probability that it is grey is $\frac{3}{4}$.

How many square are grey?

$$\text{Number of grey squares} = \frac{3}{4} \times 16$$

$$= 12$$

7 The probability of drawing a king from a handful of 12 cards is $\frac{1}{6}$.

How many kings are there in the hand?

Tip Remember that the number of outcomes = P(outcome happens) ×(number of times the experiment is repeated).

8 In a car park there is a probability of $\frac{2}{7}$ that a car picked at random is red.

There are 112 cars in the car park. How many of them are not red?

9 From a handful of cards the probability of drawing a club at random is $\frac{2}{3}$.

There are 6 clubs. How many other cards are there?

10 In a cupboard there are only history and geography textbooks.

There is a probability of $\frac{5}{8}$ that a book taken at random is on history.

There are 30 geography books. How many history books are there?

Probability trees

Suppose we have two discs, a red one marked A on one side and B on the other, and a blue one marked E on one side and F on the other.

Tossing the red disc, the probability that we get A is $\frac{1}{2}$ and the probability that we get B is also $\frac{1}{2}$.

$P(A) = \frac{1}{2}$

$P(B) = \frac{1}{2}$

Red disc

This information can be shown in the tree diagram on the right.

Suppose that the red disc showed A and we go on to toss the blue disc. The probability of getting E is $\frac{1}{2}$ and the probability of getting F is $\frac{1}{2}$. We put this information on the diagram.

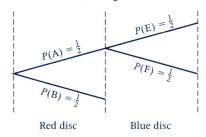

$P(E) = \frac{1}{2}$

$P(A) = \frac{1}{2}$

$P(F) = \frac{1}{2}$

$P(B) = \frac{1}{2}$

Red disc   Blue disc

We complete the diagram by considering what the probabilities are, supposing that the red disc shows a B before we toss the blue disc.

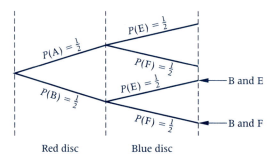

$P(E) = \frac{1}{2}$

$P(A) = \frac{1}{2}$

$P(F) = \frac{1}{2}$

$P(E) = \frac{1}{2}$ B and E

$P(B) = \frac{1}{2}$

$P(F) = \frac{1}{2}$ B and F

Red disc   Blue disc

To use the tree diagram to find the probability that we get an A and then an E, follow the path from left to right for an A on the first branch and an E on the second. The two probabilities we find there are $\frac{1}{2}$ and $\frac{1}{2}$. Multiply them together to get $\frac{1}{4}$.

The reason for multiplying the two probabilities together can be seen if we consider tossing the red disc, say 100 times. We would expect to get the side A 50 times. On half of these occasions we would expect to go on to get side E on the blue disc.

To find the probability that we get a B on the red disc and an F on the blue one, follow the B and F path and multiply the probabilities, i.e.

$$P(\text{B and F}) = \tfrac{1}{2} \times \tfrac{1}{2} = \tfrac{1}{4}$$

In general we multiply the probabilities when we follow the path along the branches of the probability tree.

Exercise 7d

Two coins are tossed, one after the other. Find the probability that

a both coins show heads

b the first shows a tail and the second a head.

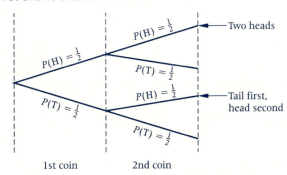

1st coin   2nd coin

a $P(\text{two heads}) = \tfrac{1}{2} \times \tfrac{1}{2} = \tfrac{1}{4}$

b $P(\text{first tail, second head}) = \tfrac{1}{2} \times \tfrac{1}{2} = \tfrac{1}{4}$

In each question, draw a probability tree to show the given information.

1 The probability that Mark gets to work on time is $\frac{7}{8}$ and the probability that he leaves work on time is $\frac{3}{5}$.

a Find the probability that he does not leave work on time.

b Copy and extend the given probability tree.

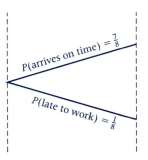

Getting to work   Leaving work

What is the probability that

c Mark gets to work on time but does not leave on time

d Mark is late for work but leaves on time?

2 When a drawing pin falls to the ground the probability that it lands point up is $\frac{1}{4}$.

a Find the probability that a pin does not land point up.

Two drawing pins fall one after the other. Find the probability that

b both drawing pins land point up

c both drawing pins do not land point up.

3 The first of two boxes of tennis balls contains one white and two yellow; the second box contains three yellow and two lime green. A ball is taken at random from each box.

Find the probability that

a both balls are yellow **b** one is white and one is lime green.

4 **a** If a die is rolled what is the probability of getting

i a six **ii** a number other than six?

b Two dice, a red and a blue, are rolled. Find the probability that

i both dice show sixes

ii the red die gives a six but the blue die does not

iii the blue die gives a six but the red die does not.

c What is the total probability that just one six appears?

5 On the way home Larry passes two sets of traffic lights. The probability that the first set are green is $\frac{3}{5}$ and the probability that the second set are green is $\frac{2}{3}$.

Find the probability that he has to stop at

a just one set of lights **b** at least one set of lights

c both sets of lights.

6 The probabilities that Alice and Bernard can solve a problem are $\frac{1}{2}$ and $\frac{2}{3}$ respectively.

Find the probability that

a both can solve it **b** just one of them can solve it

c neither can solve it.

Combining probabilities

In the worked example on page 123, we see that there are two ways in which a tail and a head can appear.

We have to follow two paths along the branches.

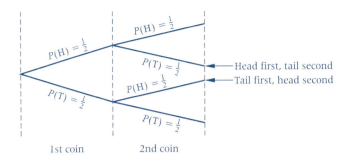

We *add* the probabilities resulting from these paths. Each new possible way of achieving the event increases the PROBABILITY.

In general:

We *multiply* the probabilities when we follow a path along the branches of the probability tree and *add* the results of following several paths.

The first of two parcels contains 3 French books and 2 German books. The second parcel contains 1 French book and 3 German books. Two books are taken at random, one from each parcel.

What is the probability that one book is French and one German?

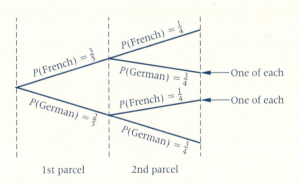

(As the order of choosing the books does not matter, there are two paths on the tree that give a French and a German book.)

$P(\text{one of each}) = \frac{3}{5} \times \frac{3}{4} + \frac{2}{5} \times \frac{1}{4}$

$\qquad\qquad\qquad = \frac{9}{20} + \frac{2}{20}$

$\qquad\qquad\qquad = \frac{11}{20}$

For each question, draw a probability tree to illustrate the given information.

1 In a group of 6 girls, four are fair and two are dark. Of 5 boys, two are fair and three are dark. One boy and one girl are picked at random. What is the probability that, of the two pupils, one is fair and one is dark?

2 In a class of 20, four are left-handed. In a second class of 24, six are left-handed. One pupil is chosen at random from each class.

What is the probability that one of the pupils is left-handed and one is not?

3 Derek and Alexis keep changing their minds about whether to send Christmas cards to each other. In any one year, the probability that Derek sends a card is $\frac{3}{4}$ and that Alexis sends one is $\frac{5}{6}$.

Find the probability that next year

a they both send cards   **b** only one of them sends a card

c neither sends a card.

What should the three answers add up to?

4 Copy the probability tree in the worked example in Exercise 7d and, by adding more branches to the right, extend it to show the following information.

Three unbiased coins are tossed, one after the other.

Find the probability that

a three heads appear   **b** three tails appear

c two heads and one tail appear in any order.

Finding probability by experiment

We have assumed that if you toss a coin it is equally likely to land head up or tail up so that $P(\text{a head}) = \frac{1}{2}$. Coins like this are called 'fair' or 'unbiased'.

Most coins are likely to be unbiased but it is not necessarily true of all coins. A particular coin may be slightly bent or even deliberately biased so that there is not an equal chance of getting a head or a tail.

The only way to find out if a particular coin is unbiased is to toss it several times and count the number of times that it lands head up.

Then for that coin

$$P(\text{a head}) \approx \frac{\text{number of heads}}{\text{total number of tosses}}$$

The approximation gets nearer to the truth as the number of tosses gets larger.

Exercise 7f

Work with a partner or collect information from the whole class.

1 Toss a 5 c coin 100 times and count the number of times it lands head up and the number of times it lands tail up.
Use tally marks, in groups of five, to count as you toss.
Find, approximately, the probability of getting a head with this coin.

2 Repeat question 1 with a 10 c coin.

3 Repeat question 1 with the 5 c coin that you used first but this time stick a small piece of plasticine on one side.

4 Choose two 5 c coins and toss them both once. What do you think is the probability of getting two heads? Now toss the two coins 100 times and count the number of times that both coins land head up together. Use tally marks to count as you go: you will need to keep two tallies, one to count the total number of tosses and one to count the number of times you get two heads. Use your results to find approximately the probability of getting two heads.

5 Take an ordinary pack of playing cards and keep them well shuffled. If the pack is cut, what do you think is the probability of getting a red card? Cut the pack 100 times and keep count, using tally marks as before, of the number of times that you get a red card. Now find an approximate value for the probability of getting a red card.

6 Using the pack of cards again, what do you think is the probability of getting a spade? Now find this probability by experiment.

7 Use an ordinary six-sided die. Toss it 25 times and keep count of the number of times that you get a six. Use your results to find an approximate value for the probability of getting a six. Now toss the die another 25 times and add the results to the last set. Use these to find again the probability of getting a six. Now do another 25 tosses and add the results to the last two sets to find another value for the probability. Carry on doing this in groups of 25 tosses until you have done 200 tosses altogether.

You know that the probability of getting a six is $\frac{1}{6}$. Now look at the sequence of results obtained from your experiment. What do you notice? (It is easier to compare your results if you use your calculator to change the fractions into decimals correct to 2 d.p.)

8 Remove all the diamonds from an ordinary pack of playing cards. Shuffle the remaining cards well and then cut the pack. What do you think is the probability of getting a black card? Shuffle and cut the pack 100 times and use the results to find approximately the probability of cutting a black card.

9 Roll two dice together 100 times. Find the total for each toss and put a mark beside a number each time you toss that number. Use a copy of the chart below.

a What is the probability of getting a total of 7?

b What is the probability of getting

i a total 2?   **ii** a total 12?

Total	Number of times
2	
3	
4	
5	
.	
.	
.	
12	

Investigation

This is an experiment to find out if you can see into the future!

You need to work in pairs and you need one coin. One of you tosses the coin and records the results; the other is the guesser.

a The guesser predicts whether the coin will land head up or tail up. The tosser then tosses the coin.

If the guesser has no psychic powers,

i what is the probability that he/she guesses the actual outcome?

ii when this experiment is repeated 100 times, about how many times do you expect the guesser to predict the actual outcome?

b Now perform the experiment described at least 100 times and record each result 'right' or 'wrong' as appropriate.

c Compare what you expected to happen with what did happen, using appropriate diagrams as illustrations. Comment on the likelihood of the guesser being able to predict which way the coin will land.

d How could you make your results more reliable?

Did you know that much of the theory of probability was the work of the French philosopher and mathematician, Blaise Pascal (1623–1662). He also constructed, between 1642 and 1645, the first calculating machines, several of which still survive. His machine would add and subtract.

IN THIS CHAPTER...

you have seen that:

- a possibility space is a rectangular array showing all the possible ways in which the events being studied can occur. It can be used to find the probability that a particular event happens

- to estimate a probability by experiment find

$$\frac{\text{the number of times the event happens}}{\text{the number of times the experiment is repeated}}$$

- you multiply the probabilities when you follow a path along the branches of a tree diagram

- you add the results of following several paths along a tree diagram.

8 CONSUMER ARITHMETIC

AT THE END OF THIS CHAPTER...

you should be able to:

1   Total supermarket and other bills, and find the change from a given amount.

2   Calculate workers' wages, commission, bonuses and tax deductions.

3   Calculate sales tax and exchange rates.

4   Calculate the amount due on telephone and electricity bills, given the necessary information.

The clock strikes 12 and both hands point upwards. How many times will the minute hand and hour hand coincide before the hands point upwards at 12 again?

BEFORE YOU START

you need to know:
✓ how to work with decimals and fractions
✓ how to use a calculator

KEY WORDS

amount, bonus, commission, compound interest, exchange rate, gross and net wage, income tax, inflation, kilowatt hour, principal, salary, sales tax, simple interest, standing charge

Shopping bills

Total the following supermarket bills. In each case find the change from a $20 note.

	$		$		$		$		$
1	.88	**2**	.62	**3**	.55	**4**	.36	**5**	1.26
	.82		.37		.43		.72		.49
	.44		.37		.43		.42		.53
	.17		.37		.27		.42		.75
	.38		.42		.64		.93		.44
	.24		.18		.59		.45		.45
	.29		.23		.19		.45		.45
	.33		1.04		.19		.37		.45
	.34		.77		.54		.37		.45
	.23		.64		.62		.85		.62
	1.29		.53		.73		4.21		.41
	.29		.22		.80		.62		.87
	.59		.22		.34		.14		.73
	.43		.22		.37		.14		.49
	.23		.89		.52		.25		.61
	.32		.73		.49		.25		.72
	.32		.32		.26		.72		.17
	.28		.32		.37		.64		.17
	.16		2.76		1.04		.45		.43
	.77		3.49		.92		.27		.56
	1.43		.23		.76		.27		.92
	.49		.42		.43		.84		.44
	.42				.52		.92		.73
	.18						.66		.84
									.44
									.62

Copy and complete the following bills: $

6 2 tins of paint at $23.75 per tin
3 brushes at $5.40 each
24 ceramic tiles at $2.15 each   _____
_____

$

7 2 kg butter at $11.60 per kg
3 litres milk at $1.20 per litre
1 jar of nutmeg jam at $3.40 each   _____
_____

$

8 7 oranges at 50 c each
8 grapefruit at 78 c each
3 lb apples at $1.50 per lb
2 lb bananas at $1.32 per lb   _____
_____

$

9 2 jars of guava jelly at $3.75 per jar
3 jars of jam at $3.50 per jar
2 jars of marmalade at $4.20 per jar
3 jars of honey at $5.10 per jar   _____
_____

$

10 2 cans ackee at $6.25 each
3 cans of grape juice at $3.45 per can
2 packets of peas at $1.59 each
1 packet of Jello at $2.37
1 bottle of bitters at $5.89   _____
_____

11 An importer of car batteries in Barbados pays the following costs:

Price of battery	$59.98
Freight	$12.00
Duty	$32.39
Consumption tax	$10.44
Stamp duty	$ 8.64

Calculate the total cost of a battery to the importer.

12 Residential lots of land in a Caribbean country are advertised as follows:

Downpayment	$1200.00
Monthly payment	$ 250.00

How much will it cost a person who pays for his land by making a downpayment followed by 24 monthly payments?

Puzzle

I have two old coins. One is marked Elizabeth I and the other George VI. A knowledgeable friend told me that one was probably genuine but that the other was definitely a fake. How could she be so certain?

Wages

Everybody who goes to work expects to get paid. Some are paid an annual amount or *salary*, but many people are paid a wage at a fixed sum per hour. There is usually an agreed length to the working week and any hours worked over and above this may be paid for at a higher rate.

If John Duffy works for 37 hours for an agreed hourly rate of $4.50, he receives payment of $4.50 × 37, i.e. $166.50. This figure is called his *gross wage* for the week. From this, deductions are made for such things as National Insurance contributions and Income Tax. After the deductions have been made he receives his *net wage* or 'take-home' pay.

All this information is gathered together by the employer on a pay slip, an example of which is given below.

STAFF No.	DATE	Basic Salary	Additional Payts. A	Deduction for Absence	Gross Pay
01035932	JAN 2005	130.34	24.44		154.78
Loan Repayts/Adv. Recovered	Vol. Dedns. B	Nat. Ins.		Income Tax	Total Deducted
		13.54		46.50	60.04
A—Acting Allow.	Commission	Bonuses	Other	Non-Taxble. Allces	NET PAY
24.44					94.74
Detail	Detail	Detail	Detail	B—Voluntary Deductions Union Dues	

Exercise 8b

Calculate the gross weekly wage for each of the following factory workers.

	Name	Number of hours worked	Hourly rate of pay
1	E. D. Nisbett	40	$5
2	A. Dexter	35	$5.50
3	T. Wilson	$38\frac{1}{2}$	$5.60
4	A. Smith	44	$5.40
5	D. Thomas	$39\frac{1}{2}$	$7.40

In the questions that follow, it is assumed that the meal breaks are unpaid.

Sally Green works a five-day week Monday to Friday. She starts work every day at 8 a.m. and finishes at 4.30 p.m. She has 1 hour off for lunch. How many hours does she work in a week? Find her gross pay if her rate is $4.92 for each hour worked.

Number of hours from 8 a.m. to 4.30 p.m. is $8\frac{1}{2}$.

She has 1 hour off for lunch, so

number of hours worked each day is $7\frac{1}{2}$

$$\text{Number of hours worked each week} = 7\tfrac{1}{2} \times 5 = 37\tfrac{1}{2}$$

$$\text{Gross pay for the week} = \$4.92 \times 37\tfrac{1}{2} = \$184.50$$

6 Edna Owen works a five-day week. She starts work each day at 7.30 a.m. and finishes at 4.15 p.m. She has 45 minutes for lunch and a 10-minute break each morning and afternoon. How long does she actually work

a in a day   **b** in a week?

If her hourly rate is $4.66, calculate her gross wage for the week.

7 Martin Jones starts work each day at 7 a.m. and finishes at 4.30 p.m. He has a 45-minute lunch break. How many hours does he work in a normal five-day week? Find his gross weekly wage if his rate of pay is $5.24 per hour.

8 Elaine Mock works 'afternoons'. She starts every day at 2 p.m. and finishes at 10.30 p.m., and is entitled to a meal break from 6 p.m. to 6.45 p.m. How many hours does she work

a in a day   **b** in a five-day week?

Calculate her gross weekly wage if she is paid $4.52 per hour.

Mary Killick gets paid $4.12 per hour for her normal working week of $37\frac{1}{2}$ hours. Any overtime is paid at time-and-a-half. Find her gross pay in a week when she works $45\frac{1}{2}$ hours.

$$\text{Basic weekly pay} = \$4.12 \times 37.5 = \$154.50$$

$$\text{Number of hours overtime} = (45\tfrac{1}{2} - 37\tfrac{1}{2}) \text{ hours} = 8 \text{ hours}$$

Because overtime is paid at time-and-a-half, that is one-and-a-half times the hourly rate, the rate of overtime pay is $4.12 \times 1.5 = \$6.18$ per hour

$$\text{Payment for overtime} = \$6.18 \times 8 = \$49.44$$

Total gross pay = basic pay + overtime pay

$$= \$154.50 + \$49.44$$

$$= \$203.94$$

9 Tom Shepherd works for a builder who pays $3.10 per hour for a basic week of 38 hours. If overtime worked is paid at time-and-a-half, how much will he earn in a week when he works for

a 38 hours **b** 48 hours **c** 50 hours?

10 Maxine Brown works in a factory where the basic hourly rate is $1.96 for a 35-hour week. Any overtime is paid at time-and-a-half. How much will she earn in a week when she works for 46 hours?

11 Walter Markland works a basic week of $37\frac{1}{2}$ hours. Overtime is paid at time-and-a-quarter. How much does he earn in a week when he works $44\frac{1}{2}$ hours if the hourly rate is $3.40?

12 Peter Ambler's time sheet showed that he worked 7 hours overtime in addition to his basic 38-hour week. If his basic hourly rate is $3.16 and overtime is paid at time-and-a-half, find his gross pay for the week.

13 During a certain week Adelle Dookham worked $8\frac{1}{2}$ hours Monday to Friday together with 4 hours on Saturday. The normal working day was 7 hours and any time worked in excess of this was paid at time-and-a-half, with Saturday working being paid at double time. Calculate her gross wage for the week if she was paid $4.32 per hour.

14 Diana Read works a basic week of 39 hours. Overtime is paid at time-and-a-half. How much does she earn in a week when she works $47\frac{1}{2}$ hours if the hourly rate is $3.64?

15 Joan Danby's pay slip showed that she worked $5\frac{1}{2}$ hours overtime in addition to her basic 37-hour week. If her basic rate of pay is $2.96 and overtime is paid at time-and-a-half, find her gross pay for the week.

16 The timesheet for Anne Stent showed that during the last week in November she worked as follows:

	Morning		Afternoon	
Day	In	Out	In	Out
Monday	7.45 a.m.	12 noon	1.00 p.m.	5.45 p.m.
Tuesday	7.45 a.m.	12 noon	1.00 p.m.	4.15 p.m.
Wednesday	7.45 a.m.	12 noon	1.00 p.m.	4.15 p.m.
Thursday	7.45 a.m.	12 noon	1.00 p.m.	4.15 p.m.
Friday	7.45 a.m.	12 noon	1.00 p.m.	4.15 p.m.

a What is the length of her normal working day?

b How many hours make up her basic working week?

c Calculate her basic weekly wage if the hourly rate is $2.84.

d How much overtime was worked?

e Calculate her gross wage if overtime is paid at time-and-a-half.

Commission and bonus incentives

Some workers, such as salesmen and representatives, are paid in a different way. They are given a fairly low basic wage but they also get commission on every order they secure. The commission is usually a percentage of the value of the order.

Other workers get paid a fixed wage plus an amount that depends on the amount of work they do.

For example, Pete gets paid $120 a week plus 40c for every article he produces after the first 30.

Exercise 8c

1 In addition to a basic weekly wage of $40, Miss Black receives a commission of 1% for selling second-hand cars. Calculate her gross wage for a week when she sells cars to the value of $50 000.

> **Tip** Find 1% of the value of the cars she sold and add this to the basic wage.

2 A salesman receives a basic wage of $50 per week plus commission at 6% on the value of the goods he sells. Find his income in a week when sales amount to $5300.

3 Tom Hannah receives a basic wage of $55 per week and receives a commission of 2% on all sales over $1000. Find his income for a week when he sells goods to the value of $8800.

4 Sue Renner receives a basic wage of $100 per week plus a commission of 2% on her sales. Find her income for a week when she sells goods to the value of $21 200.

5 Penny George is paid a basic wage of $85 per week plus a commission of 1½% on her sales over $1500. Find her income for a week when she sells goods to the value of $21 300.

6 Alan McKay is paid a basic wage of $50 per week plus a commission of 3% on all sales over $2400. Find his income for a week when he sells goods to the value of $17 400.

7 In addition to a weekly wage of $170, Olive MacCarthy receives commission of 1½% on the sales of antique furniture. Calculate her gross wage in a week when she sells furniture to the value of $15 500.

8 Don Smith receives a guaranteed weekly wage of $260 plus a bonus of 40c for every circuit board he completes each day after the first 20. During a particular week the number of boards he produced are as follows:

> **Tip** First find the total number on which the bonus is paid: 33 the first day, 28 the second, and so on. Next calculate the total bonus and add it to the gross wage.

Monday 53, Tuesday 48, Wednesday 55, Thursday 51, Friday 47.

Calculate his gross wage for the week.

9 Audley Davis gets paid 40 c for each article he completes up to 100 per day. For every article above this figure he receives 45 c. In a particular week his production figures are

Mon	Tues	Wed	Thurs	Fri
216	192	234	264	219

a How many articles does he produce in the week?

b For how many of these is he paid 40 c each?

c For how many of these is he paid 45 c each?

d Find his earnings for the week.

10 The table shows the number of electric light fittings produced by five factory workers each day for a week.

	Mon	Tues	Wed	Thurs	Fri
Ms Arnold	34	38	34	39	41
Mr Beynon	37	40	37	44	–
Miss Capstick	35	40	43	37	39
Mr Davis	42	45	40	52	46
Mrs Edmunds	39	38	37	35	42

The rate of payment is: 95 c for each fitting up to 20 per day and $1.35 for each fitting above 20 per day.

a How many fittings does each person produce in the week?

b For each person find

i how many fittings are paid at 95 c each

ii how many fittings are paid at $1.35 each.

c Find each person's income for the week.

d On which day of the week does this group of workers produce the greatest number of fittings?

Taxes

Most people who work pay tax on their income. This is known as *income tax*, and the amount varies depending on the amount earned. In general the more you earn the more you pay.

The government is forever looking for ways of extracting money from us to pay for its spending. One such way is to put a tax on almost everything that is sold. This *sales tax* is usually a fixed percentage of the selling price. It is also called value added tax (VAT) in some countries.

Exercise 8d

1 Jane Axe earns $20 000 a year. She pays tax on this at 15%.

   a How much tax must she pay?   b Find her net income.

2 Freddy Davis earns $35 000 a year. He pays tax at 18%.

   a How much tax does he pay?

   b Work out his net income   i a year   ii a week.

3 Complete the table

Name	Gross weekly pay	Tax rate	Tax due	Net weekly pay
M Davis	$820	10%		
P Evans	$1230	12%		
G Brown	$1765	18%		
A Khan	$2182	25%		

4 A CD costs $38 plus sales tax at 12%. Find

   a the sales tax to be added   b the price I must pay for the CD.

In questions 5 to 7 find the total purchase price of the item. Take the rate of sales tax as 17.5%.

5 An electric cooker marked $1600 + sales tax.

6 A calculator costing $20 + sales tax.

7 A van marked $21 000 + sales tax.

8 The price tag on a television gives $655 plus sales tax at 15%.
   What does the customer have to pay?

9 In March, Nicki looked at a camera costing $320 plus sales tax. The sales tax rate at that time was 17½%. How much would the camera have cost in March? Nicki decided to wait until June to buy the camera but by then the sales tax had been raised to 22%. How much did she have to pay?

10 An electric cooker was priced in a showroom at $1100 plus value added tax at 15%.

   a What was the price to the customer?

   Later in the year value added tax was increased to 17.5%. The showroom manager placed a notice on the cooker that read:

   Due to the increase in VAT this cooker will now cost you $1296.63

   b Was the manager correct?

   c If your answer is 'Yes', state how the manager calculated the new price. If your answer is 'No', give your reason and find the correct price.

Saving and borrowing

In Book 1 we saw that you could calculate the *simple interest*, I, on a sum of money using the formula $I = \frac{PRT}{100}$ where P is the principal, R the rate per cent each year, and T the time in years for which the principal is borrowed or invested. The formula can be arranged as $100I = PRT$, which can be used to find any one quantity if the other three are known.

Sometimes, when the interest is due, the whole amount remains invested or is not paid back.

Suppose $2000 is invested for 1 year at 8%.

The interest earned is $2000 \times 0.08 = \$160$

so the amount at the end of the year is $2160.

If this amount is invested for a year at 8% the interest due is $2160 \times 0.08 = \$172.80$

The amount now is $2160 + \$172.80 = \$2332.80$

If you invest money, do not spend the interest, and the annual rate stays the same, your money will increase by larger and larger amounts each year. The total amount by which it grows is called the *compound interest*.

Exercise 8e

1 Find the simple interest on

   a $420 invested for 3 years at 10% p.a.

   b $280 invested for 6 years at 12% p.a.

   c $834 invested for 5 years at 9% p.a.

   d $500 invested for 8 years at $12\frac{1}{2}$% p.a.

   e $726 invested for 3 years at $7\frac{1}{2}$% p.a.

2 What sum of money invested for 5 years at 12% p.a. gives $264 simple interest?

3 How long must $370 be invested at 9% p.a. simple interest to give interest of $233.10?

4 What annual rate of simple interest is necessary to give interest of $416 on a principal of $800 invested for 8 years?

5 What sum of money earns $312 simple interest if invested for 8 years at 13%?

6 Find the annual rate per cent that earns $234 simple interest when $900 is invested for $6\frac{1}{2}$ years.

7 Find the amount if $280 is invested for 5 years at 9% p.a. simple interest.

Find the compound interest on $550 invested for 2 years at 6%.

Interest for first year at 6% is 6% of the original principal.

New principal at end of first year = 100% of original principal + 6% of original principal

= 106% of original principal

= 1.06 × original principal

∴ principal at end of first year = 1.06 × $550 = $583

Similarly, new principal at end of second year = 106% of principal at beginning of second year

= 1.06 × $583 = $617.98

Compound interest = principal at end of second year − original principal

= $617.98 − $550 = $67.98

Find the compound interest on

8 $200 for 2 years at 10% p.a.

9 $300 for 2 years at 12% p.a.

10 $400 for 3 years at 8% p.a.

11 $650 for 3 years at 9% p.a.

12 $520 for 2 years at 13% p.a.

13 $624 for 3 years at 11% p.a.

14 $40 000 is invested at compound interest of 10% each year. What will it be worth in 2 years' time?

15 Brian Barnes borrows $5000 at 12% compound interest. He agrees to clear the debt at the end of 2 years. How much must he pay?

16 A postage stamp increases in value by 15% each year. If it is bought for $50, what will it be worth in 3 years' time?

17 A motorcycle bought for $1500 depreciates in value by 10% each year. Find its value after 3 years.

18 A motor car bought for £20 000 depreciates in any one year by 20% of its value at the beginning of that year. Find its value after 2 years.

Tip Find 10% of the purchase price and subtract this from the purchase price to give the value after 1 year. Now find 10% of the new value. Deduct this from the value at the end of the first year to give the value at the end of the second year and so on.

Telephone bills

The cost of a telephone call depends on three factors:

i the distance between the caller and the person being called

ii the time of day and/or the day of the week on which the call is being made,

iii the length of the call.

These three factors are put together in various ways to give metered units of time, each unit being charged at a fixed rate.

In common with gas and electricity there is usually a fixed charge each quarter in addition to the charge for the metered units.

For example, suppose that Chris Reynolds' telephone account for the last quarter showed that his telephone had been used for 546 metered units. If the standing charge was $31.00 and each unit cost 5 c, his telephone bill for the quarter can be worked out as follows:

$$\text{Cost of 546 units at 5 c per unit} = 546 \times 5\,c$$
$$= \$27.30$$
$$\text{Standing charge} = \$31$$
$$\therefore \quad \text{the telephone bill for the quarter was } \$58.30$$

In some islands, for example Barbados, telephone bills are calculated at a fixed rate. Additional charges are made only for overseas calls, which will depend on the three factors above. Details similar to the following will be shown on such a bill.

Date	Country	Area Code	Number	Mins	Amount
06/11	Tdad & Tobago	000	645-3272	7	$18.92
06/19	Tdad & Tobago	000	622-0000	12	$30.74
06/29	Jamaica	000	927-9751	8	$17.60
06/29	Jamaica	000	927-8798	4	$ 8.80
07/03	Jamaica	000	927-9751	12	$34.32
07/20	Miami	305	279-1319	1	$ 3.96
07/20	Miami	305	949-4616	9	$35.64

Balance Forward	$177.80
Payments	$177.80
Overseas Calls	$149.98
Other Charges	$ 0.00
Service Charge	$ 25.40
TOTAL DUE	$175.38

Exercise 8f

Find the quarterly telephone bill for each of the following households.

	Name	Number of units used	Standing charge	Cost per unit
1	Mrs Keeling	750	$28	5 c
2	Mr Hodge	872	$32	6 c
3	Miss Hutton	1040	$33	7 c
4	Miss Jacob	1134	$37.60	8.5 c
5	Mrs Buckley	1590	$36.80	8.3 c
6	Mr Leeson	765	$42	7.68 c
7	Mrs Solly	965	$51	10.5 c

Calculate the monthly telephone bill for each of the following people.

Name	1st 3 mins.	Each additional min.	Number of mins.	Service charge
8 Singh	$ 3.90	$1.30	70	$25.40
9 Bird	$ 6.45	$2.15	28	$30.00
10 Lee	$ 9.15	$3.05	105	$25.40

Electricity: kilowatt-hours

A kilowatt-hour is electric power of one kilowatt used for one hour. Electric companies charge one rate for the first number of kilowatt-hours and a lower rate for additional usage.

We all use electricity in some form and we know that some appliances cost more to run than others. For example, an electric fire costs much more to run than a light bulb. Electricity is sold in units called kilowatt-hours (kWh) and each appliance has a rating that tells us how many kilowatt-hours it uses each hour.

A typical rating for an electric fire is 2 kW. This tells us that it will use 2 kWh each hour, i.e. 2 units per hour. On the other hand, a light bulb can have a rating of 100 W. Because 1 kilowatt = 1000 watts (kilo means 'thousand' as we have already seen in kilometre and kilogram), the light bulb uses $\frac{1}{10}$ kWh each hour, or $\frac{1}{10}$ of a unit.

Exercise 8g

How many units (i.e. kilowatt-hours) will each of the given appliances use in 1 hour?

1 a 3 kW electric fire

2 a 100 W bulb

3 a $1\frac{1}{2}$ kW fire

4 a 1200 W hair dryer

5 a 60 W video recorder

6 a 20 W radio

7 an 8 kW cooker

8 a 2 kW dishwasher

With the help of an adult, find the rating of any of the following appliances that you might have at home. The easiest place to find this information is probably from the instructions.

9 an electric kettle

10 the refrigerator

11 the washing machine

12 the television set

13 a bedside lamp

14 the electric cooker

How many units of electricity would

15 a 2 kW fire use in 8 hours

16 a 100 W bulb use in 10 hours

17 an 8 kW cooker use in $1\frac{1}{2}$ hours

18 a 60 W bulb use in 50 hours

19 a 150 W refrigerator use in 12 hours

20 a 12 W radio use in 12 hours

21 an 8 W night bulb use in a week at 10 hours per night

22 a 5 W clock use in 1 week

For how long could the following appliances be run on one unit of electricity?

23 a 250 W bulb

24 a 2 kW electric fire

25 a 100 W television set

26 a 360 W electric drill

In the following questions assume that 1 unit of electricity costs 6 c.

How much does it cost to run

27 a 100 W bulb for 5 hours

28 a 250 W television set for 8 hours

29 a 3 W clock for 1 week

30 a 3 kW kettle for 5 min

Electricity bills

It is clear from the questions in the previous exercise that lighting from electricity is cheap but heating is expensive.

While electricity is a difficult form of energy to store, it is convenient to produce it continuously at the power stations, 24 hours a day. There are therefore times of the day when more electricity is produced than is normally required. The Electricity Boards are able to solve this problem by selling 'off-peak', or 'white meter', electricity to domestic users at a cheaper rate. Most of the electricity consumed in this way is for domestic heating.

Domestic electricity bills are calculated by charging every household a fixed amount, together with a charge for each unit used. Off-peak electricity is sold at approximately half price. The amount used is recorded on a meter, the difference between the readings at the beginning and end of a month or quarter showing how much has been used.

The following shows an electricity bill that might be received by a customer in a West Indian Island.

LIGHT & POWER COMPANY   August 2004

Meter No.	Meter reading & date		kWh Used	Fuel Cts/kWh
	Previous	Present		
F12906	09118 04-07-08	09218 04-08-08	100	5.5408

Charges	
Fixed	$20.60
Energy	5.54
Subtotal	26.14
Arrears	
TOTAL	$26.14

The above bill shows the number of kWh registered on the meter on 04-07-08 as 09118 and on 04-08-08 as 09218. The number used for the one-month period is calculated as the difference between 09118 and 09218, i.e. 100 kWh.

The cost of the energy is therefore 100 kWh at 5.5408 c per kWh = $5.54.

There is also a fixed charge of $20.60 to add, so the total is $20.60 + $5.54 = $26.14.

Sometimes a customer is unable to pay the full amount in a particular period and only pays a portion. The remainder is then added to his next bill as 'Arrears'. In the above bill, there were no arrears.

In some territories bills are sent quarterly instead of monthly.

Exercise 8h   Mrs Comerford uses 1527 units of electricity in a quarter. If the standing charge is $9.45 and each unit costs 8 c, how much does electricity cost her for the quarter?

$$\text{Cost of 1527 units at 8 c per unit} = 1527 \times 8\,c$$
$$= \$122.16$$
$$\text{Standing charge} = \$9.45$$
$$\text{Total bill} = \$131.61$$

Find the quarterly electricity bills for each of the following households:

	Name	Number of units used	Standing charge	Cost per unit
1	Mr George	500	$20	5 c
2	Mrs Newton	600	$24	5 c
3	Mr Churchman	950	$30	10 c
4	Mr Khan	750	$28	12 c
5	Mr Vincent	1427	$31.80	6.65 c
6	Mrs Jackson	684	$36	11 c
7	Mr Wilton	938	$32.80	7.36 c

Find the quarterly electricity bills for each of the following households. Assume in each case that there is a standing charge of $20, and that off-peak units are bought at half price.

	Name	Number of units used At the basic price	Off-peak	Basic cost per unit
8	Mr Bennett	1000	500	10 c
9	Miss Cann	800	600	8 c
10	Mr Hadley	640	1200	7.5 c

Exchange rates

When we shop abroad, prices quoted in the local currency often give us little idea of value so we tend to convert prices into the currency we are familiar with. To do this we need to know the exchange rate.

This tells us how many units of currency are equivalent to 1 unit of our own currency.

For example, using an exchange rate of 1 US dollar (US$1) = 0.8 euros

means that $100 = 100 \times 0.8$ euros = 80 euros

and that 100 euros $= US\$ \frac{100}{80} = US\$125$.

Exercise 8i

This table gives the equivalent of US$1 in various currencies.

US$	UK£	Barbadian $	Canadian $	Jamaican $
1	0.55	2.0	1.325	61.4

Use this table to convert

1 US$45 into Jamaican dollars

2 US$550 into Canadian dollars

3 US$400 into UK pounds

4 US$68.90 into Barbadian dollars

Use this table

a to estimate

b to calculate to the nearest whole number, the equivalent value in US dollars of

5 £100

> **Tip** The table shows that £0.55 is equivalent to US$1 so £1 is equivalent to US$$\frac{1}{0.55}$ ∴ £100 is equivalent to US$$100 \times \frac{1}{0.55}$

6 345 Barbadian dollars

7 5567 Jamaican dollars

8 500 Barbadian dollars

9 £642

10 462 Canadian dollars

11 1000 Jamaican dollars

12 £246.40

13 1188 Barbadian dollars

Use the table to convert

14 £200 into Barbadian dollars

15 500 Canadian dollars into Jamaican dollars

> **Tip** Start with £0.55 = 2 Bds, so £1 = $\frac{2}{0.55}$Bds. The equivalent of £200 will be 200 times this value.

16 3000 Bds into £s

17 275 Jamaican dollars into Barbadian dollars

18 £754 into Jamaican dollars

Investigation

Look up the meaning of 'inflation' in your dictionary, then discuss it with your teacher.

Now try the following problem.

If an item is bought for $1.00 now and inflation continues at 10% per year, how many years will it take for the price to double? Make up a table like the one below to help you.

Year	Cost
Now	$1.00
1	$1.00 + (0.10 \times 1.00) = 1.1 \times 1.00 = \$1.10$
2	$1.10 + (0.10 \times 1.10) = 1.1 \times 1.10 = \$1.21$

Compute the remaining costs.

Farmer Giles has a problem. He has 4 hens and they laid 4 eggs in 4 days. He buys another 8 hens and has an order for 36 eggs. How long will it take 12 hens to lay 36 eggs?

IN THIS CHAPTER...

you have seen that:

- wages can be checked when you have all the relevant details

- a kWh is 1 kilowatt used for 1 hour

- bills for domestic utilities usually include a fixed charge and a charge for the number of units used

- you earn money (interest) when you invest money and have to pay when you borrow it

- there are deductions made from earnings for tax and possibly other things

- sales tax is added to the selling price of an article at a rate fixed by the government

- exchange rates are used to convert from one currency to another

9 VECTORS

AT THE END OF THIS CHAPTER...

you should be able to:

1   Differentiate between a vector and a scalar.

2   Represent a vector by a straight line.

3   Write a vector as an ordered pair in a column.

4   Find the end point/starting point of a vector represented in column form given the starting/end point.

5   Classify vectors as equal, parallel or opposite.

6   Find the vector sum/difference of two or more vectors given in column form.

Who was the mathematician of the third century BC who was also an astronomer? Here are some clues:

He knew that the Earth was round and made a good measurement of the distance around it.

BEFORE YOU START

you need to know:
✓ how to plot points on a grid
✓ what coordinates are
✓ how to add and subtract negative numbers

KEY WORDS

coordinates, equal vectors, magnitude, negative vectors, ordered pair, parallel vectors, scalar, vector

If you arranged to meet your friend 3 km from your home, this information would not be enough to ensure that you both went to the same place. You would also need to know which way to go.

Two pieces of information are required to describe where one place is in relation to another: the distance and the direction. Quantities that have both *size* (magnitude) and *direction* are called *vectors*.

A quantity that has magnitude but not direction is called a *scalar*. For example, the amount of money in your pocket or the number of pupils in your school are scalar quantities. On the other hand, the velocity of a hurricane, which states the speed and the direction in which it is moving, is a vector quantity.

Exercise 9a

State whether the following sentences refer to vector or scalar quantities:

1 There are 24 pupils in my class.

2 To get to school I walk $\frac{1}{2}$ km due north.

3 There are 11 players in a cricket team.

4 John walked at 6 km per hour.

5 The vertical cliff face is 50 m high.

6 Give other examples of

    a vector quantities   **b** scalar quantities.

Representing vectors

Because a vector has both size and direction we can represent a vector by a straight line and indicate its direction with an arrow. For example

We use **a**, **b**, **c**, ... to name the vectors.

When writing by hand it is difficult to write **a**, which is in heavy type, so we use $\underline{a}$.

In the diagram on the right, the movement along **a** corresponds to 4 across and 2 up and we can write

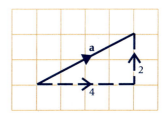

$$\mathbf{a} = \begin{pmatrix} 4 \\ 2 \end{pmatrix}$$

The vector **b** can be described as 8 across and 4 down. As with coordinates, which we looked at in Book 1, we use negative numbers to indicate movement down or movement to the left.

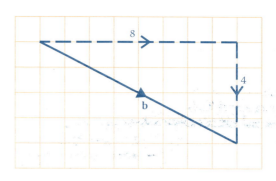

Therefore $\quad \mathbf{b} = \begin{pmatrix} 8 \\ -4 \end{pmatrix}$

Notice that the top number represents movement across and that the bottom number represents movement up or down.

Exercise 9b

Write the following vectors in the form $\begin{pmatrix} p \\ q \end{pmatrix}$:

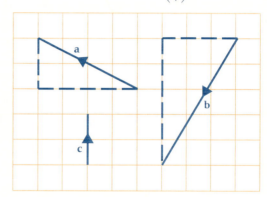

To move from the start to the end of **a**, you go 4 squares back (to the left) and 2 squares up: $\mathbf{a} = \begin{pmatrix} -4 \\ 2 \end{pmatrix}$.

For **b** you need to go 3 squares back and 5 squares down: $\mathbf{b} = \begin{pmatrix} -3 \\ -5 \end{pmatrix}$.

For **c** you do not need to go across, but you go 2 squares up: $\mathbf{c} = \begin{pmatrix} 0 \\ 2 \end{pmatrix}$.

Write the following vectors in the form $\begin{pmatrix} p \\ q \end{pmatrix}$:

1

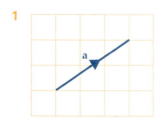

> **Tip** Move in the direction of the arrow and remember that the top number gives the distance across and the bottom number gives the distance up or down.

2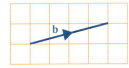

4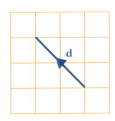

6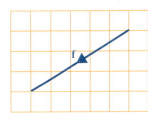

3

5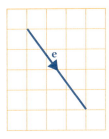

7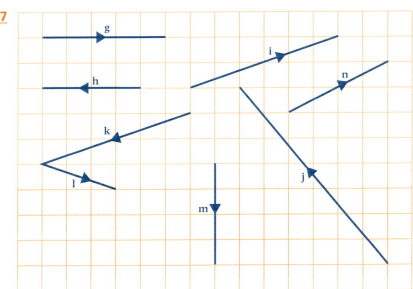

On squared paper draw the following vectors. Label each vector with its letter and an arrow:

8 $\mathbf{a} = \begin{pmatrix} 3 \\ 5 \end{pmatrix}$

11 $\mathbf{d} = \begin{pmatrix} 6 \\ -12 \end{pmatrix}$

14 $\mathbf{g} = \begin{pmatrix} 6 \\ 10 \end{pmatrix}$

9 $\mathbf{b} = \begin{pmatrix} -4 \\ -3 \end{pmatrix}$

12 $\mathbf{e} = \begin{pmatrix} -4 \\ 3 \end{pmatrix}$

15 $\mathbf{h} = \begin{pmatrix} -1 \\ -5 \end{pmatrix}$

10 $\mathbf{c} = \begin{pmatrix} 2 \\ -4 \end{pmatrix}$

13 $\mathbf{f} = \begin{pmatrix} -2 \\ 5 \end{pmatrix}$

16 $\mathbf{i} = \begin{pmatrix} -6 \\ 2 \end{pmatrix}$

17 What do you notice about the vectors in questions **8** and **14**, and in questions **10** and **11**?

Exercise **9c**

In each of the following questions you are given a vector followed by the coordinates of its starting point. Find the coordinates of its other end, or end point:

$\begin{pmatrix} 5 \\ 2 \end{pmatrix}$, (2, 1)

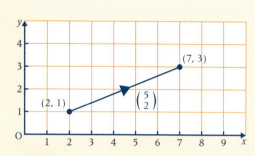

The coordinates of its other end are (7, 3).

1 $\begin{pmatrix} 3 \\ 3 \end{pmatrix}$, (4, 1)

5 $\begin{pmatrix} 5 \\ 2 \end{pmatrix}$, (3, −1)

9 $\begin{pmatrix} 4 \\ 3 \end{pmatrix}$, (−2, −3)

2 $\begin{pmatrix} 3 \\ 1 \end{pmatrix}$, (−2, −3)

6 $\begin{pmatrix} 4 \\ -2 \end{pmatrix}$, (4, 2)

10 $\begin{pmatrix} 5 \\ -3 \end{pmatrix}$, (2, −1)

3 $\begin{pmatrix} -6 \\ 2 \end{pmatrix}$, (3, 5)

7 $\begin{pmatrix} -3 \\ 4 \end{pmatrix}$, (2, −4)

11 $\begin{pmatrix} -5 \\ 2 \end{pmatrix}$, (−4, −3)

4 $\begin{pmatrix} -4 \\ -3 \end{pmatrix}$, (5, −2)

8 $\begin{pmatrix} -6 \\ -6 \end{pmatrix}$, (−3, −2)

12 $\begin{pmatrix} -4 \\ -2 \end{pmatrix}$, (−3, −1)

In each of the following questions a vector is given followed by the coordinates of its end point. Find the coordinates of its starting point:

$\begin{pmatrix} 6 \\ 4 \end{pmatrix}$, (8, 6)

From the end of the vector, you need to go backwards, i.e. 6 units left and 4 units down to get to the start of the vector.

The coordinates of the vector's starting point are (2, 2).

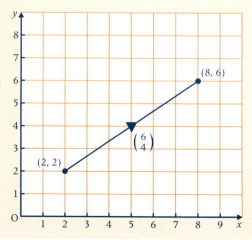

13 $\begin{pmatrix} 10 \\ 2 \end{pmatrix}$, (4, 1)

17 $\begin{pmatrix} -3 \\ 4 \end{pmatrix}$, (−2, 1)

Tip Remember that the point given is the end of the vector so you have to go in the opposite direction to get to its start point.

14 $\begin{pmatrix} 5 \\ -1 \end{pmatrix}$, (3, −4)

18 $\begin{pmatrix} -6 \\ -3 \end{pmatrix}$, (−5, 2)

15 $\begin{pmatrix} -5 \\ -2 \end{pmatrix}$, (−2, −4)

19 $\begin{pmatrix} 4 \\ -2 \end{pmatrix}$, (−3, 2)

21 $\begin{pmatrix} 1 \\ 4 \end{pmatrix}$, (−5, −2)

16 $\begin{pmatrix} 8 \\ 6 \end{pmatrix}$, (6, 3)

20 $\begin{pmatrix} -2 \\ 6 \end{pmatrix}$, (−3, −4)

22 $\begin{pmatrix} 2 \\ -3 \end{pmatrix}$, (1, 7)

Capital letter notation

In the diagram A and B are two points.

We can denote the vector from A to B

as $\overrightarrow{AB}$ where $\overrightarrow{AB} = \begin{pmatrix} 7 \\ 1 \end{pmatrix}$.

Similarly $\overrightarrow{BC} = \begin{pmatrix} -4 \\ 6 \end{pmatrix}$.

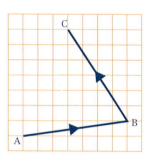

Exercise 9d   Write down the vector $\overrightarrow{AB}$ where A is (2, 4), B is (3, 9).

Plot the start and end points of the vector. Then you can see that you need to go 1 forward and 5 up.

Vector $\overrightarrow{AB}$ is $\begin{pmatrix} 1 \\ 5 \end{pmatrix}$.

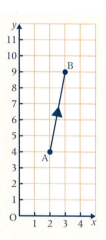

Write down the vector $\overrightarrow{AB}$ where:

1 A is (1, 4), B is (7, 6)

2 A is (−3, 4), B is (2, 3)

3 A is (7, 3), B is (1, 2)

4 A is (−1, 4), B is (5, 9)

5 A is (2, 1), B is (−3, −5)

6 A is (3, 0), B is (5, −2)

7 A is (6, 3), B is (4, 1)

8 A is (−1, −3), B is (−5, −8)

9 A is (2, 6), B is (2, −6)

10 A is (2, −3), B is (4, 5)

Equal vectors, parallel vectors and negative vectors

In the diagram, we can see that

$$\mathbf{a} = \begin{pmatrix} 5 \\ 2 \end{pmatrix} \quad \text{and} \quad \mathbf{b} = \begin{pmatrix} 5 \\ 2 \end{pmatrix}$$

so we say that $\mathbf{a} = \mathbf{b}$

The lines representing **a** and **b** are parallel and equal in length.

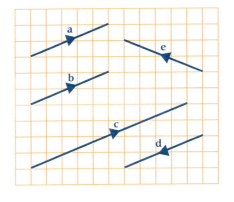

Now $\quad \mathbf{c} = \begin{pmatrix} 10 \\ 4 \end{pmatrix} \quad \text{and} \quad \mathbf{a} = \begin{pmatrix} 5 \\ 2 \end{pmatrix}$

so $\quad \mathbf{c} = 2\mathbf{a}$

This time **a** and **c** are parallel but **c** is twice the size of **a**.

If we look at **a** and **d** we have

$$\mathbf{d} = \begin{pmatrix} -5 \\ -2 \end{pmatrix} \quad \text{and} \quad \mathbf{a} = \begin{pmatrix} 5 \\ 2 \end{pmatrix}$$

Although **d** and **a** are parallel and the same size, they are in opposite directions,

so we say that $\quad \mathbf{d} = -\mathbf{a}$

Now $\quad \mathbf{e} = \begin{pmatrix} -5 \\ 2 \end{pmatrix}$

and we can see from the diagram that although **e** is the same size as **a**, **b** and **d**, it is *not* parallel to them or to **c**. So **e** is not equal to any of the other vectors.

Exercise 9e

1 Write the vectors **a**, **b**, **c**, **d** and **e** in the form $\begin{pmatrix} p \\ q \end{pmatrix}$.

What is the relationship between

a **a** and **b**　　d **a** and **e**

b **a** and **c**　　e **b** and **e**

c **a** and **d**　　f **d** and **c**?

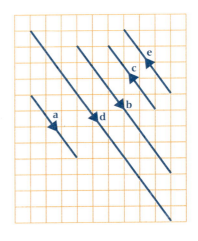

2

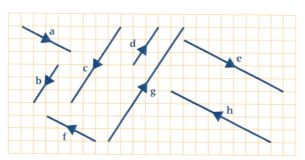

Find as many relationships as you can between the vectors in the diagram.

3 If $\mathbf{a} = \begin{pmatrix} 4 \\ 6 \end{pmatrix}$ draw diagrams to represent $\mathbf{a}$, $2\mathbf{a}$, $-\mathbf{a}$, $\frac{1}{2}\mathbf{a}$.

Now write the vectors $2\mathbf{a}$, $-\mathbf{a}$, $\frac{1}{2}\mathbf{a}$ in the form $\begin{pmatrix} p \\ q \end{pmatrix}$.

4 If $\mathbf{b} = \begin{pmatrix} -2 \\ 4 \end{pmatrix}$ draw diagrams to represent $\mathbf{b}$, $-\mathbf{b}$, $2\mathbf{b}$, $-2\mathbf{b}$.

Now write the vectors $-\mathbf{b}$, $2\mathbf{b}$, $-2\mathbf{b}$ in the form $\begin{pmatrix} p \\ q \end{pmatrix}$.

5 If $\mathbf{c} = \begin{pmatrix} 5 \\ -4 \end{pmatrix}$ draw diagrams to represent $\mathbf{c}$, $2\mathbf{c}$, $-\mathbf{c}$, $3\mathbf{c}$.

Now write the vectors $2\mathbf{c}$, $-\mathbf{c}$, $3\mathbf{c}$ in the form $\begin{pmatrix} p \\ q \end{pmatrix}$.

6 If $\mathbf{d} = \begin{pmatrix} -3 \\ -6 \end{pmatrix}$ draw diagrams to represent $\mathbf{d}$, $-\mathbf{d}$, $2\mathbf{d}$, $-2\mathbf{d}$.

Now write the vectors $-\mathbf{d}$, $2\mathbf{d}$, $-2\mathbf{d}$ in the form $\begin{pmatrix} p \\ q \end{pmatrix}$.

7 If $\mathbf{e} = \begin{pmatrix} 5 \\ 1 \end{pmatrix}$ write, in the form $\begin{pmatrix} p \\ q \end{pmatrix}$, the vectors $2\mathbf{e}$, $-\mathbf{e}$, $3\mathbf{e}$ and $-4\mathbf{e}$.

8 If $\mathbf{f} = \begin{pmatrix} -2 \\ 0 \end{pmatrix}$ write, in the form $\begin{pmatrix} p \\ q \end{pmatrix}$, the vectors $3\mathbf{f}$, $-2\mathbf{f}$, $5\mathbf{f}$ and $-4\mathbf{f}$.

9 If $\mathbf{g} = \begin{pmatrix} 6 \\ -4 \end{pmatrix}$ write, in the form $\begin{pmatrix} p \\ q \end{pmatrix}$, the vectors $-\mathbf{g}$, $3\mathbf{g}$, $\frac{1}{2}\mathbf{g}$, and $-2\mathbf{g}$.

10 If $\mathbf{h} = \begin{pmatrix} -6 \\ -20 \end{pmatrix}$ write, in the form $\begin{pmatrix} p \\ q \end{pmatrix}$, the vectors $3\mathbf{h}$, $-4\mathbf{h}$, $\frac{1}{2}\mathbf{h}$, and $-5\mathbf{h}$.

Addition of vectors

The displacement from A to B followed by the displacement from B to C is equivalent to the displacement from A to C, so we write

$$\overrightarrow{AB} + \overrightarrow{BC} = \overrightarrow{AC}$$

or

$$\begin{pmatrix} 7 \\ 1 \end{pmatrix} + \begin{pmatrix} 3 \\ 6 \end{pmatrix} = \begin{pmatrix} 10 \\ 7 \end{pmatrix}$$

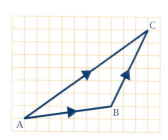

To add two vectors we add the corresponding numbers in the ordered pairs that represent them.

Exercise 9f

If $\mathbf{a} = \begin{pmatrix} 2 \\ 4 \end{pmatrix}$ and $\mathbf{b} = \begin{pmatrix} 6 \\ -8 \end{pmatrix}$ find $\mathbf{a} + \mathbf{b}$ and illustrate on a diagram.

$$\mathbf{a} + \mathbf{b} = \begin{pmatrix} 2 \\ 4 \end{pmatrix} + \begin{pmatrix} 6 \\ -8 \end{pmatrix} = \begin{pmatrix} 2+6 \\ 4+(-8) \end{pmatrix}$$

$$= \begin{pmatrix} 8 \\ -4 \end{pmatrix}$$

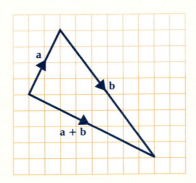

In questions **1** to **10** find $\mathbf{a} + \mathbf{b}$ and illustrate on a diagram:

1 $\mathbf{a} = \begin{pmatrix} 6 \\ 4 \end{pmatrix}$, $\mathbf{b} = \begin{pmatrix} 1 \\ -5 \end{pmatrix}$

6 $\mathbf{a} = \begin{pmatrix} 4 \\ 0 \end{pmatrix}$, $\mathbf{b} = \begin{pmatrix} 0 \\ 3 \end{pmatrix}$

2 $\mathbf{a} = \begin{pmatrix} -3 \\ 4 \end{pmatrix}$, $\mathbf{b} = \begin{pmatrix} -5 \\ -2 \end{pmatrix}$

7 $\mathbf{a} = \begin{pmatrix} -2 \\ -4 \end{pmatrix}$, $\mathbf{b} = \begin{pmatrix} -4 \\ -2 \end{pmatrix}$

3 $\mathbf{a} = \begin{pmatrix} 5 \\ 0 \end{pmatrix}$, $\mathbf{b} = \begin{pmatrix} 2 \\ -4 \end{pmatrix}$

8 $\mathbf{a} = \begin{pmatrix} 5 \\ 2 \end{pmatrix}$, $\mathbf{b} = \begin{pmatrix} -2 \\ 4 \end{pmatrix}$

4 $\mathbf{a} = \begin{pmatrix} -4 \\ 3 \end{pmatrix}$, $\mathbf{b} = \begin{pmatrix} 6 \\ 3 \end{pmatrix}$

9 $\mathbf{a} = \begin{pmatrix} 2 \\ 6 \end{pmatrix}$, $\mathbf{b} = \begin{pmatrix} 5 \\ 2 \end{pmatrix}$

5 $\mathbf{a} = \begin{pmatrix} 5 \\ 3 \end{pmatrix}$, $\mathbf{b} = \begin{pmatrix} 5 \\ -3 \end{pmatrix}$

10 $\mathbf{a} = \begin{pmatrix} 5 \\ 1 \end{pmatrix}$, $\mathbf{b} = \begin{pmatrix} 1 \\ -5 \end{pmatrix}$

In questions **11** to **20** find the following vectors:

11 $\begin{pmatrix} 2 \\ 6 \end{pmatrix} + \begin{pmatrix} 4 \\ 3 \end{pmatrix}$

16 $\begin{pmatrix} 1 \\ 4 \end{pmatrix} + \begin{pmatrix} -3 \\ 6 \end{pmatrix}$

12 $\begin{pmatrix} 5 \\ 5 \end{pmatrix} + \begin{pmatrix} 2 \\ 6 \end{pmatrix}$

17 $\begin{pmatrix} 2 \\ 1 \end{pmatrix} + \begin{pmatrix} -4 \\ -5 \end{pmatrix}$

13 $\begin{pmatrix} 4 \\ 9 \end{pmatrix} + \begin{pmatrix} 3 \\ 1 \end{pmatrix}$

18 $\begin{pmatrix} -7 \\ 2 \end{pmatrix} + \begin{pmatrix} 2 \\ -4 \end{pmatrix}$

14 $\begin{pmatrix} 6 \\ 2 \end{pmatrix} + \begin{pmatrix} 4 \\ -2 \end{pmatrix}$

19 $\begin{pmatrix} -5 \\ 9 \end{pmatrix} + \begin{pmatrix} -3 \\ -4 \end{pmatrix}$

15 $\begin{pmatrix} 5 \\ 9 \end{pmatrix} + \begin{pmatrix} -6 \\ 2 \end{pmatrix}$

20 $\begin{pmatrix} 2 \\ 4 \end{pmatrix} + \begin{pmatrix} -2 \\ -4 \end{pmatrix}$

Order of addition

If $\mathbf{a} = \begin{pmatrix} 4 \\ 2 \end{pmatrix}$ and $\mathbf{b} = \begin{pmatrix} 2 \\ 3 \end{pmatrix}$

$$\mathbf{a} + \mathbf{b} = \begin{pmatrix} 4 \\ 2 \end{pmatrix} + \begin{pmatrix} 2 \\ 3 \end{pmatrix} = \begin{pmatrix} 4+2 \\ 2+3 \end{pmatrix} = \begin{pmatrix} 6 \\ 5 \end{pmatrix}$$

and $$\mathbf{b} + \mathbf{a} = \begin{pmatrix} 2 \\ 3 \end{pmatrix} + \begin{pmatrix} 4 \\ 2 \end{pmatrix} = \begin{pmatrix} 2+4 \\ 3+2 \end{pmatrix} = \begin{pmatrix} 6 \\ 5 \end{pmatrix}$$

i.e. $$\mathbf{a} + \mathbf{b} = \mathbf{b} + \mathbf{a}$$

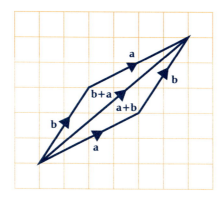

i.e. the order in which you do the addition does not matter.

This means that vector addition is commutative.

Addition of more than two vectors

If $\mathbf{a} = \begin{pmatrix} 4 \\ 1 \end{pmatrix}$, $\mathbf{b} = \begin{pmatrix} 4 \\ 4 \end{pmatrix}$ and $\mathbf{c} = \begin{pmatrix} -2 \\ 4 \end{pmatrix}$ we can find $\mathbf{a} + \mathbf{b} + \mathbf{c}$ by adding the corresponding numbers in the ordered pairs.

$$\mathbf{a} + \mathbf{b} + \mathbf{c} = \begin{pmatrix} 4 \\ 1 \end{pmatrix} + \begin{pmatrix} 4 \\ 4 \end{pmatrix} + \begin{pmatrix} -2 \\ 4 \end{pmatrix} = \begin{pmatrix} 6 \\ 9 \end{pmatrix}$$

Again the order of addition does not matter, as you can see from the diagrams below.

i.e. $$\mathbf{a} + \mathbf{b} + \mathbf{c} = \mathbf{a} + \mathbf{c} + \mathbf{b}$$

There are other possible orders in which we could add $\mathbf{a}$, $\mathbf{b}$ and $\mathbf{c}$.

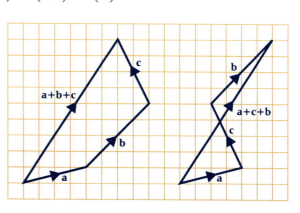

Exercise **9g**

1 If $\mathbf{a} = \begin{pmatrix} 2 \\ 3 \end{pmatrix}$, $\mathbf{b} = \begin{pmatrix} 5 \\ 2 \end{pmatrix}$ and $\mathbf{c} = \begin{pmatrix} 3 \\ 4 \end{pmatrix}$ find:

a $\mathbf{a} + \mathbf{b}$ **c** $\mathbf{b} + \mathbf{c}$ **e** $2\mathbf{a}$ **g** $\mathbf{a} + \mathbf{b} + \mathbf{c}$

b $\mathbf{b} + \mathbf{a}$ **d** $\mathbf{c} + \mathbf{b}$ **f** $3\mathbf{a}$ **h** $\mathbf{c} + \mathbf{b} + \mathbf{a}$

2 If $\mathbf{a} = \begin{pmatrix} 5 \\ -2 \end{pmatrix}$, $\mathbf{b} = \begin{pmatrix} -2 \\ 4 \end{pmatrix}$ and $\mathbf{c} = \begin{pmatrix} -5 \\ -3 \end{pmatrix}$ find:

a $\mathbf{a} + \mathbf{b}$ **c** $\mathbf{a} + \mathbf{c}$ **e** $3\mathbf{b}$

b $\mathbf{b} + \mathbf{a}$ **d** $\mathbf{c} + \mathbf{a}$ **f** $4\mathbf{c}$

3 If $\mathbf{a} = \begin{pmatrix} 3 \\ 2 \end{pmatrix}$, $\mathbf{b} = \begin{pmatrix} -3 \\ 2 \end{pmatrix}$ and $\mathbf{c} = \begin{pmatrix} 5 \\ 6 \end{pmatrix}$ find:

a $\mathbf{a} + \mathbf{b} + \mathbf{c}$ **b** $2\mathbf{a} + \mathbf{b} + 3\mathbf{c}$ **c** $\mathbf{a} + 2\mathbf{b} + 3\mathbf{c}$

4 If $\mathbf{a} = \begin{pmatrix} -6 \\ -2 \end{pmatrix}$, $\mathbf{b} = \begin{pmatrix} 4 \\ -3 \end{pmatrix}$ and $\mathbf{c} = \begin{pmatrix} -5 \\ 3 \end{pmatrix}$ find:

a $2\mathbf{a} + 2\mathbf{b} + 3\mathbf{c}$ **b** $\mathbf{a} + 5\mathbf{b} + 2\mathbf{c}$

Subtraction of vectors

If $\mathbf{a} = \begin{pmatrix} 4 \\ 3 \end{pmatrix}$ and $\mathbf{b} = \begin{pmatrix} 2 \\ -5 \end{pmatrix}$

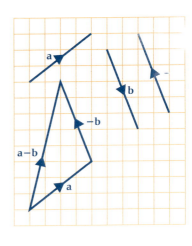

we know that $\mathbf{a} - \mathbf{b} = \mathbf{a} + (-\mathbf{b})$

$$= \begin{pmatrix} 4 \\ 3 \end{pmatrix} + \begin{pmatrix} -2 \\ 5 \end{pmatrix}$$

$$= \begin{pmatrix} 2 \\ 8 \end{pmatrix}$$

But this result is also given by

$$\begin{pmatrix} 4 \\ 3 \end{pmatrix} - \begin{pmatrix} 2 \\ -5 \end{pmatrix}$$

$$\therefore \quad \mathbf{a} - \mathbf{b} = \begin{pmatrix} 4 \\ 3 \end{pmatrix} - \begin{pmatrix} 2 \\ -5 \end{pmatrix} = \begin{pmatrix} 4 & -2 \\ 3 - (-5) \end{pmatrix} = \begin{pmatrix} 2 \\ 8 \end{pmatrix}$$

Therefore to subtract vectors we subtract the corresponding numbers in the ordered pairs.

Note that $\qquad \mathbf{b} - \mathbf{a} = \begin{pmatrix} 2 \\ -5 \end{pmatrix} - \begin{pmatrix} 4 \\ 3 \end{pmatrix} = \begin{pmatrix} -2 \\ -8 \end{pmatrix}$

and this is *not* the same as $\mathbf{a} - \mathbf{b}$.

So in subtraction, the order *does* matter.

Exercise 9h

$$a = \begin{pmatrix} 3 \\ 6 \end{pmatrix}, \; b = \begin{pmatrix} -2 \\ 3 \end{pmatrix}$$

$$a - b = \begin{pmatrix} 3 \\ 6 \end{pmatrix} - \begin{pmatrix} -2 \\ 3 \end{pmatrix} = \begin{pmatrix} 3 - (-2) \\ 6 - 3 \end{pmatrix} = \begin{pmatrix} 5 \\ 3 \end{pmatrix}$$

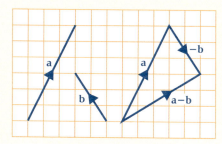

As **b** goes 2 back and 3 up,
$-b$ goes 2 forward and 3 down.

In questions **1** to **4**, find $a - b$ and draw diagrams to represent **a**, **b** and $a - b$:

1 $a = \begin{pmatrix} 2 \\ 5 \end{pmatrix}, \; b = \begin{pmatrix} -3 \\ 2 \end{pmatrix}$

2 $a = \begin{pmatrix} -3 \\ 4 \end{pmatrix}, \; b = \begin{pmatrix} -3 \\ -2 \end{pmatrix}$

3 $a = \begin{pmatrix} 6 \\ 3 \end{pmatrix}, \; b = \begin{pmatrix} 4 \\ -1 \end{pmatrix}$

4 $a = \begin{pmatrix} -4 \\ -3 \end{pmatrix}, \; b = \begin{pmatrix} 1 \\ -4 \end{pmatrix}$

Find the following vectors:

5 $\begin{pmatrix} 10 \\ 3 \end{pmatrix} - \begin{pmatrix} 8 \\ 2 \end{pmatrix}$

6 $\begin{pmatrix} 6 \\ 4 \end{pmatrix} - \begin{pmatrix} 3 \\ 2 \end{pmatrix}$

7 $\begin{pmatrix} 9 \\ 12 \end{pmatrix} - \begin{pmatrix} -2 \\ 3 \end{pmatrix}$

8 $\begin{pmatrix} 4 \\ 5 \end{pmatrix} - \begin{pmatrix} -1 \\ -3 \end{pmatrix}$

9 $\begin{pmatrix} 6 \\ 2 \end{pmatrix} - \begin{pmatrix} 9 \\ 4 \end{pmatrix}$

10 $\begin{pmatrix} 3 \\ 5 \end{pmatrix} - \begin{pmatrix} 10 \\ 2 \end{pmatrix}$

11 $\begin{pmatrix} -2 \\ 7 \end{pmatrix} - \begin{pmatrix} -4 \\ 3 \end{pmatrix}$

12 $\begin{pmatrix} -5 \\ -2 \end{pmatrix} - \begin{pmatrix} -6 \\ -1 \end{pmatrix}$

13 $\begin{pmatrix} 4 \\ 3 \end{pmatrix} + \begin{pmatrix} 2 \\ 3 \end{pmatrix} + \begin{pmatrix} -1 \\ 4 \end{pmatrix}$

14 $\begin{pmatrix} 5 \\ 2 \end{pmatrix} + \begin{pmatrix} -3 \\ -4 \end{pmatrix} + \begin{pmatrix} 2 \\ -3 \end{pmatrix}$

15 $\begin{pmatrix} 7 \\ 6 \end{pmatrix} - \begin{pmatrix} 2 \\ 3 \end{pmatrix} - \begin{pmatrix} 1 \\ 4 \end{pmatrix}$

16 $\begin{pmatrix} 3 \\ 1 \end{pmatrix} - \begin{pmatrix} 4 \\ 2 \end{pmatrix} - \begin{pmatrix} -3 \\ -4 \end{pmatrix}$

17 $\begin{pmatrix} 3 \\ 2 \end{pmatrix} + \begin{pmatrix} -4 \\ 3 \end{pmatrix} - \begin{pmatrix} 2 \\ -6 \end{pmatrix}$

18 $\begin{pmatrix} -5 \\ -2 \end{pmatrix} - \begin{pmatrix} 4 \\ -3 \end{pmatrix} + \begin{pmatrix} -2 \\ 6 \end{pmatrix}$

19 If $a = \begin{pmatrix} 6 \\ 2 \end{pmatrix}$, and $b = \begin{pmatrix} 5 \\ 0 \end{pmatrix}$ find: **a** $a - b$ **b** $b - a$

20 If $a = \begin{pmatrix} -4 \\ 2 \end{pmatrix}, \; b = \begin{pmatrix} 2 \\ 6 \end{pmatrix}$ and $c = \begin{pmatrix} 5 \\ 9 \end{pmatrix}$ find:

a $a - b$ **b** $b - c$ **c** $c - b$

21 If $a = \begin{pmatrix} 5 \\ 4 \end{pmatrix}, \; b = \begin{pmatrix} 2 \\ 6 \end{pmatrix}$ and $c = \begin{pmatrix} -3 \\ -1 \end{pmatrix}$ find:

a $2a - b$ **b** $3b - c$ **c** $2a - 5b$ **d** $a + b - c$ **e** $b - c - a$

Q: What do you get if you cross a mountain climber with a vector?

A: Nothing, a mountain climber is a scalar.

Did you know that M.A.T.H.S. stands for Madness Attacking The Head Slowly?

IN THIS CHAPTER...

you have seen that:

- a vector has size (magnitude or length) and direction

- a scalar has size only

- a vector can be represented by an ordered pair of numbers written vertically. The top number gives movement across and the bottom number gives movement up or down

- vectors can be added by adding the top numbers in the ordered pairs together and the bottom numbers in the ordered pairs together

- a vector can be subtracted by subtracting the top number in the ordered pair and subtracting the bottom number in the ordered pair

REVIEW TEST 1 – CHAPTERS 1 TO 9

Answer questions **1** to **11** by choosing the letter for the correct answer.

1 In standard form, $0.007\,32 =$

   A 0.0732×10^{-1} **B** 0.732×10^{-2} **C** 7.32×10^{-3} **D** 73.2×10^{-4}

2 To two significant figures, $6.7483 =$

   A 6.7 **B** 6.74 **C** 6.75 **D** 6.8

3 What percentage of 14 is 42 ?

   A $\frac{1}{3}\%$ **B** 3% **C** $33\frac{1}{3}\%$ **D** 300%

4 Which of the following is/are true for the diagonals of a rhombus ?

   I They bisect the angles.

   II They bisect each other.

   III They are perpendicular.

   A I and II only   **B** II and III only   **C** I and III only   **D** I, II and III

5 If the bearing of Q from P is $135°$, the bearing of P from Q is

   A $45°$ **B** $135°$ **C** $225°$ **D** $315°$

6 When simplified, $8 - 3(x - 2) =$

   A $5x - 10$ **B** $14 - 3x$ **C** $6 - 3x$ **D** $5x - 2$

7 Given $p = q - 3r$, if $q = 5$, $r = -2$, then $p =$

   A 11 **B** 9 **C** 1 **D** -4

8 If two die are tossed together what is the probability that the total shown is 5 ?

   A $\frac{4}{36}$ **B** $\frac{5}{36}$ **C** $\frac{4}{6}$ **D** $\frac{5}{6}$

9 At what rate per cent per annum simple interest will \$200 amount to \$240 in 2 years ?

   A 10% **B** 20% **C** 40% **D** 50%

10 Given $v = u - gt$, the formula with t as the subject is $t =$

   A $\dfrac{(u - v)}{g}$ **B** $\dfrac{(v - u)}{g}$ **C** $\dfrac{v - u}{g}$ **D** $\dfrac{u - v}{g}$

11 The value of $-9 + 3 - (-4)$ is

   A -10 **B** 2 **C** -2 **D** -16

12 a Express $6\,000\,000 \div 2000$ in standard form.

b A boy saved $7.00 out of each $10 that he earned. What per cent of his earnings did he save?

If he saved $105 in all, what did he earn?

13

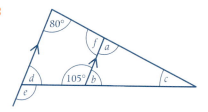

Find the size of each angle marked with a letter.

14 Without using a protractor, construct a quadrilateral LMNO with $LM = 8.4\,cm$, $MN = 5.2\,cm$, the angle $OLM = 60°$ and ON parallel to LM.

15 Solve **a** $17 - 3x = 8 + 3x$ **b** $3 - 4x = -7 - 2x$

16 Two four-sided dice, one red and one blue, with faces labelled 1, 2, 3 and 4 are rolled together. Draw up a possibility space showing all the combinations in which the dice may land.
How many outcomes are there?
What is the probability that the number on the red minus the number on the blue is 1?

17 After depositing his savings in the bank for 3 years, a man received $150 simple interest. The rate of interest was 5 per cent per annum. What sum did he deposit?

18 A bag contains red and blue discs. When a disc is drawn the probability that it is red is $\frac{1}{3}$.

a Find the probability that it is blue.

One disc is drawn and replaced. A second disc is now drawn.

Show this information on a tree diagram and use it to find the probability that

b both discs are red

c one is red and the other is blue.

19 Find the compound interest on $750 invested for 2 years at 8%.

20 A table and four chairs are priced $550 + sales tax at 15%. How much do they cost?

21 If $\mathbf{a} = \begin{pmatrix} 4 \\ -3 \end{pmatrix}$, $\mathbf{b} = \begin{pmatrix} -2 \\ -1 \end{pmatrix}$ and $\mathbf{c} = \begin{pmatrix} -5 \\ 2 \end{pmatrix}$

find

a $2\mathbf{a} + \mathbf{b}$ **b** $\mathbf{a} - 2\mathbf{b}$ **c** $3\mathbf{a} + 2\mathbf{b} - \mathbf{c}$

10 REFLECTIONS AND TRANSLATIONS

AT THE END OF THIS CHAPTER...

you should be able to:

1 Identify shapes that have axes of symmetry.

2 Complete drawings of shapes, given their axes of symmetry.

3 Draw in axes of symmetry for given shapes.

Next time you go to an electrical appliance store, look at the trademark on a Westinghouse Electrical Corporation appliance, or to a Shell Service Station, look at the Shell trademark. These trademarks both have line symmetry.

BEFORE YOU START

you need to know:

✓ how to plot points on a set of x and y axes
✓ what a vector is and how to describe it
✓ the equations of lines parallel to the x and y axes

KEY WORDS

axis of symmetry, bilateral symmetry, corresponding vertices, displacement, image, invariant line, invariant point, line symmetry, mapped, midpoint, mirror line, object, parallel, perpendicular, perpendicular bisector, right-angle, translation, reflection, vector.

Line symmetry

  **CODE**

As we saw in Book 1, shapes like these are *symmetrical*. They have line symmetry (or bilateral symmetry); the dotted line is the *axis of symmetry* because if the shape were folded along the dotted line, one half of the drawing would fit exactly over the other half.

Exercise 10a

1 Which of the following shapes have an axis of symmetry?

a   b   c

Copy the following drawings on squared paper and complete them so that the dotted line is the axis of symmetry.

2 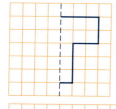   4    6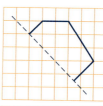

3    5 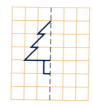   7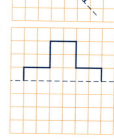

Two or more axes of symmetry

    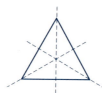   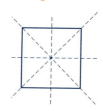

Shapes can have more than one axis of symmetry. In the drawings above, the axes are shown by dotted lines and it is clear that the first shape has two axes of symmetry, the second has three and the third has four.

Exercise 10b

Sketch or trace the shapes in questions **1** to **12**. Mark in the axes of symmetry and say how many there are. (Some shapes may have no axis of symmetry.)

1   **5**   **9**

2   **6**   **10**

3   **7**    **11**

4   **8**   **12**

Copy and complete the following drawings on squared paper. The dotted lines are the axes of symmetry.

13 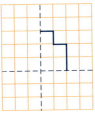   **15** 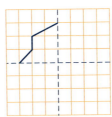   **17**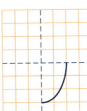

14 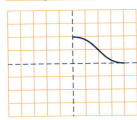   **16** 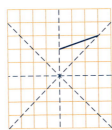   **18**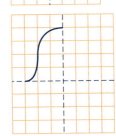

19 Draw, on squared paper or on plain paper, shapes of your own with more than one axis of symmetry.

 Puzzle

Show how sixteen counters can be arranged in ten rows with exactly four counters in each row.

Reflections

Consider a piece of paper, with a drawing on it, lying on a table. Stand a mirror upright on the paper and the reflection can be seen as in the picture.

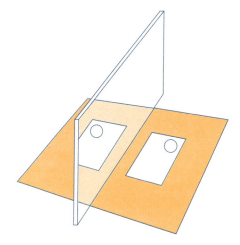

If we did not know about such things as mirrors, we might imagine that there were two pieces of paper lying on the table like this:

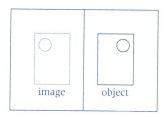

The *object* and the *image* together form a symmetrical shape and the *mirror line* is the axis of symmetry.

Exercise 10c

In this exercise it may be helpful to use a small rectangular mirror, or you can use tracing paper to trace the object and turn the tracing paper over, to find the shape of the image.

Copy the objects and mirror lines (indicated by dotted lines) on to squared paper and draw the image of each object.

1 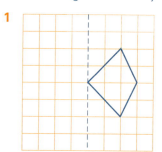 **2** 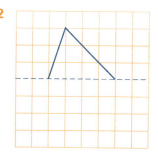 **3**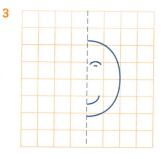

4 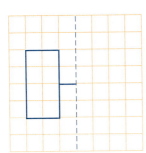   **5** 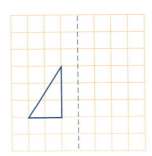   **6**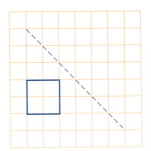

Copy triangle ABC and the mirror line on to squared paper. Draw the image. Label the corresponding vertices (corners) of the image A′, B′, C′.

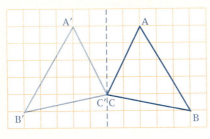

(In this case C and C′ are the same point.)

In each of the following questions, copy the object and the mirror line on to squared paper. Draw the image. Label the vertices of the object A, B, C, etc. and label the corresponding vertices of the image A′, B′, C′, etc.

7 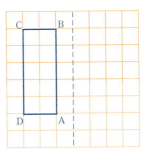   **8** 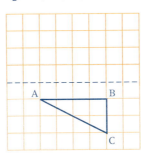   **9**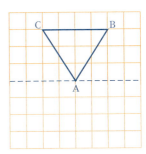

In mathematical reflection, though not in real life, the object can cross the mirror line.

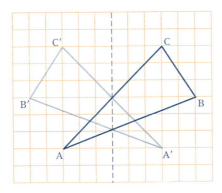

10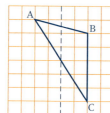

13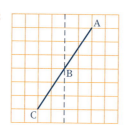

16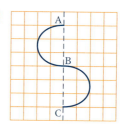

11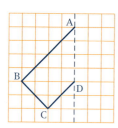

14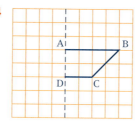

17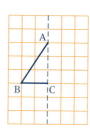

12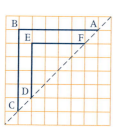

15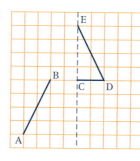

18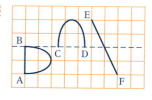

19 Which points in questions **7** to **18** are labelled twice? What is special about their positions?

20 In the diagram for question **10**, join A and A'.

   a Measure the distances of A and A' from the mirror line. What do you notice?

   b At what angle does the line AA' cut the mirror line?

21 Repeat question **20** on other suitable diagrams, in each case joining each object point to its image point. What conclusions do you draw?

In questions **22** to **25** use 1 cm to 1 unit.

22 Draw axes, for x from -5 to 5 and for y from 0 to 5. Draw triangle ABC by plotting A(1, 2), B(3, 2) and C(3, 5). Draw the image A'B'C' when ABC is reflected in the y-axis.

23 Draw axes, for x from 0 to 5 and for y from -2 to 2. Draw triangle PQR where P is (1, -1), Q is (5, -1) and R is (4, 0). Draw the image P'Q'R' when $\triangle$PQR is reflected in the x-axis.

24 Draw axes for x and y from -5 to 1. Draw rectangle WXYZ: W is (-3, -1), X(-3, -2), Y(-5, -2) and Z(-5, -1). Draw the mirror line $y = x$. Draw the image W'X'Y'Z' when WXYZ is reflected in the mirror line.

25 Draw axes for x and y from -1 to 9. Plot the points A(2, 1), B(5, 1), C(7, 3) and D(4, 3). Draw the parallelogram ABCD and its image by reflection in the line $y = x$.

26 Draw axes for x and y from -6 to 8. Draw triangle ABC where A is $(-6, -2)$, B is $(-3, -4)$ and C is $(-2, -1)$. Draw the following images of triangle ABC:

  a triangle $A_1B_1C_1$ by reflection in the y-axis

  b triangle $A_2B_2C_2$ by reflection in the line $y = -x$ (this is the straight line through the points $(2, -2)$, $(-4, 4)$)

  c triangle $A_3B_3C_3$ by reflection in the x-axis

  d triangle $A_4B_4C_4$ by reflection in the line $x = -1$

Invariant points

A point that is its own image, i.e. such that the object point and its image are in the same place, is called an *invariant point*. The previous examples showed that, with reflection, the invariant points lie on the mirror line. The mirror line is an *invariant line*.

Finding the mirror line

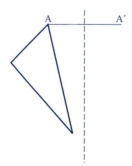
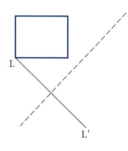

We can see from these diagrams, and from the work in the previous exercise, that the object and image points are at equal distances from the mirror line, and the lines joining them (e.g. AA′ and LL′) are perpendicular (at right angles) to the mirror line.

Exercise **10d**   Find the mirror line if △A′B′C′ is the image of △ABC.

The mirror line is halfway between an object point and its image and perpendicular to the line through them.

So the mirror line is halfway between B and B′ and perpendicular to the line BB′. Check that it also goes through the midpoint of CC′.

The mirror line is the line $x = 1$

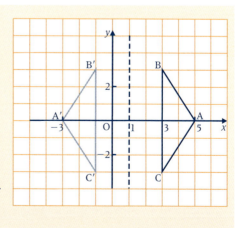

Copy the diagrams in questions **1** to **4** and draw in the mirror lines.

1

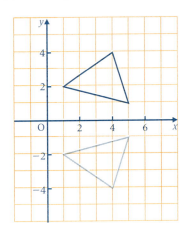

3

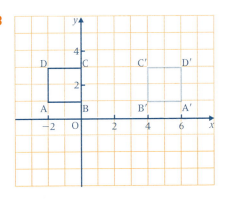

2

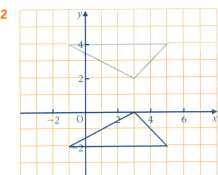

4

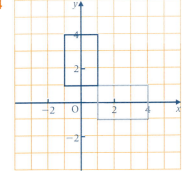

Draw axes for x and y from -5 to 5 for each of questions **5** to **8**.

5 Draw square PQRS: P(1, 1), Q(4, 1), R(4, 4), S(1, 4). Draw square P′Q′R′S′: P′(−2, 1), Q′(−5, 1), R′(−5, 4), S′(−2, 4). Draw the mirror line so that P′Q′R′S′ is the reflection of PQRS and write down its equation.

6 Draw △XYZ: X(2, 1), Y(4, 4), Z(−2, 4), and △X′Y′Z′: X′(2, 1), Y′(4, −2), Z′(−2, −2). Draw the mirror line so that △X′Y′Z′ is the reflection of △XYZ and write down its equation. Are there any invariant points? If there are, name them.

7 Draw △ABC: A(−2, 0), B(0, 2), C(−3, 3), and △PQR: P(3, −1), Q(4, −4), R(1, −3). Draw the mirror line so that △PQR is the reflection of △ABC. Which point is the image of A? Are there any invariant points? If there are, name them.

8 Draw lines AB and PQ: A(2, −1), B(4, 4), P(−2, −1), Q(−5, 4). Is PQ a reflection of AB? If it is, draw the mirror line. If not, give a reason.

If A'B'C' is the reflection of ABC,
draw the mirror line.

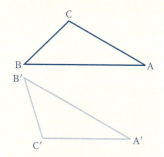

(Join AA' and BB' and find their midpoints,
marking them P and Q. Then PQ is the mirror line.)

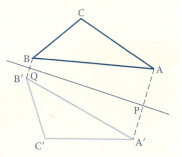

Whenever you attempt to draw a mirror line in this way, always check that the
mirror line is at right angles to AA' and BB'. If it is not, then A'B'C' cannot be a
reflection of ABC.

9 Trace the diagrams and draw the mirror lines.

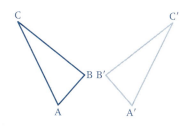
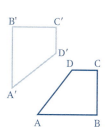
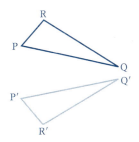

10 Draw axes for x and y from -4 to 5. Draw △ABC: A(3, 1), B(4, 5), C(1, 4), and
△A'B'C': A'(0, −2), B'(−4, −3), C'(−3, 0). Draw the mirror line so that △A'B'C' is
the image of △ABC.

11 Draw axes for x and y from -4 to 4. Draw lines AB and PQ: A(−4, 3), B(0, 4),
P(1, −2), Q(2, 2). Draw the mirror line so that AB is the image of PQ.

12 Draw axes for x and y from -3 to 5. Draw △XYZ: X(3, 2), Y(5, 2), Z(3, 5), and
△LMN: L(0, −3), M(0, −1), N(−3, −1). Draw the mirror line so that △LMN is the
image of △XYZ.

Construction of the mirror line

If we have only one point and its image, and we cannot use squares to guide us, we can use the fact that the mirror line goes through the midpoint of AA′ and is perpendicular to AA′. The mirror line is therefore the perpendicular bisector of AA′ and can be constructed with compasses.

Exercise 10e

1 On plain paper mark two points P and P′ about 10 cm apart in the middle of the page and construct the perpendicular bisector of PP′. Join PP′ and check that it is cut in half by the line you have constructed and that the two lines cut at right angles. Are we correct in saying that P′ is the reflection of P in the constructed line?

2 On squared paper draw axes for x and y from −5 to 5, using 1 cm to 1 unit. A is the point (5, 2) and A′ is the point (−3, −3). Construct the mirror line so that A′ is the reflection of A.

3 Draw axes for x and y from −1 to 8, using 1 cm to 1 unit. B is the point (−1, 0) and B′ is the point (6, 3). Construct the mirror line so that B′ is the reflection of B.

Other transformations

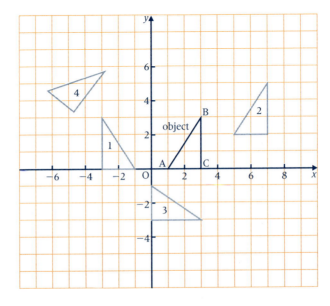

Imagine a triangle ABC cut out of card and lying in the position shown. We can reflect △ABC in the *y*-axis by picking up the card, turning it over and putting it down again in position 1.

Starting again from its original position, we can change its position by sliding the card over the surface of the paper to position 2, 3 or 4. Some of these movements can be described in a simple way, some are more complicated.

Translations

Consider the movements in the diagram: All these movements are of the same type. The side AB remains parallel to the *x*-axis in each case and the triangle continues to face in the same direction. This type of movement is called a *translation*.

Although not a reflection we still use the words *object* and *image*.

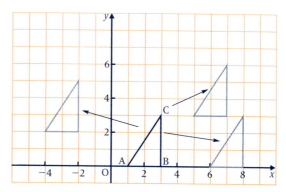

*Exercise **10f***

1 In the following diagram, which images of △ABC are given by translations?

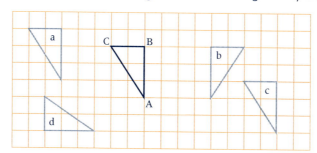

2 In the following diagram, which images of △ABC are given by a translation, which by a reflection and which by neither?

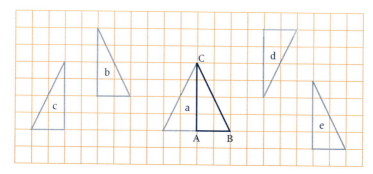

3 Repeat question **2** with the diagram at the top of page 170.

Descriptions of translations

Draw sketches to illustrate the following translations:

1 An object is translated 6 cm to the left.
2 An object is translated 4 units parallel to the *x*-axis to the right.
3 An object is translated 3 m due north.
4 An object is translated 5 km south-east.
5 An object is translated 3 units parallel to the *x*-axis to the right and then 4 units parallel to the *y*-axis upwards.

Using vectors to describe translations

The translation in question **5**, Exercise 10g, can be given more briefly in vector form as $\begin{pmatrix} 3 \\ 4 \end{pmatrix}$.

In Book 1 we saw that the top number gives the displacement parallel to the *x*-axis and the lower number gives the displacement parallel to the *y*-axis.

$\begin{pmatrix} 1 \\ 2 \end{pmatrix}$ $\begin{pmatrix} 3 \\ -2 \end{pmatrix}$ 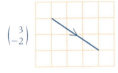 $\begin{pmatrix} -4 \\ 3 \end{pmatrix}$ 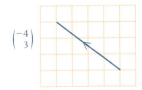 $\begin{pmatrix} -2 \\ -1 \end{pmatrix}$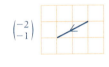

If the top number is negative, the displacement is to the left and if the lower number is negative, the displacement is downwards.

Consider the diagram:

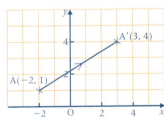

$$\overrightarrow{AA'} = \begin{pmatrix} 5 \\ 3 \end{pmatrix}$$

A′ is the image of A under the translation described by the vector $\begin{pmatrix} 5 \\ 3 \end{pmatrix}$.

A is *mapped* to A′ by the translation described by the vector $\begin{pmatrix} 5 \\ 3 \end{pmatrix}$.

Find the images of the points given in questions **1** to **10** under the translations described by the given vectors.

> **Tip** Draw the point (3, 1) then move 4 units to the right and 2 units up.

 1 (3, 1), $\begin{pmatrix} 4 \\ 2 \end{pmatrix}$ **2** (4, 5), $\begin{pmatrix} 2 \\ 4 \end{pmatrix}$

3 $(-2, 4)$, $\begin{pmatrix} 4 \\ 3 \end{pmatrix}$

4 $(3, 2)$, $\begin{pmatrix} -2 \\ 3 \end{pmatrix}$

Tip $\begin{pmatrix} -2 \\ 3 \end{pmatrix}$ means 2 units to the left and 3 units up.

5 $(4, 5)$, $\begin{pmatrix} -3 \\ -2 \end{pmatrix}$

7 $(-6, -3)$, $\begin{pmatrix} 4 \\ 1 \end{pmatrix}$

9 $(3, -2)$, $\begin{pmatrix} 6 \\ -4 \end{pmatrix}$

6 $(4, -4)$, $\begin{pmatrix} 2 \\ -3 \end{pmatrix}$

8 $(1, 1)$, $\begin{pmatrix} -5 \\ -3 \end{pmatrix}$

10 $(7, 4)$, $\begin{pmatrix} -5 \\ -4 \end{pmatrix}$

In questions **11** to **16**, find the vectors describing the translations that map A to A′.

11 A$(1, 2)$, A′$(5, 3)$

13 A$(-4, -3)$, A′$(0, 0)$

15 A$(-3, -4)$, A′$(-5, -6)$

12 A$(3, 8)$, A′$(2, 9)$

14 A$(-2, 6)$, A′$(2, 6)$

16 A$(4, -2)$, A′$(5, -1)$

In questions **17** to **19**, the given point A′ is the image of an object point A under the translation described by the given vector. Find A.

17 A′$(7, 9)$, $\begin{pmatrix} 2 \\ 3 \end{pmatrix}$

18 A′$(0, 6)$, $\begin{pmatrix} 2 \\ 3 \end{pmatrix}$

19 A′$(-3, -2)$, $\begin{pmatrix} 1 \\ 3 \end{pmatrix}$

Investigation

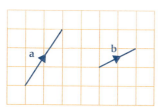

This diagram shows the vectors $\mathbf{a} = \begin{pmatrix} 2 \\ 3 \end{pmatrix}$ and $\mathbf{b} = \begin{pmatrix} 2 \\ 1 \end{pmatrix}$.

The point $(6, 7)$ can be reached from the origin by adding $\mathbf{a}$ and $\mathbf{b}$, and then adding $\mathbf{a}$ to the result. This is shown in the diagram.

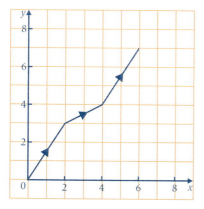

a In how many different ways can these vectors be combined to get from the origin to the point $(6, 7)$? Show them on a diagram.

b The vector in the direction opposite to $\mathbf{a}$ is called $-\mathbf{a}$ so as

$\mathbf{a} = \begin{pmatrix} 2 \\ 3 \end{pmatrix}$, $-\mathbf{a} = \begin{pmatrix} -2 \\ -3 \end{pmatrix}$. Similarly $-\mathbf{b} = \begin{pmatrix} -2 \\ -1 \end{pmatrix}$.

Investigate how many of the points on this grid it is possible to get to from the origin if combinations of $\mathbf{a}$, $\mathbf{b}$, $-\mathbf{a}$ and $-\mathbf{b}$ are allowed.

A translation moves each point of an object the same distance in the same direction.

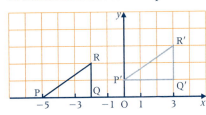

$$\overrightarrow{PP'} = \begin{pmatrix} 5 \\ 1 \end{pmatrix} \quad \overrightarrow{RR'} = \begin{pmatrix} 5 \\ 1 \end{pmatrix} \quad \overrightarrow{QQ'} = \begin{pmatrix} 5 \\ 1 \end{pmatrix}$$

i.e. $\overrightarrow{PP'} = \overrightarrow{QQ'} = \overrightarrow{RR'}$

Exercise 10i

1   Given the following diagrams, find the vectors $\overrightarrow{AA'}$, $\overrightarrow{BB'}$ and $\overrightarrow{CC'}$. Are they all equal? Is the transformation a translation?

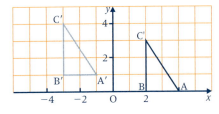

2   Given the following diagrams, find the vectors $\overrightarrow{LL'}$, $\overrightarrow{MM'}$ and $\overrightarrow{NN'}$. Are they all equal? Is the transformation a translation?

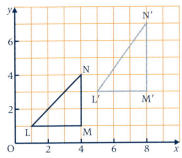

3   Find the vector that describes the translation mapping A to A', B to B' and C to C'.

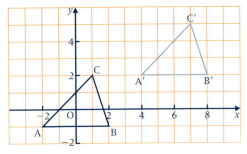

4   Give the vectors describing the translations that map

   a   △ABC to △PQR

   b   △PQR to △ABC.

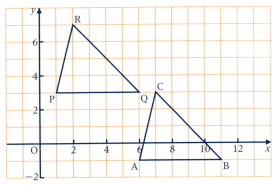

5 Give the vectors describing the translations that map

    a △ABC to △PQR

    b △ABC to △LMN

    c △XYZ to △ABC

    d △ABC to △ABC

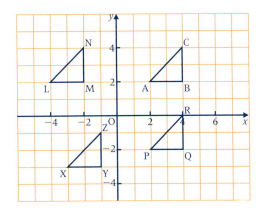

6 Draw axes for x and y from -4 to 5. Draw the following triangles:

    △ABC with A(2, 2), B(4, 2), C(2, 5);
    △PQR with P(1, −2), Q(3, −2), R(1, 1);
    △XYZ with X(−3, 1), Y(−1, 1), Z(−3, 4).

    Give the vectors describing the translations that map

    a △ABC to △PQR   **b** △PQR to △ABC

    c △PQR to △XYZ   **d** △ABC to △ABC

7 Draw axes for x and y from 0 to 9. Draw △ABC with A(3, 0), B(3, 3), C(0, 3) and △A′B′C′ with A′(8, 2), B′(8, 5), C′(5, 5).

    Is △A′B′C′ the image of △ABC under a translation? If so, what is the vector describing the translation?

    Join AA′, BB′ and CC′. What type of quadrilateral is AA′B′B? Give reasons. Name other quadrilaterals of the same type in the figure.

8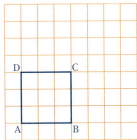

    a Square ABCD is translated parallel to AB a distance equal to AB. Sketch the diagram and draw the image of ABCD.

    b Square ABCD is translated parallel to AC a distance equal to AC. Sketch the diagram and draw the image of ABCD.

9 Draw axes for x and y from -2 to 7. Draw △ABC with A(−2, 5), B(1, 3), C(1, 5).

Translate △ABC using the vector $\begin{pmatrix} 5 \\ 1 \end{pmatrix}$. Label this image $A_1B_1C_1$. Then translate

△$A_1B_1C_1$ using the vector $\begin{pmatrix} -1 \\ -3 \end{pmatrix}$. Label this new image $A_2B_2C_2$.

Give the vectors describing the translations that map

    a △ABC to △$A_2B_2C_2$ **b** △$A_2B_2C_2$ to △ABC   **c** △$A_2B_2C_2$ to △$A_1B_1C_1$

Tangrams

What is a tangram?

A tangram is a puzzle composed of seven parts called tans. The tans are formed by cutting a square and its interior into five triangles, a square and a parallelogram.

Measure all angles of the tans. What do you find?

Find the areas of a large triangle and a small triangle. What is the relationship?

Can you find other relationships?

Tangram play began in China and was introduced into the western world during the 1800s. John Q. Adams and Edgar A. Poe are said to have enjoyed tangrams.

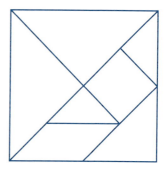

The object of the puzzle is to put the seven tans together to form outlines of all sorts. Use all the pieces, and do not overlap the tans.

It is said that tangrams contain serious as well as playful mathematics. But, what doesn't?

IN THIS CHAPTER...

you have seen that:

- when an object is reflected in a mirror line, the object and the image are symmetrical about the mirror line

- the mirror line is the perpendicular bisector of the line joining a point on the object to the corresponding point on the image

- a translation moves an object without turning it or reflecting it

- a translation can be described by a vector.

11 ROTATIONS

Did you know that 50 is the same as 20?

Trace this **50** now turn the tracing paper over and turn it round.

BEFORE YOU START

you need to know:
- ✓ how to recognise shapes with line symmetry
- ✓ how to mark scales and plot points on a set of axes
- ✓ how to use ruler and compasses to draw the perpendicular bisector of a line
- ✓ how to use a protractor to measure angles

Rotational symmetry

Some shapes have a type of symmetry different from line symmetry.

a 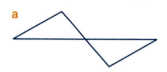 b  c

These shapes do not have an axis of symmetry but can be turned or rotated about a centre point and still look the same. Such shapes are said to have *rotational symmetry*. The centre point is called the *centre of rotation*.

Exercise 11a

1 a  b 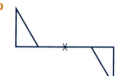 c

Trace each of the shapes above, then turn the tracing paper about the centre of rotation (put a compass point or a pencil point in the centre). Turn until the traced shape fits over the original shape again. In each case state through what fraction of a complete turn the shape has been rotated.

2 Which of the following shapes have rotational symmetry?

a  b

c  d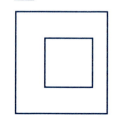

Order of rotational symmetry

If the smallest angle that a shape needs to be turned through is a third of a complete turn to fit, then it will need two more such turns to return it to its original position. So, starting from its original position, it takes three turns, each one-third of a revolution, to return it to its starting position.

It has *rotational symmetry of order 3*.

Exercise 11b   Give the order of rotational symmetry of the following shape.

You need to decide whether turning this shape about its centre through some angle leaves it looking unchanged to the eye. If so, how big is this angle? How many times can you do this before getting back to the starting position?

The smallest angle turned through is a right angle or one-quarter of a complete turn.

The shape has rotational symmetry of order 4.

1   Give the orders of rotational symmetry of the shapes in Exercise 11a, question **1**.

2   Give the orders of rotational symmetry, if any, of the shapes in Exercise 11a, question **2**.

Copy and complete the diagram, given that there is rotational symmetry of order 4.

If the order of rotational symmetry is to be 4 you must turn the given shape about the cross (×) through one-quarter of a turn, i.e. through 90°. Repeat this until you get back to the starting position.

Each of the diagrams in questions **3** to **8** has rotational symmetry of the order given and **X** marks the centre of rotation.

Copy and complete the diagrams. (Tracing paper may be helpful.)

3

Rotational symmetry of order 4

4

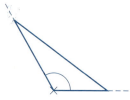

Rotational symmetry of order 3

5

Rotational symmetry of order 2

6

Rotational symmetry of order 4

7

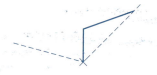

Rotational symmetry of order 3

8

Rotational symmetry of order 2

<u>9</u> In questions **3** to **8**, give the size of the angle, in degrees, through which each shape is turned.

Exercise 11c

Some shapes have both line symmetry and rotational symmetry:

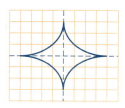

Two axes of symmetry
Rotational symmetry of order 2

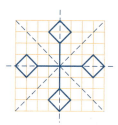

Four axes of symmetry
Rotational symmetry of order 4

Which of the following shapes have

a rotational symmetry only

b line symmetry only

c both?

> **Tip** You can spot line symmetry by seeing if any line acts as a mirror; you can spot rotational symmetry if you can rotate the shape about a point that leaves it looking unchanged.

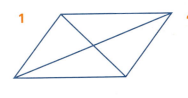 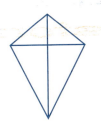 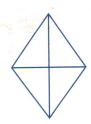

  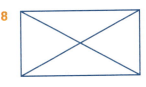

10 Make up three shapes that have rotational symmetry only. Give the order of symmetry and the angle of turn, in degrees.

11 Make up three shapes with line symmetry only. Give the number of axes of symmetry.

12 Make up three shapes that have both line symmetry and rotational symmetry.

13 The capital letter **X** has line symmetry (two axes) and rotational symmetry (of order 2). Investigate the other letters of the alphabet.

 ## Puzzle

A dried-out well is 12 metres deep. A snail starts from the bottom and tries to climb out. It climbs up 3 m every night and falls back 2 m every day. How long will it take the snail to climb out?

Transformations: rotations

a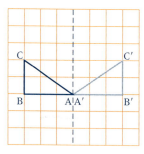
b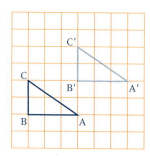
c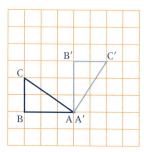

So far, in transforming an object we have used reflections, as in a, and translations, as in b, but for c we need a rotation.

In this case we are rotating △ABC about A through 90° clockwise (@). We could also say △ABC was rotated through 270° anticlockwise ('). Clockwise rotation is sometimes referred to as positive rotation or rotating through a positive angle. Anticlockwise rotation is rotating through a negative angle.

For a rotation of 180° we do not need to say whether it is clockwise or anticlockwise.

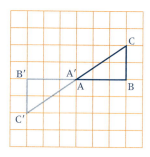

Exercise 11d

Give the angle of rotation when △ABC is mapped to △A'B'C'.

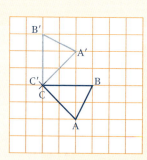

If you rotate △ABC anticlockwise about C it can be turned to position A'B'C'. You must now measure the angle it has turned through. Compare the position of one side of the object triangle with the corresponding side of the image triangle.

The angle of rotation is 90° anticlockwise.

In questions **1** to **4**, give the angle of rotation when △ABC is mapped to △A'B'C'.

1

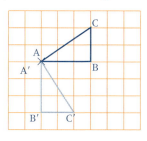

3

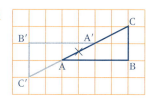

Tip Don't forget to say, where appropriate, whether the rotation is clockwise or anticlockwise.

2

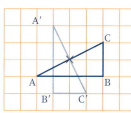

4

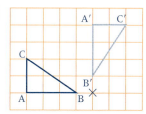

In questions **5** to **10**, state the centre of rotation and the angle of rotation. △ABC is the object in each case.

5

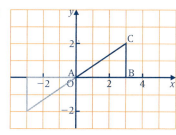

8

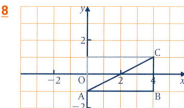

6

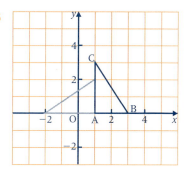

9

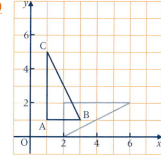

7

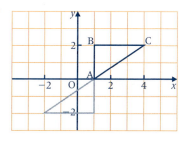

10

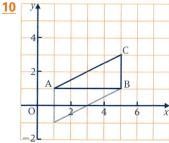

183

Copy the diagrams in questions 11 to 18, using 1 cm to 1 unit.
Find the images of the given objects under the rotations described.

Tip Be careful to do the rotation in the right direction.

11

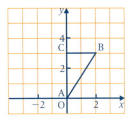

Centre of rotation (0, 0)
Angle of rotation 90° anticlockwise

12

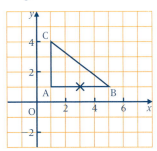

Centre of rotation (3, 1)
Angle of rotation 180°

13

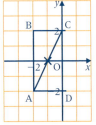

Centre of rotation (−1, 0)
Angle of rotation 180°

14

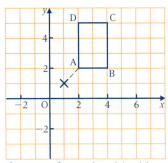

Centre of rotation (1, 1)
Angle of rotation 180°

(As the centre of rotation is not a point on the object, join it to A first.)

15

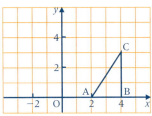

Centre of rotation (0, 0)
A negative angle of rotation of 90°

Tip A negative angle of rotation means anticlockwise.

16

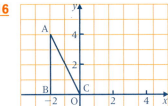

Centre of rotation (2, 0)
A positive angle of rotation of 90°

17

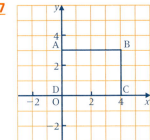

Centre of rotation (2, 0)
Angle of rotation of 90° anticlockwise

18

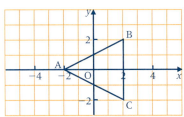

Centre of rotation (0, 0)
Angle of rotation of 180°

19 △ABC is rotated about O through 180° to give the image, △A′B′C′. Copy and complete the diagram, using 1 cm to 1 unit.

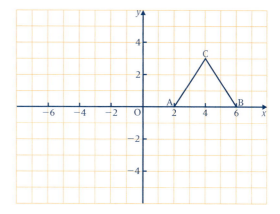

a What is the shape of the path traced out by C as it moves to C′?

b Measure OC and OC′. How do they compare?

Repeat with OB and OB′.

20 Why is the direction of rotation sometimes not given?

21 Draw the diagram accurately. Then draw accurately, using a protractor, the image of △ABC under a rotation of 60° anticlockwise about O.

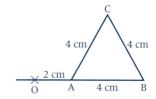

Finding the centre of rotation by construction

As we have seen we can often spot the centre of rotation just by looking at the diagram but sometimes it is not obvious.

In such cases we can use the fact that an object point A and its image point A′ are the same distance from the centre.

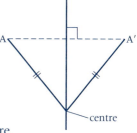

So the centre lies on the perpendicular bisector of AA′.

It also lies on the perpendicular bisector of BB′.

Therefore the point P, where these two bisectors meet, is the centre.

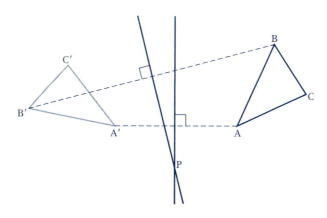

(The perpendicular bisector of CC′ will also go through P.)

Exercise 11e

1

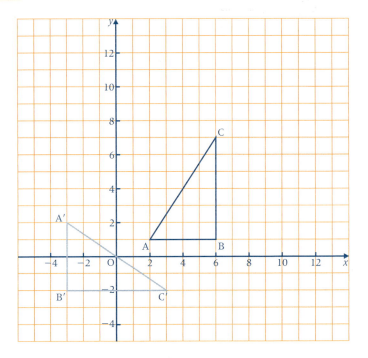

a Copy the diagram, drawing axes for *x* and *y* from −5 to 12. Use 1 cm to 1 unit. △A′B′C′ is the image of △ABC under a rotation.

> **Tip** You need a *sharp* pencil and a good pair of compasses for this work.

b Construct the perpendicular bisectors of AA′ and BB′.

c Mark the centre of rotation, P (that is, the point where the two perpendicular bisectors meet).

d Check that it is the centre by using tracing paper and the point of your compasses.

e Join BP and B′P. Measure BP̂B′. What is the angle of rotation?

2 Draw axes for *x* and *y* from −5 to 10, using 1 cm to 1 unit. Draw △ABC with A(−1, 8), B(5, 4), C(−1, 1) and △A′B′C′ with A′(4, 1), B′(0, −5), C′(−3, 1).

Repeat **b** to **e** in question 1.

3 Draw axes for *x* and *y* from −5 to 10, using 1 cm to 1 unit. Draw △ABC with A(−4, −2), B(2, −2), C(−4, 4) and △A′B′C′ with A′(4, 0), B′(4, 6), C′(−2, 0).

Repeat **b** to **e** in question 1.

Finding the angle of rotation

Having found the centre of rotation, the angle of rotation can be found by joining both an object point and its image to the centre.

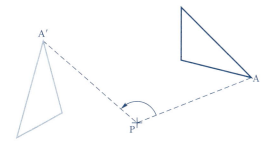

In the diagram above, A′ is the image of A and P is the centre of rotation.

Join both A and A′ to P. AP̂A′ is the angle of rotation.

In this case the angle of rotation is 120° anticlockwise.

Exercise 11f

Trace each of the diagrams and, by drawing in the necessary lines, find the angle of rotation when △ABC is rotated about the centre P to give △A′B′C′.

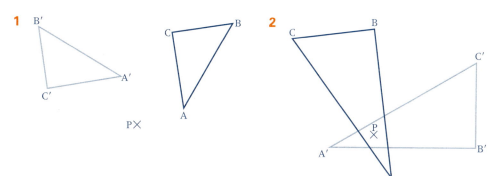

> **Tip** Remember that when you measure an angle, put the 0° line on the protractor over one arm of the angle, then count the number of degrees to turn to the other arm.

 Investigation

Create three shapes, none of which has appeared in this chapter, such that

- one has both line symmetry and rotational symmetry
- another has line symmetry but not rotational symmetry
- and a third has rotational symmetry but not line symmetry.

Mixed questions on reflections, translations and rotations

Name the transformation, describing it fully, if the grey triangle is the image of the black one.

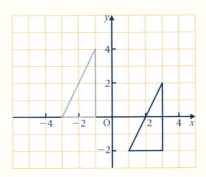

The transformation is a translation given by the vector $\begin{pmatrix} -4 \\ 2 \end{pmatrix}$

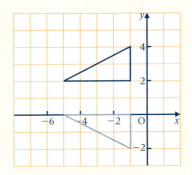

The transformation is a reflection in the line $y = 1$

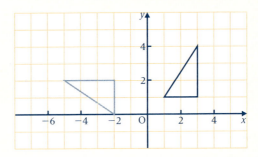

The transformation is a rotation about $(0, -1)$ through an angle of $90°$ anticlockwise

Mixed questions on reflections, translations and rotations

Name the transformations in questions **1** to **10**, describing them fully. The black shape is the object, the grey shape is the image. Describe a translation by a vector; for a reflection give the mirror line and for a rotation give the centre of rotation and the angle turned through.

1

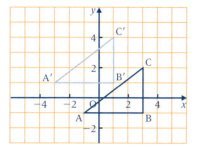

5

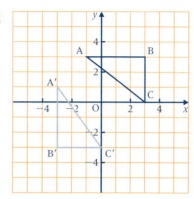

2

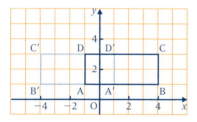

6

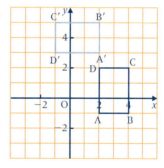

3

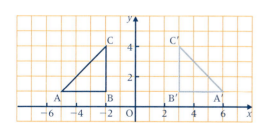

7

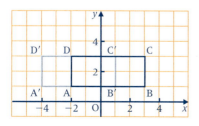

4

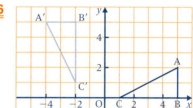

8

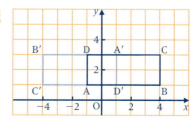

9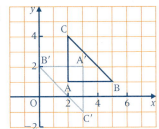

10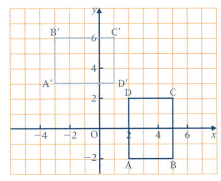

Sometimes we do not know which point is the image of a particular object point. In such cases there could be more than one possible transformation.

(Remember that a rotation of 90° anticlockwise is the same as a rotation of 270° clockwise. Do not give these as two independent transformations.)

11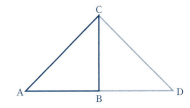

Name and describe two possible transformations that will map the object △ABC to the image △BCD.

12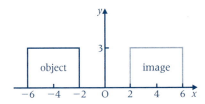

Name and describe three possible transformations that will map the object to the image.

13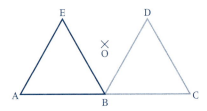

Name and describe four possible transformations that will map the left-hand triangle to the right-hand triangle.

Mixed questions on reflections, translations and rotations

14

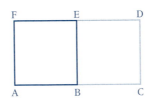

Name and describe five possible transformations that will map the left-hand square to the right-hand square.

15 Copy the diagram using 1 cm to 1 unit.

Reflect $\triangle ABC$ in the line $x = 1$ to give $\triangle A_1B_1C_1$.

Then reflect $\triangle A_1B_1C_1$ in the line $x = -1$ to give $\triangle A_2B_2C_2$.

What single transformation will map $\triangle ABC$ to $\triangle A_2B_2C_2$?

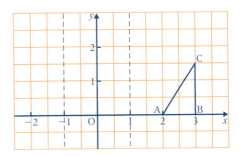

16 Copy the diagram in question **15** again but draw the axis for x from -5 to 7. Repeat the two reflections but use the line $x = -1$ first and $x = 1$ second. What single transformation is needed this time?

17 A car is turning a corner and two of its positions are shown. Trace the drawing, allowing plenty of space above and below, and find the centre of the turning circle.

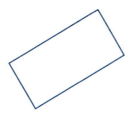
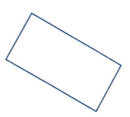

18 Look at the diagram below. Taking one of the shapes as the object, what types of transformations will map it to other shapes in the diagram?

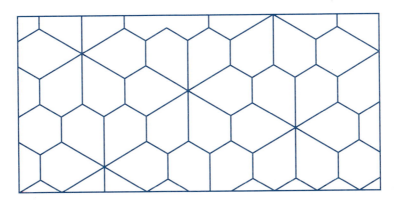

19 Draw axes for x and y from -5 to 5, using 1 cm to 1 unit. Draw lines AB and BC with A(2, 2), B(5, 2) and C(3, 0). Draw the images of ABC under the reflections in the four lines $x = 0$, $x = y$, $y = 0$ and $y = -x$. Draw the images of ABC under the three rotations about O through angles of 90°, 180° and 270° anticlockwise. (The seven images of △ABC, together with △ABC itself, form an eight-pointed star.)

20 Draw axes for x and y from -5 to 5, using 1 cm to 1 unit. Draw △ABC with A(2, 1), B(4, 1) and C(4, 2).

a Reflect △ABC in the line $y = x$ to produce the image △$A_1B_1C_1$. Then rotate △$A_1B_1C_1$ through 180° about O to produce △$A_2B_2C_2$. What single transformation will map △ABC to △$A_2B_2C_2$?

b Rotate △ABC through 180° about O then reflect the image in the line $y = x$. Is the final image the same as △$A_2B_2C_2$?

c Try other pairs of reflections and rotations, starting a fresh diagram where necessary. In each case find the single transformation which is equivalent to the pair. Does the order in which you do the transformations matter? Are the single transformations themselves all reflections or rotations?

Rolling stones gather no moss. What about rolling wheels?

Place a mark on a circular object and roll it along a straight line. As the circle rolls along the line the mark will trace out a curve called a *cycloid*. This name was given by Galileo, who admired this curve – he thought it looked like the arches of a bridge.

IN THIS CHAPTER...

you have seen that:

- a shape has rotational symmetry if it can be rotated about a point to a different position but still look the same

- the order of rotational symmetry is the number of times the shape can be rotated and still look the same until it returns to its original position

- when rotational symmetry is described as turning about a point through a given angle you must say whether the turning is clockwise or anticlockwise

- you can find the centre of rotation by drawing the perpendicular bisectors of lines joining points in the object to the corresponding points in the image

- you can measure the angle of rotation with a protractor.

12

CIRCLES: CIRCUMFERENCE AND AREA

AT THE END OF THIS CHAPTER...

you should be able to:

1   State the relationship between the circumference and diameter of a circle.

2   Calculate the circumference of a circle given its diameter or radius.

3   Solve problems involving the calculation of circumferences of circles.

4   Calculate the radius of a circle of given circumference.

5   Calculate the area of a circle of given radius.

6   Calculate the area of a sector of a circle of given angle.

MATHS IS OUT THERE

Circumference means 'the perimeter of a circle'. The word comes from the Latin *circumferre* – 'to carry around'. The symbol π is the first letter of the Greek word for circumference – *perimetron*. In Germany, π is identified as the ludolphine number, because of the work of Ludolph van Ceulen, who tried to find a better estimate of the number.

BEFORE YOU START

you need to know:
✓   the meaning of significant figures
✓   how to change the subject of a formula
✓   the units of area

KEY WORDS

annulus, circle, circumference, cone, cylinder, diameter, perimeter, pi (π), quadrant, radius, rectangle, revolution, sector, semicircle, square

Diameter, radius and circumference

When you use a pair of compasses to draw a circle, the place where you put the point is the *centre* of the circle. The length of the line that the pencil draws is the *circumference* of the circle.

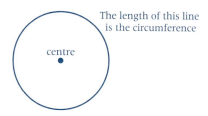

The length of this line is the circumference

centre

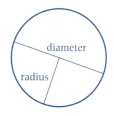

diameter

radius

Any straight line joining the centre to a point on the circumference is a *radius*.

A straight line across the full width of a circle (i.e. going through the centre) is a *diameter*.

The diameter is twice as long as the radius. If *d* stands for the length of a diameter and *r* stands for the length of a radius, we can write this as a formula:

$$d = 2r$$

Exercise 12a

In questions **1** to **6**, write down the length of the diameter of the circle whose radius is given

1

6 cm

2

5 m

3 15 mm

4 3.5 cm

5 1 km

6 4.6 cm

7 For this question you will need some thread and a cylinder (e.g. a tin of soup, a soft drink can, the cardboard tube from a roll of kitchen paper).

Measure across the top of the cylinder to get a value for the diameter. Wind the thread 10 times round the can. Measure the length of thread needed to do this and then divide your answer by 10 to get a value for the circumference. If *C* stands for the circumference and *d* for the length of the diameter, find, approximately, the value of $C \div d$.

(Note that you can also use the label from the cylindrical tin. If you are careful you can reshape it and measure the diameter and then unroll it to measure the circumference.)

8 Compare the results from the whole class for the value of $C \div d$.

Introducing π

From the last exercise you will see that, for any circle,

$$circumference \approx 3 \times diameter$$

The number that you have to multiply the diameter by to get the circumference is slightly larger than 3.

This number is unlike any number that you have met so far. It cannot be written down exactly, either as a fraction or as a decimal: as a fraction it is approximately, but *not* exactly, $\frac{22}{7}$; as a decimal it is approximately 3.142, correct to 3 decimal places.

Now with a computer to do the arithmetic we can find its value to as many decimal places as we choose: it is a never-ending, never-repeating decimal fraction. To sixty decimal places, the value of this number is

3.141592653589793238462643383279502884197169399375105820974944 ...

Because we cannot write it down exactly we use the Greek letter π (pi) to stand for this number. Then we can write a formula connecting the circumference and diameter of a circle in the form $C = \pi d$. But $d = 2r$ so we can rewrite this formula as

$$C = 2\pi r$$

where C = circumference and r = radius

The symbol π was first used by an English writer, William Jones, in 1706. It was later adopted in 1737 by Euler.

Calculating the circumference

Exercise 12b Using 3.142 as an approximate value for π, find the circumference of a circle of radius 3.8 m.

Remember that to get your answer correct to 3 s.f. you must work to 4 s.f.

3.8 m

Using $C = 2\pi r$

with $\pi = 3.142$ and $r = 3.8$

gives $C = 2 \times 3.142 \times 3.8$

$= 23.9$ to 3 s.f.

Circumference $= 23.9$ m to 3 s.f.

Using 3.142 as an approximate value for π and giving your answers correct to 3 s.f., find the circumference of a circle of radius:

1	2.3 m	**4**	53 mm	**7**	36 cm	**10**	0.014 km	**13**	1.4 m
2	4.6 cm	**5**	8.7 m	**8**	4.8 m	**11**	7 cm	**14**	35 mm
3	2.9 cm	**6**	250 mm	**9**	1.8 m	**12**	28 mm	**15**	5.6 cm

For questions **16** to **23** you can use $C = 2\pi r$ or $C = \pi d$.

Read the question carefully before you decide which one to use.

Using $\pi \approx 3.14$ and giving your answer correct to 2 s.f., find the circumference of a circle of:

16 radius 154 mm

17 diameter 28 cm

18 diameter 7.7 m

19 radius 210 mm

20 radius 34.6 cm

21 diameter 511 mm

22 diameter 630 cm

23 diameter 9.1 m

In early times, $\sqrt{10}$ was used as an approximation for π.

 # Investigation

Count Buffon's experiment

Count Buffon was an 18th-century scientist who carried out many probability experiments. The most famous of these is his 'Needle Problem'. He dropped needles on to a surface ruled with parallel lines and considered the drop successful when a needle fell between two lines. His amazing discovery was that the number of successful drops divided by the number of unsuccessful drops was an expression involving π.

You can repeat his experiment and get a good approximation for the value of π from it:

Take a matchstick or a similar small stick and measure its length. Now take a sheet of paper measuring about $\frac{1}{2}$ m each way and fill the sheet with a set of parallel lines whose distance apart is equal to the length of the stick. With the sheet on the floor drop the stick on to it from a height of about 1 m. Repeat this about a hundred times and keep a tally of the number of times the stick touches or crosses a line and of the number of times it is dropped. Then find the value of

$$\frac{2 \times \text{number of times it is dropped}}{\text{number of times it crosses or touches a line}}$$

Problems

Exercise 12c

Use the value of π on your calculator and give your answers correct to 3 s.f.

Find the perimeter of the given semicircle.

(The prefix 'semi' means half.)

Remember that perimeter means distance all round, so you need to find half the circumference then add the length of the straight edge.

8 m

The complete circumference of the circle is $2\pi r$

The curved part of the semicircle is $\frac{1}{2} \times 2\pi r$

$$= \frac{1}{2} \times 2 \times \pi \times 4\,\text{m}$$

$$= 12.57\,\text{m (correct to 4 s.f.)}$$

The perimeter = curved part + straight edge

$$= (12.57 + 8)\,\text{m}$$

$$= 20.57\,\text{m}$$

$$= 20.6\,\text{m} \quad \text{to 3 s.f.}$$

Find the perimeter of each of the following shapes:

1   4 cm

2

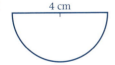

3 cm

Tip (This is called a *quadrant*: it is one quarter of a circle.)

3

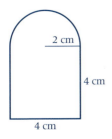

2 cm

4 cm

4 cm

4

5 cm

120°

Tip This is one-third of a circle because 120° is $\frac{1}{3}$ of 360°.

5

10 cm

45°

Tip A 'slice' of a circle is called a *sector*. $\frac{45}{360} = \frac{1}{8}$, so this sector is $\frac{1}{8}$ of a circle.

6

7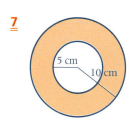

8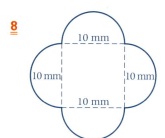

9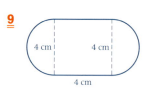

10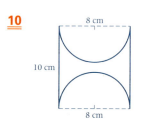

> **Tip** The perimeter is the distance round all the edges. The region between two concentric circles is called an *annulus*.

Exercise 12d

Use the value of π on your calculator and give your answers correct to 3 s.f.

A circular flower bed has a diameter of 1.5 m. A metal edging is to be placed round it. Find the length of edging needed and the cost of the edging if it is sold by the metre (i.e. you can only buy a whole number of metres) and costs 60 c a metre.

First find the circumference of the circle, then how many metres you need.

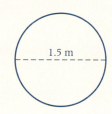

Using $C = \pi d$,

$$C = \pi \times 1.5$$

$$= 4.712\ldots$$

Length of edging needed $= 4.71\,\text{m}$ to 3 s.f.

(Note that if you use $C = 2\pi r$, you must remember to halve the diameter.)

As the length is 4.71 m we have to buy 5 m of edging.

$$\text{Cost} = 5 \times 60\,\text{c}$$

$$= 300\,\text{c} \quad \text{or} \quad \$3.00$$

1 Measure the diameter, in millimetres, of a 25 c coin. Use your measurement to find the circumference of a 25 c coin.

2 Repeat question **1** with a 10 c coin and a 5 c coin.

3 A circular tablecloth has a diameter of 1.4 m. How long is the hem of the cloth?

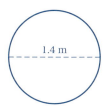

4 A rectangular sheet of metal measuring 50 cm by 30 cm has a semicircle of radius 15 cm cut from each short side as shown. Find the perimeter of the shape that is left.

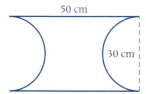

5 A bicycle wheel has a radius of 28 cm. What is the circumference of the wheel?

6 How far does a bicycle wheel of radius 28 cm travel in one complete revolution? How many times will the wheel turn when the bicycle travels a distance of 352 m?

7 A cylindrical tin has a radius of 2 cm. What length of paper is needed to put a label on the tin if the edges just meet?

8 A square sheet of metal has sides of length 30 cm. A quadrant (one quarter of a circle) of radius 15 cm is cut from each of the four corners. Sketch the shape that is left and find its perimeter.

9 A boy flies a model aeroplane on the end of a wire 10 m long. If he keeps the wire horizontal, how far does his aeroplane fly in one revolution?

10 If the aeroplane described in question 9 takes 1 second to fly 10 m, how long does it take to make one complete revolution? If the aeroplane has enough power to fly for 1 minute, how many turns can it make?

11 A cotton reel has a diameter of 2 cm. There are 500 turns of thread on the reel. How long is the thread?

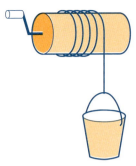

12 A bucket is lowered into a well by unwinding rope from a cylindrical drum. The drum has a radius of 20 cm and with the bucket out of the well there are 10 complete turns of the rope on the drum. When the rope is fully unwound the bucket is at the bottom of the well. How deep is the well?

13 A garden hose is 100 m long. For storage it is wound on a circular hose reel of diameter 45 cm. How many turns of the reel are needed to wind up the hose?

14 The cage that takes miners up and down the shaft of a coal mine is raised and lowered by a rope wound round a circular drum of diameter 3 m. It takes 10 revolutions of the drum to lower the cage from ground level to the bottom of the shaft. How deep is the shaft?

 # Investigation

Ken entered a 50 km sponsored cycle ride. He wondered how many pedal strokes he made. The diameter of each wheel is 70 cm.

a Investigate this problem if one pedal stroke gives one complete turn of the wheels.

b What happens if Ken uses a gear that gives two turns of the wheel for each pedal stroke?

c Find out how the gears on a racing bike affect the ratio of the number of pedal strokes to the number of turns of the wheels.

Discuss the assumptions made in order to answer parts **a** and **b**.

Write a short report on how these assumptions affect the reasonableness of your answers.

Finding the radius of a circle given the circumference

If a circle has a circumference of 24 cm, we can find its radius from the formula $C = 2\pi r$ either by using the formula as it stands,

i.e. $\qquad 24 = 2 \times 3.142 \times r$

and solving this equation for r

or by first making r the subject of $C = 2\pi r$ as follows

Finding the radius of a circle given the circumference

Divide both sides by 2 and π
$$C = 2 \times \pi \times r$$

$$\frac{C}{1} \times \frac{1}{2 \times \pi} = \frac{\cancel{2}}{1} \times \frac{\cancel{\pi}}{1} \times \frac{r}{1} \times \frac{1}{\cancel{2} \times \cancel{\pi}}$$

$$\frac{C}{2\pi} = r$$

i.e.
$$r = \frac{C}{2\pi}$$

Exercise 12e

Use the value of π on your calculator and give your answers correct to 3 s.f.

> The circumference of a circle is 36 m. Find the radius of this circle.
>
> *Either:*   Using $C = 2\pi r$ gives
>
> $$36 = 2 \times \pi \times r$$
> $$36 = 6.283 \times r$$
>
> (Writing down the first 4 digits in the calculator display)
>
> $$\frac{36}{6.283} = r \quad (\text{dividing both sides by } 6.283)$$
> $$= 5.729 \ldots$$
> $$r = 5.73 \quad \text{to 3 s.f.}$$
>
> *Or:*   Using $r = \dfrac{C}{2\pi}$ gives
>
> $$r = \frac{\cancel{36}^{\,18}}{\cancel{2}_{\,1} \times \pi}$$
> $$= 5.729 \ldots$$
> $$= 5.73 \quad \text{to 3 s.f.}$$
>
> Therefore the radius is 5.73 cm to 3 s.f.

Find the radius of the circle whose circumference is:

1 44 cm	**3** 550 m	**5** 462 mm	**7** 36.2 mm	**9** 582 cm
2 121 mm	**4** 275 cm	**6** 831 cm	**8** 391 m	**10** 87.4 m

11 Find the diameter of a circle whose circumference is 52 m.

12 A roundabout at a major road junction is to be built. It has to have a minimum circumference of 188 m. What is the corresponding minimum diameter?

13 A bicycle wheel has a circumference of 200 cm. What is the radius of the wheel?

14 A car has a turning circle whose circumference is 63 m. What is the narrowest road that the car can turn round in without going on the pavement?

15 When the label is taken off a tin of soup it is found to be 32 cm long.
If there was an overlap of 1 cm when the label was on the tin, what is the radius of the tin?

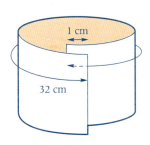

16 The diagram shows a quadrant of a circle. If the curved edge is 15 cm long, what is the length of a straight edge?

17 A tea cup has a circumference of 24 cm. What is the radius of the cup? Six of these cups are stored edge to edge in a straight line on a shelf. What length of shelf do they occupy?

18 Make a cone from a sector of a circle as follows:

On a sheet of paper draw a circle of radius 8 cm. Draw two radii at an angle of 90°. Make a tab on one radius as shown. Cut out the larger sector and stick the straight edges together. What is the circumference of the circle at the bottom of the cone?

19 A cone is made by sticking together the straight edges of the sector of a circle, as shown in the diagram.

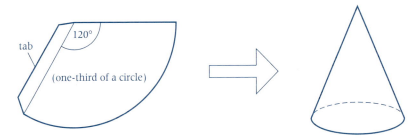

The circumference of the circle at the bottom of the finished cone is 10 cm. What is the radius of the circle from which the sector was cut?

20 The shape in the diagram on the right is made up of a semicircle and a square.
Find the length of a side of this square.

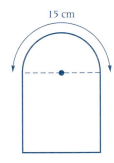

21 The curved edge of a sector of angle 60° is 10 cm.
Find the radius and the perimeter of the sector.

 **Puzzle**

What is the exact time after 1 o'clock when the minute hand of a clock is immediately over the hour hand?

The area of a circle

The formula for finding the area of a circle is

$$A = \pi r^2$$

You can see this if you cut a circle up into sectors and place the pieces together as shown to get a shape which is roughly rectangular. Consider a circle of radius r whose circumference is $2\pi r$.

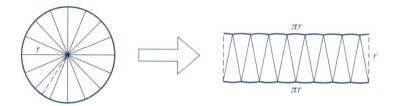

$$\text{Area of circle} = \text{area of 'rectangle'}$$
$$= \text{length} \times \text{width}$$
$$= \pi r \times r = \pi r^2$$

Exercise 12f

Use the value of π on your calculator and give your answers correct to 3 s.f.

Find the area of a circle of radius 2.5 cm.

Using $A = \pi r^2$

with $r = 2.5$

gives $A = \pi \times (2.5)^2$

$\qquad = 19.63 \ldots$

$\qquad = 19.6$ to 3 s.f.

Area is 19.6 cm² to 3 s.f.

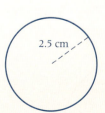

2.5 cm

Find the areas of the following circles:

1
4 cm

4
10 mm

7
3.8 m

Tip Be careful!

2
8 cm

5
3.5 m

8
3.5 km

3
5 m

6
60 cm

9
80 m

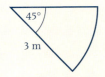
45°
3 m

This is a *sector* of a circle. Find its area.

If we find 45° as a fraction of 360° this will tell us what fraction this sector is of a circle.

$$\frac{\overset{9}{\cancel{45}}}{\underset{72}{\cancel{360}}} = \frac{\overset{1}{\cancel{9}}}{\underset{8}{\cancel{72}}} = \frac{1}{8}$$

$\therefore$ area of sector $= \frac{1}{8}$ of area of circle of radius 3 m

$$\text{Area of sector} = \frac{1}{8} \text{ of } \pi r^2$$

$$= \frac{1}{8} \times \pi \times 9 \,\text{m}^2$$

$$= 3.534 \,\text{m}^2$$

$$= 3.53 \,\text{m}^2 \quad \text{to 3 s.f.}$$

Find the areas of the following shapes:

10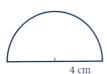
4 cm

12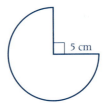
5 cm

11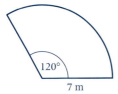
120°
7 m

13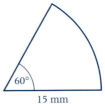
60°
15 mm

14

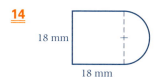

18 mm

18 mm

15

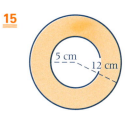

5 cm
12 cm

16

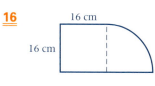

16 cm

16 cm

17

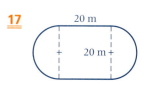

20 m

20 m

18

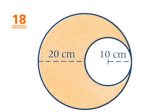

20 cm   10 cm

19

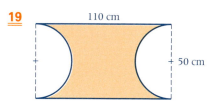

110 cm

50 cm

20

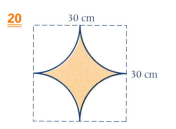

30 cm

30 cm

Problems

Exercise 12g

Use the value of π on your calculator and make a rough sketch to illustrate each problem. Give your answers to 3 s.f.

A circular table has a radius of 75 cm. Find the area of the table top. The top of the table is to be varnished. One tin of varnish covers 4 m². Will one tin be enough to give the table top three coats of varnish?

The area to varnish is three times the area of the top of the table.

75 cm

Area of table top is πr^2

$= \pi \times 75 \times 75 \text{ cm}^2$

$= 17671.4 \ldots$

$= 17\,670 \text{ cm}^2$ to 4 s.f.

$= 17\,670 \div 100^2 \text{ m}^2$

$= 1.767 \text{ m}^2$ to 4 s.f.

(To give an answer correct to 3 s.f. work to 4 s.f.)

For three coats, enough varnish is needed to cover

$$3 \times 1.767\,m^2 = 5.30\,m^2 \text{ to 3 s.f.}$$

So one tin of varnish is not enough.

1 The minute hand on a clock is 15 cm long. What area does it pass over in 1 hour?

2 What area does the minute hand described in question 1 cover in 20 minutes?

3 The diameter of a 25 c coin is 25 mm. Find the area of one of its flat faces.

4 The hour hand of a clock is 10 cm long. What area does it pass over in 1 hour?

5 A circular lawn has a radius of 5 m. A bottle of lawn weedkiller says that the contents are sufficient to cover 50 m². Is one bottle enough to treat the whole lawn?

6 The largest possible circle is cut from a square of paper 10 cm by 10 cm. What area of paper is left?

7 Circular place mats of diameter 8 cm are made by stamping as many circles as possible from a rectangular strip of card measuring 8 cm by 64 cm. How many mats can be made from the strip of card and what area of card is wasted?

8 A wooden counter top is a rectangle measuring 280 cm by 45 cm. There are three circular holes in the counter, each of radius 10 cm. Find the area of the wooden top.

9 The surface of the counter top described in question 8 is to be given four coats of varnish. If one tin covers 3.5 m², how many tins will be needed?

10 Take a cylindrical tin of food with a paper label:

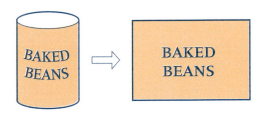

Measure the diameter of the tin and use it to find the length of the label. Find the area of the label. Now find the total surface area of the tin (two circular ends and the curved surface).

Investigation

In addition to using the formula $A = \pi r^2$ to calculate the area of a circle, there are other methods that provide reasonably good results. One such method is by weighing.

This method requires floor tiles, linoleum, or other material of measurable weight that can be easily cut. Trace the circle whose area is required on the tile and cut out the resulting circular region.

Use the remaining tiles to cut out three 10 cm squares, five 10 cm by 1 cm rectangles, and ten 1 cm squares. You will need more cutouts if the circle is large. These rectangular cutouts will be used as weights.

Place the circular cutout in the scale pan and use the rectangular pieces as weights. When the scale balances, remove the rectangular pieces and find their total area. This area will be the area of the circular cutout.

Why does this method of weighing to find area make sense?

Mixed exercises

Use the value of π on your calculator. Give your answers to 3 s.f.

Exercise *12h*

1 Find the circumference of a circle of radius 2.8 mm.

2 Find the radius of a circle of circumference 60 m.

3 Find the circumference of a circle of diameter 12 cm.

4 Find the area of a circle of radius 2.9 m.

5 Find the area of a circle of diameter 25 cm.

6 Find the perimeter of the quadrant in the diagram.

8 mm

7 Find the area of the sector in the diagram.

45°

4.5 cm

Exercise *12i*

1 Find the circumference of a circle of diameter 20 m.

2 Find the area of a circle of radius 12 cm.

3 Find the radius of a circle of circumference 360 cm.

4 Find the area of a circle of diameter 8 m.

5 Find the diameter of a circle of circumference 280 mm.

6 Find the perimeter of the sector in the diagram.

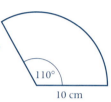

7 Find the area of the shaded part of the diagram.

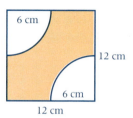

Exercise 12j

1 Find the area of a circle of radius 2 km.

2 Find the circumference of a circle of radius 49 mm.

3 Find the radius of a circle of circumference 88 m.

4 Find the area of a circle of diameter 14 cm.

5 Find the area of a circle of radius 3.2 cm.

6

An ornamental pond in a garden is a rectangle with a semicircle on each short end. The rectangle measures 5 m by 3 m and the radius of each semicircle is 1 m. Find the area of the pond.

Over the centuries mathematicians have spent a lot of time trying to find the true value of π. The ancient Chinese used 3. Three is also the value given in the Old Testament (1 Kings 7:23). The Egyptians (*c.* 1600 BC) used $4 \times \left(\frac{8}{9}\right)^2$. Archimedes (*c.* 225 BC) was the first person to use a sound method for finding its value and a mathematician called Van Ceulen (1540–1610) spent most of his life finding it to 35 decimal places !

IN THIS CHAPTER...

you have seen that:

● for any circle the circumference divided by the diameter gives a fixed value; this value is denoted by π and its approximate value is 3.142

● you can find the circumference of a circle using either the formula $C = 2\pi r$ or the formula $C = \pi d$ when you know the radius or diameter of the circle

● you can use the formula $A = \pi r^2$ to find the area of a circle

● a sector of a circle is shaped like a slice of cake; the fraction that the angle at the point is of 360° gives the fraction that its area is of the area of the circle.

13 ENLARGEMENTS

What is enlarged when it is cut at both ends? A ditch.

BEFORE YOU START

you need to know:
- ✓ the sum of the three angles in a triangle
- ✓ how to find one quantity as a fraction of another
- ✓ how to draw x and y axes and plot points

KEY WORDS

centre of enlargement, enlargement, fractional scale factor, guideline, image, object, negative scale factor, reflection, rotation, scale factor, translation

Enlargements

All the transformations we have used so far (i.e. reflections, translations and rotations) have moved the object and perhaps turned it over to produce the image, but its shape and size have not changed. Next we come to a transformation which keeps the shape but alters the size.

Think of the picture thrown on the screen when a slide projector is used.

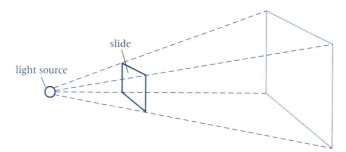

The picture on the screen is the same as that on the slide but it is very much bigger.

We can use the same idea to enlarge any shape.

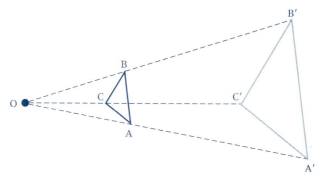

△A′B′C′ is the image of △ABC under an *enlargement, centre O*.

O is the *centre of enlargement*.

We call the dotted lines *guidelines*.

Centre of enlargement

In all these questions, one triangle is an enlargement of the other.

Exercise *13a*

1   Copy the diagram overleaf using 1 cm to 1 unit. Draw P′P, Q′Q and R′R and continue all three lines until they meet.
The point where the lines meet is called the centre of enlargement.
Give the coordinates of the centre of enlargement.

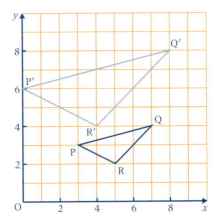

Repeat question **1** using the diagrams in questions **2** and **3**.

2

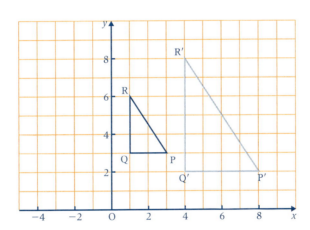

3

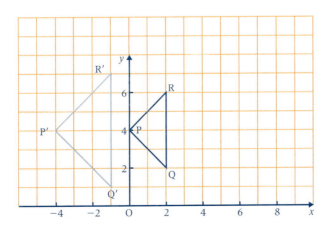

4 In questions **1** to **3**, name pairs of lines that are parallel.

5 Draw axes for x and y from 0 to 9 using 1 cm as 1 unit.
Draw △ABC: A(2, 3), B(4, 1), C(5, 4).
Draw △A'B'C': A'(2, 5), B'(6, 1), C'(8, 7).
Draw A'A, B'B and C'C and extend these lines until they meet.

   a Give the coordinates of the centre of enlargement.

   b Measure the sides and angles of the two triangles. What do you notice?

6 Repeat question **5** with △ABC: A(8, 4), B(6, 6), C(6, 4) and △A'B'C': A'(6, 2), B'(0, 8), C'(0, 2)

7 Draw axes for x and y from 0 to 10 using 1 cm as 1 unit.
Draw △XYZ with X(8, 2), Y(6, 6) and Z(5, 3) and △X'Y'Z' with X'(6, 2), Y'(2, 10) and Z'(0, 4). Find the centre of enlargement and label it P.
Measure PX, PX', PY, PY', PZ, PZ'. What do you notice?

The centre of enlargement can be anywhere, including a point inside the object or a point on the object.

The centres of enlargement in the diagrams below are marked with a cross.

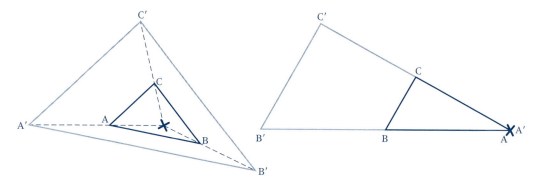

Exercise 13b

1 Copy the diagram using 1 cm as 1 unit. Draw A'A, B'B and C'C and extend the lines until they meet. Give the coordinates of the centre of enlargement.

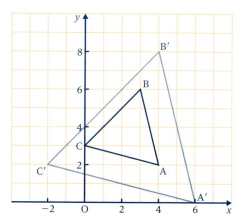

2 In the diagram on the right, which point is the centre of enlargement?

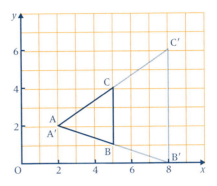

3 Draw axes for x and y from −3 to 10 using 1 cm as 1 unit. Draw △ABC with A(4, 0), B(4, 4) and C(0, 2). Draw △A′B′C′ with A′(5, −2), B′(5, 6) and C′(−3, 2). Find the coordinates of the centre of enlargement.

4 Repeat question **3** with A(1, 4), B(5, 2), C(5, 5) and A′(−3, 6), B′(9, 0), C′(9, 9).

Scale factors

If we measure the lengths of the sides of the two triangles PQR and P′Q′R′ and compare them, we find that the lengths of the sides of △P′Q′R′ are three times those of △PQR.

We say that △P′Q′R′ is the image of △PQR under an enlargement, centre O, with *scale factor 3*.

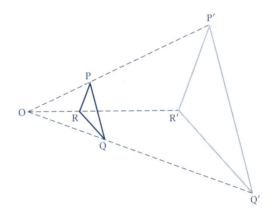

Finding an image under enlargement

If we measure OR and OR′ in the diagram above, we find R′ is three times as far from O as R is. This enables us to work out a method for enlarging an object with a given centre of enlargement (say O) and a given scale factor (say 3).

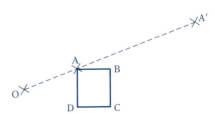

Measure OA. Multiply it by 3. Mark A′ on the guideline three times as far from O as A is.

$$OA' = 3 \times OA$$

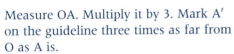

Repeat for B and the other vertices of ABCD.

Then A′B′C′D′ is the image of ABCD. To check, measure A′B′ and AB. A′B′ should be three times as large as AB.

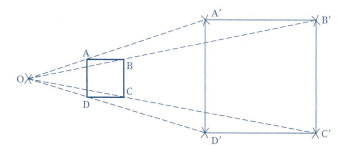

Exercise *13c*

1   Copy the diagram using 1 cm as 1 unit. P is the centre of enlargement. Draw the image of △ABC under an enlargement scale factor 2.

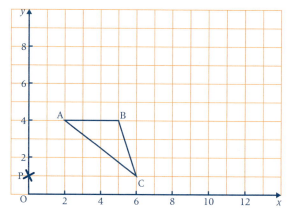

2   Repeat question **1** using this diagram.

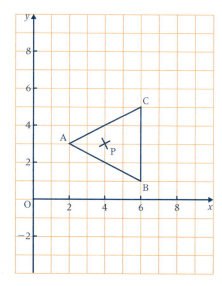

In questions **3** to **6**, draw axes for *x* and *y* from 0 to 10, using 1 cm as 1 unit. In each case, find the image A′B′C′ of △ABC using the given enlargement. Check by measuring the lengths of the sides of the two triangles.

3 △ABC: A(3, 3), B(6, 2), C(5, 6).

Enlargement with centre (5, 4), and scale factor 2.

4 △ABC: A(1, 2), B(3, 2), C(1, 5).

Enlargement with centre (0, 0) and scale factor 2.
What do you notice about the coordinates of A′ compared with those of A?

5 △ABC: A(2, 1), B(4, 1), C(3, 4).

Enlargement with centre (1, 1) and scale factor 3.

6 △ABC: A(1, 2), B(7, 2), C(1, 6).

Enlargement with centre (1, 2) and scale factor $1\frac{1}{2}$.

7 On plain paper, mark a point P near the left-hand edge. Draw a small object (a pin man perhaps, or a square house) between P and the middle of the page. Using the method of enlargement, draw the image of the object with centre P and scale factor 2.

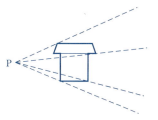

8 Repeat question **7** with other objects and other scale factors. Think carefully about the space you will need for the image.

9 Draw axes for *x* and *y* from 0 to 10 using 1 cm as 1 unit. Draw △ABC with A(2, 2), B(5, 1) and C(3, 4). Taking the origin as the centre of enlargement and a scale factor of 2, draw the image of △ABC by counting squares and without drawing the guidelines.

10 Draw axes for *x* and *y* from 0 to 8 using 1 cm as 1 unit. Draw △ABC with A(1, 2), B(5, 2) and C(2, 5). Taking (3, 2) as the centre of enlargement and a scale factor of 2, draw the image △ABC by counting squares and without drawing the guidelines.

 Puzzle

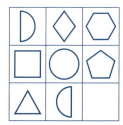 Which of these goes into the blank box?

Fractional scale factors

We can reverse the process of enlargement and shrink or reduce the object, producing a smaller image. If the lengths of the image are one-third of the lengths of the object then the scale factor is $\frac{1}{3}$.

There is no satisfactory word to cover both enlargement and shrinking (some people use 'dilation' and some 'scaling') so *enlargement* tends to be used for both. An enlargement may therefore be defined as a transformation which maps an object onto an image of similar shape. If the scale factor is less than 1, the image is smaller than the object. If the scale factor is greater than 1, the image is larger than the object.

Exercise **13d**

In questions **1** to **4**, △A'B'C' is the image of △ABC. Give the centre of enlargement and the scale factor.

1

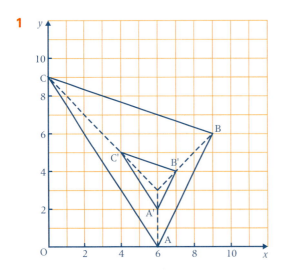

> **Tip** To find the scale factor, find the length of a side on the image as a fraction of the length of the corresponding side on the object.

2

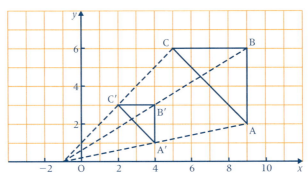

3 Draw axes for *x* and *y* from −2 to 8 using 1 cm as 1 unit. Draw △ABC with A(−1, 4), B(5, 1) and C(5, 7), and △A'B'C' with A'(2, 4), B'(4, 3) and C'(4, 5).

4 Draw axes for *x* and *y* from 0 to 9 using 1 cm as 1 unit. Draw △ABC with A(1, 2), B(9, 2) and C(9, 6), and △A′B′C′ with A′(1, 2), B′(5, 2) and C′(5, 4).

In questions **5** and **6**, draw axes for *x* and *y* from −1 to 11 using 1 cm as 1 unit. Find the image of △ABC under the given enlargement.

5 △ABC: A(9, 1), B(11, 5), C(7, 7). Centre (−1, 1), scale factor $\frac{1}{2}$.

6 △ABC: A(4, 0), B(10, 9), C(1, 6). Centre (4, 3), scale factor $\frac{1}{3}$.

Negative scale factors

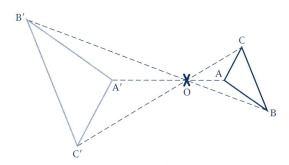

As you can see in the diagram above it is possible to produce an image twice the size of the object by drawing the guidelines backwards rather than forwards from the centre O. To show that we are going the opposite way we say that the scale factor is −2.

The image is the same shape but has been rotated through a half turn compared with the image produced by a scale factor of +2.

The following diagrams show enlargements with scale factors of −3 and +3.

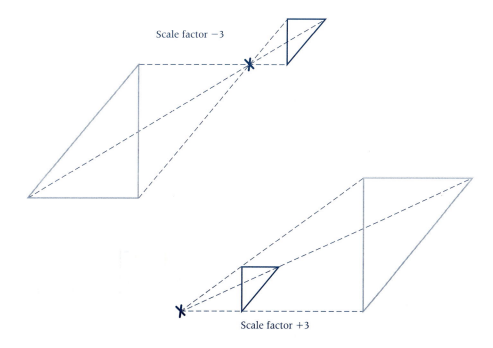

Scale factor −3

Scale factor +3

Exercise 13e

In questions **1** and **2** give the centre of enlargement and the scale factor.

1

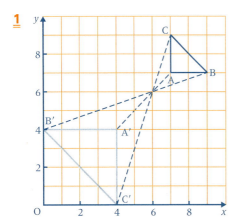

Tip The size of the scale factor is given by:

Distance from the centre of enlargement to a point on the image

Distance from the centre of enlargement to the corresponding point on the object

The scale factor is negative when the image is rotated with respect to the object.

2

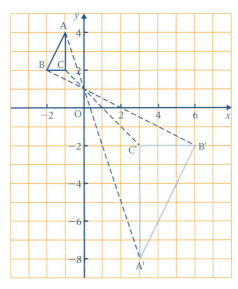

Copy the diagram in questions **3** to **6** using 1 cm to 1 unit. Find the centre of enlargement and the scale factor.

3

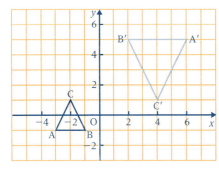

4

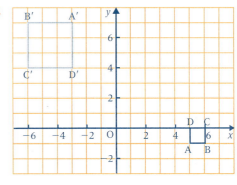

5

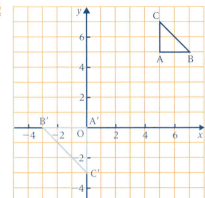

6

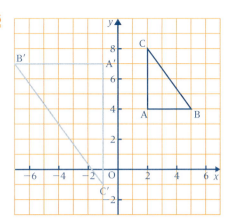

In questions **7** to **9**, draw axes for x and y from -6 to 6. Draw the object and image and find the centre of enlargement and the scale factor.

7　Object: △ABC with A(6, −1), B(4, −3), C(4, −1)
　　Image: △A′B′C′ with A′(−3, 2), B′(1, 6), C′(1, 2)

8　Object: Square ABCD with A(1, 1), B(5, 1), C(5, −3), D(1, −3)
　　Image: Square A′B′C′D′ with A′(−2, 2), B′(−4, 2), C′(−4, 4), D′(−2, 4)

9　Object: △ABC with A(2, 3), B(4, 3), C(2, 6)
　　Image: △A′B′C′ with A′(2, 3), B′(−4, 3), C′(2, −6)

10

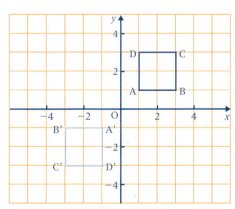

　　a　If A′B′C′D′ is the image of ABCD under enlargement, give the centre and the scale factor.

　　b　What other transformation would map ABCD to A′B′C′D′ ?

11 On plain paper, draw an object such as a pin man in the top left-hand corner. Mark the centre of enlargement somewhere between the object and the centre of the page. By drawing guidelines, draw the image with a scale factor of −2.

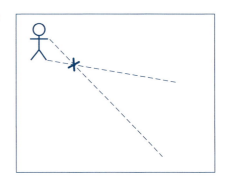

In questions **12** and **13**, copy the diagrams and find the images of the triangles using P as the centre of enlargement and a scale factor of −2.

12

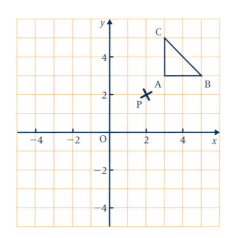

13

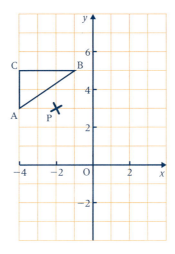

14 Draw axes for x from −10 to 4 and for y from −2 to 2.
Draw △ABC with A(2, 1), B(4, 1) and C(2, 2).
If the centre of enlargement is (1, 1) and the scale factor is −3, find the image of △ABC.

Investigation

Daniel designs packaging. The sketch shows a design that will fold to make a package suitable for sending books of various sizes by post. The design can be produced in several different sizes if it is enlarged by a given scale factor.

The basic rectangle is divided into 6 smaller rectangles, marked A, B, C, D, E and F on the sketch. It is cut along the heavy lines and fold marks are pressed into it along the broken lines.

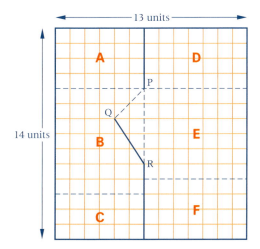

By folding along PQ and PR, you will find that rectangle B now fits over rectangle E and this acts as the base of the package. The remaining 4 rectangles fold up and over allowing books of different thicknesses to be packed.

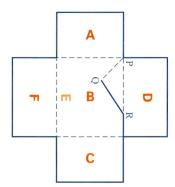

a On 5 mm squared paper, draw an enlargement of the diagram with a scale factor of 2.

b Cut it out and fold it to show how it is used.

    c If you have a much larger sheet of paper or card, repeat parts **a** and **b** using a larger scale factor, for example, 4.

    d **i** Can the fold PQ be drawn at any angle?

        ii Can the cut QR be made at any angle?

        Give reasons for your answers.

    f What is the ratio of the length of the original rectangle to its breadth?

        If you alter this ratio, can you still make a satisfactory package? Investigate.

    f Can you improve on the design?

Time for your library search!

Einstein (1879–1955); C. Goldbach (1690–1764).

IN THIS CHAPTER...

you have seen that:

- to find the centre of enlargement, you can draw guide lines between corresponding points on the object and the image

- you can find the scale factor by comparing lengths of corresponding sides on the object and the image

- to find an image of a given shape under an enlargement you need the scale factor and centre of enlargement

- a scale factor less than 1 gives an image smaller than the object whereas a scale factor larger than 1 gives an image larger than the object

- a negative scale factor gives an image that has been rotated through a half turn with respect to the object.

14 SIMILAR FIGURES

AT THE END OF THIS CHAPTER...

you should be able to:

1   State whether or not given figures are similar.

2   Determine if two triangles are similar or not, by comparing the sizes of their angles or their sides.

3   Write down equal ratios in two similar triangles.

4   Calculate the length of one side of a triangle from necessary data on a pair of similar triangles.

5   Use the scale factor for enlarging one triangle into another, to find a missing length.

Q: What do you call a perfect angle?

A: Acute angle.

BEFORE YOU START

you need to know:
✓ the angle sum of a triangle
✓ how to find one quantity as a fraction of another
✓ how to solve equations involving fractions
✓ the meaning of ratio

KEY WORDS

alternate angles, enlargement, ratio, scale factor, similar, vertically opposite angles

Similar figures

Two figures are similar if they are the same shape but not necessarily the same size. One figure is an enlargement of the other.

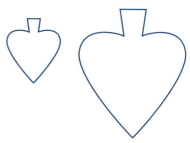

One may be turned round compared with the other.

One figure may be turned over compared with the other.

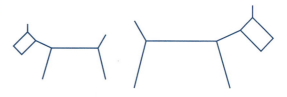

The following figures are not similar although their angles are equal.

Exercise *14a*

State whether or not the pairs of figures in questions **1** to **10** are similar.

1

3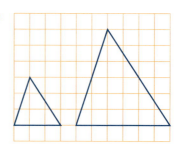

2

4

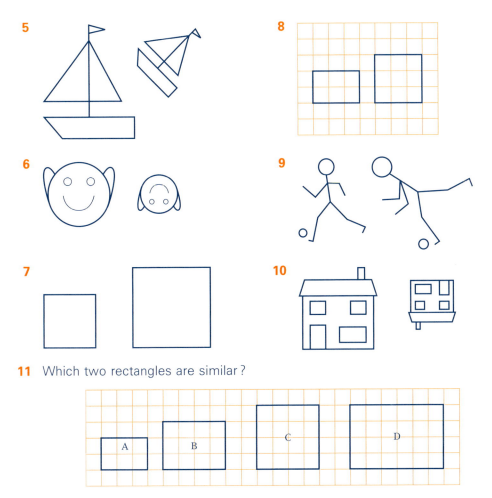

11 Which two rectangles are similar?

12 Draw your own pairs of figures and state whether or not they are similar. (The second figure may be turned round or over or both, compared with the first.)

Similar triangles

Some of the easiest similar figures to deal with are triangles. This is because only a small amount of information is needed to prove them to be similar.

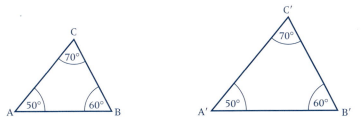

In these triangles the corresponding angles are equal and so the triangles are the same shape. One triangle is an enlargement of the other. These triangles are *similar*.

Exercise **14b**

1 Draw the following triangles accurately:

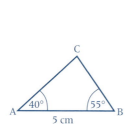 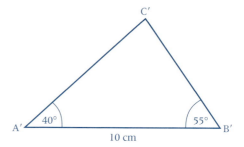

a Are the triangles similar?

b Measure the remaining sides.

c Find $\dfrac{A'B'}{AB}$, $\dfrac{B'C'}{BC}$ and $\dfrac{C'A'}{CA}$ (as decimals if necessary)

d What do you notice about the answers to part **c**?

Repeat question **1** for the pairs of triangles in questions **2** to **5**.

2

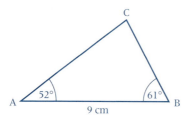 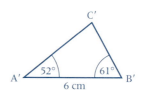

3

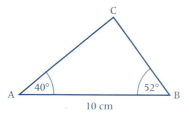 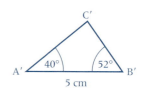

4

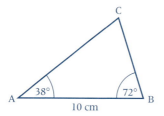 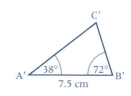

5

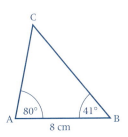

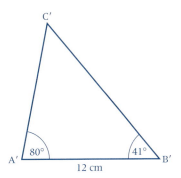

Sketch the following pairs of triangles and find the sizes of the missing angles. In each question state whether the two triangles are similar. (One triangle may be turned round or over compared with the other.)

6

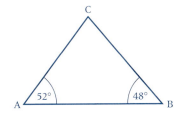

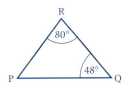

7

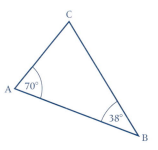

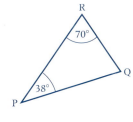

8

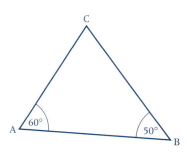

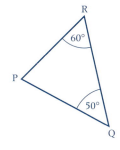

9

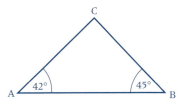

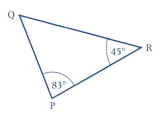

Corresponding vertices

These two triangles are similar and we can see that X corresponds to A, Y to B and Z to C.

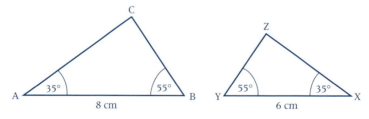

We can write $\triangle$s $\dfrac{\text{ABC}}{\text{XYZ}}$ are similar

Make sure that X is written below A, Y below B and Z below C.

The pairs of corresponding sides are in the same ratio,

that is $\qquad \dfrac{\text{AB}}{\text{XY}} = \dfrac{\text{BC}}{\text{YZ}} = \dfrac{\text{CA}}{\text{ZX}}$

Exercise 14c State whether triangles ABC and PQR are similar and, if they are, give the ratios of the sides.

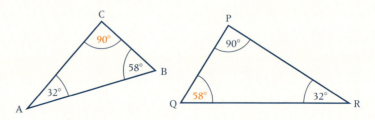

First find the third angle in each triangle

$$\hat{\text{Q}} = 58° \quad (\text{angles of a triangle})$$

and $\qquad\qquad\qquad \hat{\text{C}} = 90°$

Now you can see that the triangles are similar and that A corresponds to R, and so on,

so $\qquad\qquad\qquad \triangle$s $\dfrac{\text{ABC}}{\text{RQP}}$ are similar

and $\qquad\qquad\qquad \dfrac{\text{AB}}{\text{RQ}} = \dfrac{\text{BC}}{\text{QP}} = \dfrac{\text{CA}}{\text{PR}}$

In questions **1** to **8**, state whether the two triangles are similar and, if they are, give the ratios of the sides.

1

C

42° 65° B

A

P 42°

R

65°

Q

2

C

41° B

39°

A

P 39°

R

41°

Q

3

C

56°

A 45° B

R

45° Q

89°

P

4

C

71°

58°

A B

P

Q 58°

51°

R

5 Use the triangles given in question **6** of Exercise 14b.

6 Use the triangles given in question **7** of Exercise 14b.

7 Use the triangles given in question **8** of Exercise 14b.

8 Use the triangles given in question **9** of Exercise 14b.

Puzzle

There are 13 stations on a railway line. All tickets are printed with the name of the station you board the train and the station you leave the train. How many different tickets are needed?

Finding a missing length

Exercise 14d State whether the two triangles are similar. If they are, find AB.

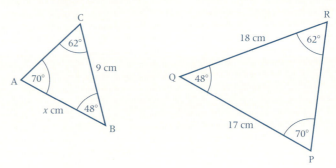

First find the third angle in each triangle.

$$\hat{C} = 62° \quad \text{and} \quad \hat{Q} = 48° \quad (\text{angles of a triangle})$$

$$\text{so} \quad \triangle s \; \frac{ABC}{PQR} \; \text{are similar} \quad \text{and} \quad \frac{AB}{PQ} = \frac{BC}{QR} = \frac{CA}{RP}$$

Ring the side you want and the sides you know: $\dfrac{\widehat{AB}}{\widehat{PQ}} = \dfrac{\widehat{BC}}{\widehat{QR}} = \dfrac{CA}{RP}$

Now substitue the values $\qquad \dfrac{x}{17} = \dfrac{9}{18}$

$$\cancel{17} \times \frac{x}{\cancel{17}} = \frac{\cancel{9}^{1}}{\cancel{18}_{2}} \times 17$$

$$x = \frac{17}{2} = 8.5$$

$$AB = 8.5 \, \text{cm}$$

In questions **1** to **4**, state whether the pairs of triangles are similar. If they are, find the required side.

1 Find PR.

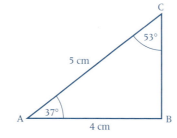
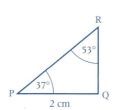

2 Find QR.

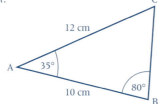
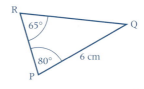

3 Find BC.

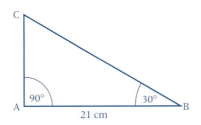 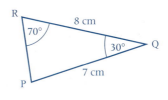

4 Find PR.

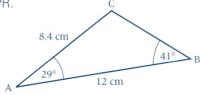 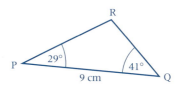

In some cases we do not need to know the sizes of the angles as long as we know that pairs of angles are equal. (Two pairs only are needed as the third pair must then be equal.)

In △s ABC and DEF, $\hat{A} = \hat{E}$ and $\hat{B} = \hat{D}$. AB = 4 cm, DE = 3 cm and AC = 6 cm. Find EF.

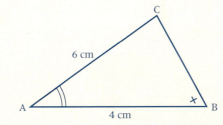 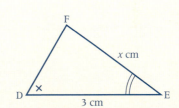

△s $\begin{matrix}\text{EDF}\\\text{ABC}\end{matrix}$ are similar because they are equiangular

(we put the triangle with the unknown side on top)

$$\frac{FE}{CA} = \frac{ED}{AB} = \frac{DF}{BC}$$

Ringing the sides known and wanted gives $\frac{\widehat{FE}}{\widehat{CA}} = \frac{\widehat{ED}}{\widehat{AB}} = \frac{DF}{BC}$

Substituting values gives $\frac{x}{6} = \frac{3}{4}$

$$^1\cancel{6} \times \frac{x}{\cancel{6}} = \frac{3}{\cancel{4}} \times \cancel{6}^3$$
$$_1_2$$

$$x = \frac{9}{2} = 4.5$$

so $\qquad$ EF = 4.5 cm

5 In △s ABC and XYZ, $\hat{A} = \hat{X}$ and $\hat{B} = \hat{Y}$.
AB = 6 cm, BC = 5 cm and XY = 9 cm.
Find YZ.

> **Tip** Draw the triangles and mark the equal angles and the lengths of the sides that are given. Label the side you have to find x cm.

6 In △s ABC and PQR, $\hat{A} = \hat{P}$ and $\hat{C} = \hat{R}$.
AB = 10 cm, PQ = 12 cm and QR = 9 cm.
Find BC.

7 In △s ABC and DEF, $\hat{A} = \hat{E}$ and $\hat{B} = \hat{F}$. AB = 3 cm, EF = 5 cm and AC = 5 cm.
Find DE.

8 In △s ABC and PQR, $\hat{A} = \hat{Q}$ and $\hat{C} = \hat{R}$. AC = 8 cm, BC = 4 cm and QR = 9 cm.
Find PR.

a Show that triangles ABC and CDE are similar.

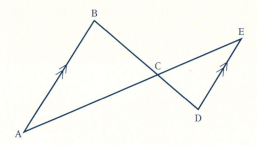

b Given that AC = 15 cm, CE = 9 cm and DE = 8 cm, find AB.

a $\hat{A} = \hat{E}$ (alternate angles, AB ∥ DE)

$\hat{B} = \hat{D}$ (alternate angles, AB ∥ DE)

(Or we could use $B\hat{C}A = E\hat{C}D$ as these are vertically opposite angles.)

so △s $\dfrac{ABC}{EDC}$ are similar.

b Sketch the diagram and mark the equal angles and the lengths of the sides that are given, and the side you need to find.

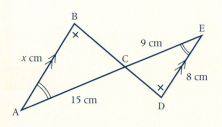

$$\frac{AB}{ED} = \frac{BC}{DC} = \frac{CA}{CE}$$

$$\frac{x}{8} = \frac{15}{9}$$

$$\overset{1}{\cancel{8}} \times \frac{x}{\cancel{8}_{1}} = \frac{\overset{5}{\cancel{15}}}{\cancel{9}_{3}} \times 8$$

$$x = \frac{40}{3}$$

$$= 13\tfrac{1}{3}$$

AB = $13\tfrac{1}{3}$ cm, or 13.3 cm correct to 3 s.f.

9

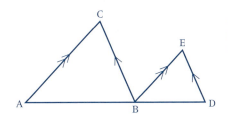

Tip Draw your own diagram.

a Show that △s ABC and BDE are similar.

b If AB = 6 cm, BD = 3 cm and DE = 2 cm, find BC.

10 **a** Show that △s ABC and CDE are similar.

b If AB = 7 cm, BC = 6 cm, AC = 4 cm and CE = 6 cm, find CD and DE.

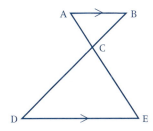

11 **a** ABCD is a square. EF is at right angles to BD. Show that △s ABD and DEF are similar.

b If AB = 10 cm, DB = 14.2 cm and DF = 7.1 cm, find EF.

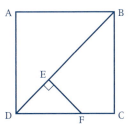

12

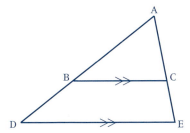

a Show that △s ABC and ADE are similar. (Notice that Â is *common* to both triangles.)

b If AB = 10 cm, AD = 15 cm, BC = 12 cm and AC = 9 cm, find DE, AE and CE.

 Puzzle

Arrange four 9s to make 100.

Using the scale factor to find the missing length

Sometimes the scale factor for enlarging one triangle into the other is very obvious and we can make use of this to save ourselves some work.

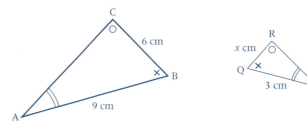

The two triangles above are similar and we can see that the scale factor for 'enlarging' the first triangle into the second is $\frac{1}{3}$. We can say straightaway that x is $\frac{1}{3}$ of 6.

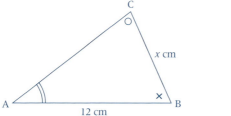 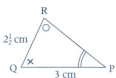

If we wish to find a length in the first triangle, we use the scale factor for enlarging the second triangle into the first.

The scale factor is 4 so $x = 4 \times 2\frac{1}{2} = 10$

Exercise 14e   Find QR.

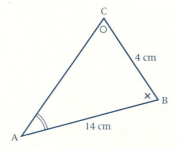 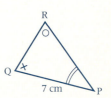

$\triangle$s $\dfrac{PQR}{ABC}$ are similar

QP and BA are corresponding sides and QP $= \frac{1}{2}$ AB

So the scale factor is $\frac{1}{2}$

QR and BC are corresponding sides

∴   QR $= \frac{1}{2} \times 4\,cm = 2\,cm$

Using the scale factor to find the missing length

 Tip Identify the corresponding sides and use them to find the scale factor.

1 Find BC.

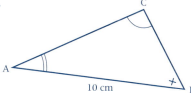
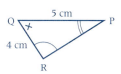

2 Find PR.

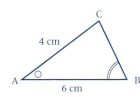
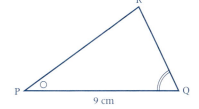

3 Find PR.

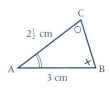
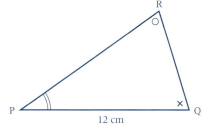

4 Find XY.

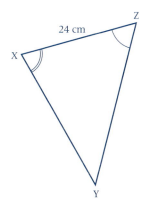
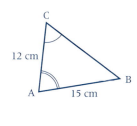

5 Find LN.

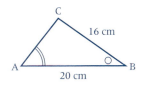
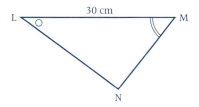

Corresponding sides

If the three pairs of sides of two triangles are in the same ratio, then the triangles are similar and their corresponding angles are equal.

When finding the ratio of three sides give the ratio as a whole number or as a fraction in its lowest terms.

Exercise 14f State whether triangles ABC and PQR are similar. Say which angle, if any, is equal to $\hat{A}$.

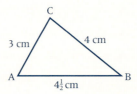 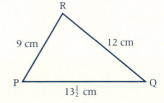

Start with the shortest side of each triangle: $\dfrac{PR}{AC} = \dfrac{9}{3} = 3$

Now the longest sides: $\dfrac{PQ}{AB} = \dfrac{13\frac{1}{2}}{4\frac{1}{2}} = \dfrac{27}{9} = 3$

Lastly the third sides $\dfrac{QR}{BC} = \dfrac{12}{4} = 3$

so $$\dfrac{PR}{AC} = \dfrac{PQ}{AB} = \dfrac{QR}{BC}$$

∴ $\triangle$s $\dfrac{PQR}{ABC}$ are similar

and $$\hat{P} = \hat{A}$$

State whether the following pairs of triangles are similar. In each case say which angle, if any, is equal to $\hat{A}$.

1

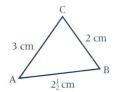 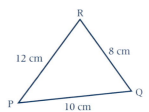

2

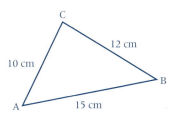

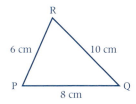

3

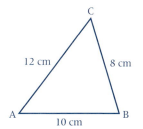

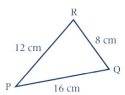

4

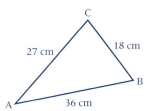

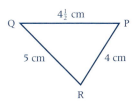

5

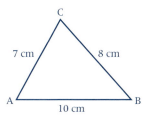

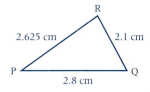

6

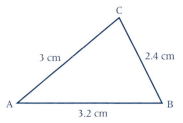

7 Are the triangles ABC and ADE similar?

Which angles are equal?

What can you say about lines BC and DE?

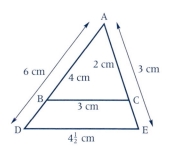

One pair of equal angles and two pairs of sides

The third possible set of information about similar triangles concerns a pair of angles and the sides containing them.

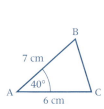 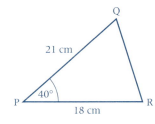 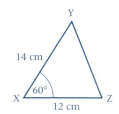

$$\frac{PR}{AC} = \frac{18}{6} = 3 \quad \text{and} \quad \frac{PQ}{AB} = \frac{21}{7} = 3$$

i.e. $\qquad \dfrac{PR}{AC} = \dfrac{PQ}{AB}$

and $\qquad \hat{A} = \hat{P}$

so $\qquad \triangle\text{s} \dfrac{ABC}{PQR}$ are similar

We can see that $\triangle PQR$ is an enlargement of $\triangle ABC$ and that the scale factor is 3.

$\left(\text{It is given by } \dfrac{PQ}{AB}.\right)$

On the other hand, $\triangle XYZ$ is a different shape from the other two and is not similar to either of them, even though two pairs of sides are in the same ratio, because the angles between the pairs of sides are not the same.

Exercise 14g State whether triangles ABC and PQR are similar. If they are, find PQ.

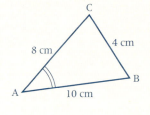 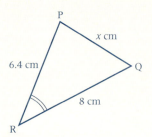

$\hat{A} = \hat{R}$ so compare the ratios of the arms containing these angles.

$$\frac{RP}{AC} = \frac{6.4}{8} = 0.8 \quad (\text{comparing the two shorter sides})$$

$$\frac{RQ}{AB} = \frac{8}{10} = 0.8 \quad (\text{comparing the other two arms})$$

$\therefore \qquad \dfrac{RP}{AC} = \dfrac{RQ}{AB} \qquad \text{and} \qquad \hat{A} = \hat{R}$

so $\qquad \triangle\text{s} \dfrac{RQP}{ABC}$ are similar

Now, $\dfrac{PQ}{CB} = \dfrac{RQ}{AB}$ *or:*   BC is half AC   so   PQ is half PR

$\therefore$ PQ = 3.2 cm

$$\dfrac{x}{4} = \dfrac{8}{10}$$

$$\cancel{4}^{1} \times \dfrac{x}{\cancel{4}_{1}} = \dfrac{\cancel{8}^{4}}{\cancel{10}_{5}} \times 4$$

$$x = 3.2$$

$$PQ = 3.2\,\text{cm}$$

State whether the following pairs of triangles are similar. If they are, find the missing lengths.

1

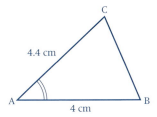

2

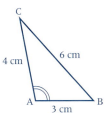

3

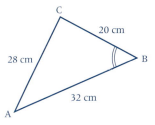

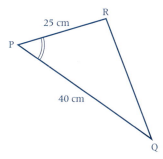

4

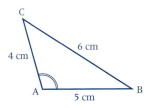

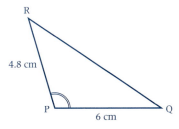

5

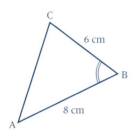
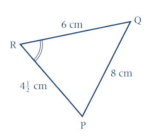

6 In △s ABC and PQR, $\widehat{A} = \widehat{P}$, AB = 8 cm, BC = 8.5 cm, CA = 6.5 cm, PQ = 4.8 cm and PR = 3.9 cm. Find QR.

7 In △s PQR and XYZ, $\widehat{P} = \widehat{X}$, PQ = 4 cm, PR = 3 cm, QR = $2\frac{1}{4}$ cm, XY = $5\frac{1}{3}$ cm and XZ = 4 cm. Find ZY.

Summary: similar triangles

If two triangles are the same shape (but not necessarily the same size) they are said to be *similar*. This word, when used in mathematics, means that the triangles are *exactly* the same shape and not vaguely alike, as two sisters may be.

One triangle may be turned over or round compared with the other.

Pairs of corresponding sides are in the same ratio. This ratio is the *scale factor* for the enlargement of one triangle into the other.

To check that two triangles are similar we need to show *one* of the three following sets of facts:

a The angles of one triangle are equal to the angles of the other (as in Exercise 14c)

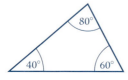

b The three pairs of corresponding sides are in the same ratio (as in Exercise 14f)

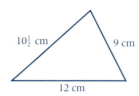
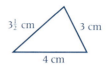

c There is one pair of equal angles and the sides containing the known angles are in the same ratio (as in Exercise 14g).

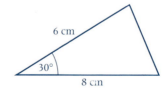

Mixed exercise

Exercise *14h*

State whether or not the pairs of triangles in questions **1** to **10** are similar, giving your reasons. If they are similar, find the required side or angle.

1 Find BC.

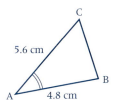
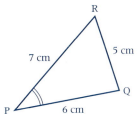

2 Find QR.

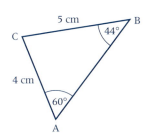
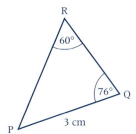

3 Find $\hat{Q}$.

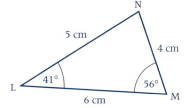
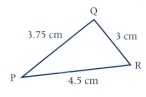

4 Find FE.

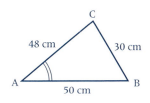
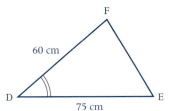

5 Find $\hat{P}$.

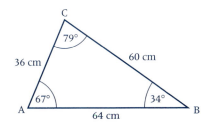
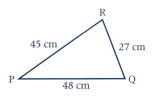

6 Find $\hat{Q}$.

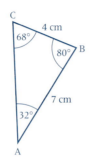

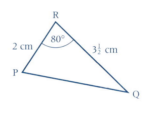

7 Find YZ.

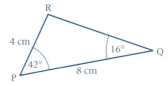

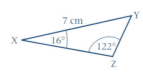

8 Find AC.

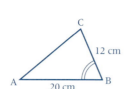

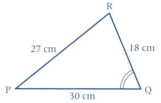

9 **a** Show that △s ABC and ADE are similar.

   b AB = 3.6 cm, AD = 4.8 cm and AE = 4.2 cm.

   Find AC and CE.

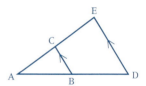

10 **a** Show that △s ABC and DEF are similar.

   b AB = 40 cm, BC = 52 cm and DE = 110 cm.
   Find EF.

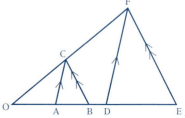

11 In the figure below there are three overlapping triangles.

   a Show that △s ABC and ABD are similar.

   b Show that △s ABC and BDC are similar.

   c Are △s ABD and BDC similar?

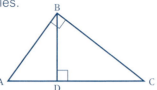

12 A pole, AB, 2 m high, casts a shadow,
AC, that is 3 m long.
Another pole, PQ, casts a shadow
15 m long. How high is the second pole?

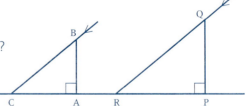

13 The shadow of a 1 m stick held upright on the ground is 2.4 m long. How long a shadow would be cast by an 8 m telegraph pole?

14 A slide measures 1.8 cm by 2.4 cm. A picture 90 cm by 120 cm is cast on the screen. On the slide, a house is 1.2 cm high. How high is the house in the picture on the screen?

 # Investigation

Draw a quadrilateral ABCD with four unequal sides. Mark the midpoints of the sides AB, BC, CD and DA with the letters P, Q, R and S in that order.

Join P, Q, R and S to give a new quadrilateral. Are the two quadrilaterals similar?

Investigate what happens if you repeat this with other quadrilaterals, including the special quadrilaterals this time.

An engineer had to measure the height of a flag pole. He had a tape measure but couldn't keep the tape along the pole. A mathematician comes along and offers to solve the problem; he removes the pole from its hole and lays it on the ground and measures it easily. When he leaves, the engineer says "Just like a mathematician! I need to know the height, and he gives me the length!"

IN THIS CHAPTER...

you have seen that:

- two shapes are similar if one is an enlargement of the other

- to prove that two triangle are similar you need to show that either

  - the triangles have the same angles or

- all three sides of each triangle are in the same ratio or

- one pair of angles are equal and the sides round those angles are in the same ratio.

15 VOLUMES: CONSTANT CROSS-SECTION

MATHS IS OUT THERE

Q How many calories are there in a slice of apple pi?

A 3.142

Volume of a cuboid

Reminder: We find the volume of a cuboid (that is, a rectangular block) by multiplying length by width by height.

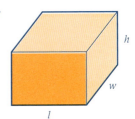

i.e.   volume = length × width × height

or $V = l \times w \times h$

Remember that the measurements must all be in the same units before they are multiplied together.

Exercise 15a

1 Find the volume of a cuboid of length 9 cm, width 6 cm and height 4 cm.

> **Tip** Use $V = lwh$ where $l = 9$, $w = 6$ and $h = 4$. Remember to give the units in your answer.

2 Find the volume of a cuboid of length 12 m, width 8 m and height 4.5 m.

3 Find the volume of a cuboid of length 300 cm, width 20 cm and height 30 cm.

4 Find the height of a cuboid of length 6.2 cm, width 3.4 cm and height 5 cm.

Find the volumes of the following cuboids, changing the units first if necessary. Do *not* draw a diagram.

	Length	Width	Height	Volume units
5	32 m	5 mm	10 mm	mm³
6	$3\frac{1}{4}$ cm	4 cm	$4\frac{1}{2}$ cm	cm³
7	1.4 cm	0.9 cm	0.32 cm	mm³
8	9.2 m	3 m	1.8 m	m³
9	0.02 cm	0.04 cm	0.01 cm	cm³
10	6.2 mm	32 mm	20 mm	cm³
11	$7\frac{1}{2}$ cm	$2\frac{1}{2}$ cm	6 cm	cm³
12	4.2 cm	3 cm	0.15 cm	cm³
13	7.2 cm	3.6 cm	5 cm	cm³
14	5.6 m	7 m	3.4 m	m³
15	72.3 mm	50 mm	40 mm	cm³
16	0.48 m	3.2 m	0.15 m	cm³

Investigation

A rectangular piece of steel sheet measuring 20cm by 14cm is to be used to make an open rectangular box. The diagram shows one way of doing this. The four corner squares are removed and the sides folded up. The four vertical seams at the corners are then sealed.

a   Copy and complete the following table, which gives the measurements of the base and the capacity of the box when squares of different sizes are removed from the corners. Continue to add numbers to the first column in the table as long as it is reasonable to do so.

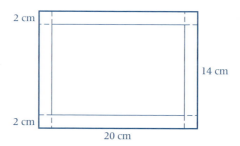

These numbers should follow the pattern indicated.

Length of edge of square (cm)	Measurements of base (cm)	Capacity of box (cm³)
0.5	13 × 19	0.5 × 13 × 19 = 123.5
1	12 × 18	1 × 12 × 18 = 216
1.5		
2		
2.5		
3		

b   What is the last number you entered in the first column of the table? Justify your choice.

c   What size of square should be removed to give the largest capacity recorded in the table?

d   Investigate whether you can find a number that you have not already entered in the first column that gives a larger capacity than any value you have found so far.

In Book 1 we saw that:

For a rectangle,   area = length × breadth

i.e. $A = l \times b$

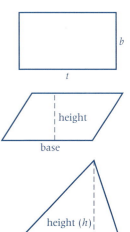

For a parallelogram,   area = base × height

For a triangle,   area = $\frac{1}{2}$ (base × perpendicular height)

i.e. $A = \frac{1}{2}bh$

The area of a compound shape is found by breaking it up into basic shapes and adding or subtracting their areas

e.g.

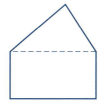

a rectangle + a triangle

a square − a triangle

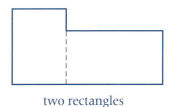

two rectangles

Volumes of solids with uniform cross-sections

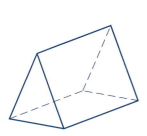

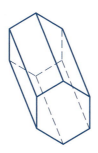

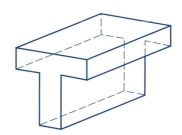

When we cut through any one of the solids above, parallel to the ends, we always get the same shape as the end. This shape is called the cross-section.

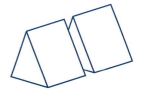

As the cross-section is the same shape and size wherever the solid is cut, the cross-section is said to be *uniform* or *constant*. These solids are also called *prisms* and we can find the volumes of some of them.

First consider a cuboid (which can also be thought of as a rectangular prism).

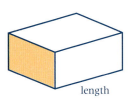

Volume = length × width × height

= (width × height) × length

= area of shaded end × length

= area of cross-section × length

Now consider a triangular prism. If we enclose it in a cuboid we can see that its volume is half the volume of the cuboid.

Volume = ($\frac{1}{2}$ × width × height) × length

= area of shaded triangle × length

= area of cross-section × length

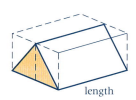

This is true of any prism so that

Volume of a prism = area of cross-section × length

Exercise 15b

Find the volume of the solid below.

Tip To find the volume you need first to find the area of the cross-section.

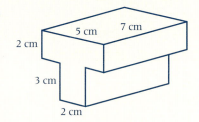

Draw the cross-section, then divide it into two rectangles.

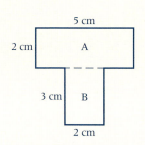

Area of A = 2 × 5 cm² = 10 cm²

Area of B = 2 × 3 cm² = 6 cm²

Area of cross-section = 16 cm²

Volume = area × length

= 16 × 7 cm³

= 112 cm³

Volumes of solids with uniform cross-sections

Find the volumes of the following prisms. Draw a diagram of the cross-section but do *not* draw a picture of the solid.

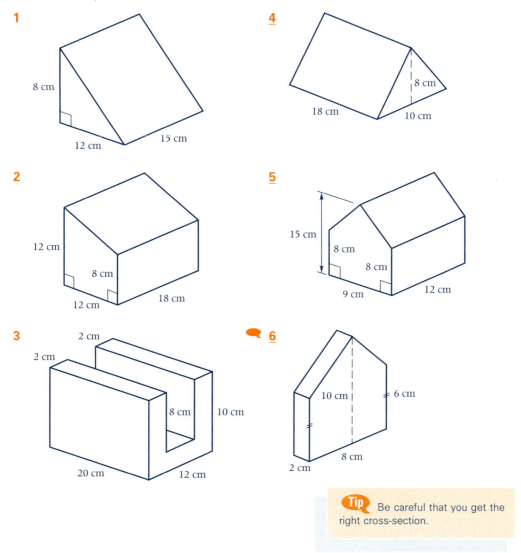

1

8 cm

12 cm   15 cm

4

18 cm   8 cm   10 cm

2

12 cm

8 cm

12 cm   18 cm

5

15 cm   8 cm   8 cm

9 cm   12 cm

3

2 cm   2 cm

8 cm   10 cm

20 cm   12 cm

6

10 cm   6 cm

2 cm   8 cm

> **Tip** Be careful that you get the right cross-section.

The following two solids are standing on their ends so the vertical measurement is the length.

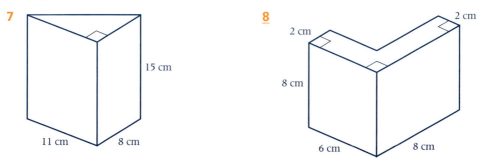

7

15 cm

11 cm   8 cm

8

2 cm   2 cm

8 cm

6 cm   8 cm

249

In questions **9** to **10**, the cross-sections of the prisms and their lengths are given. Find their volumes.

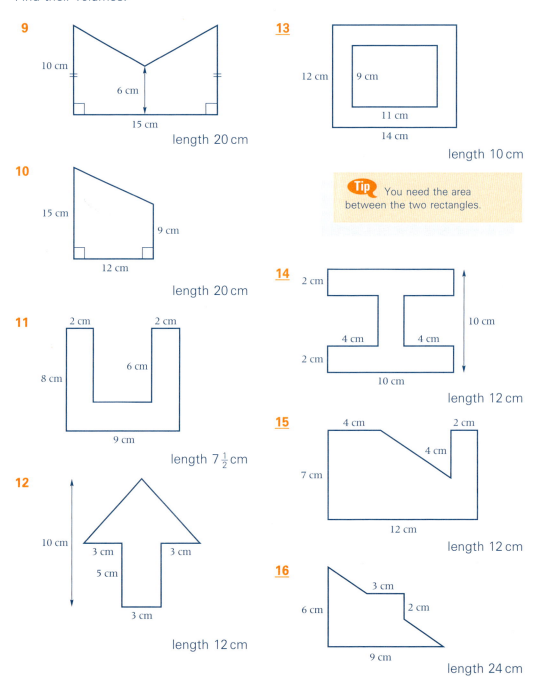

9

10 cm

6 cm

15 cm

length 20 cm

10

15 cm

9 cm

12 cm

length 20 cm

11

2 cm   2 cm

6 cm

8 cm

9 cm

length 7½ cm

12

10 cm

3 cm   3 cm

5 cm

3 cm

length 12 cm

13

12 cm   9 cm

11 cm

14 cm

length 10 cm

Tip You need the area between the two rectangles.

14

2 cm

4 cm   4 cm

10 cm

2 cm

10 cm

length 12 cm

15

4 cm   2 cm

4 cm

7 cm

12 cm

length 12 cm

16

3 cm

6 cm   2 cm

9 cm

length 24 cm

17 A tent is in the shape of a triangular prism. Its length is 2.4 m, its height 1.8 m and the width of the triangular end is 2.4 m. Find the volume enclosed by the tent.

18

The area of the cross-section of the given solid is 42 cm² and the length is 32 cm. Find its volume.

19

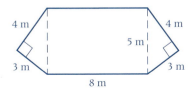

A solid of uniform cross-section is 12 m long. Its cross-section is shown in the diagram. Find its volume.

 ## Puzzle

Betty has a bag of identical triangular wooden blocks.

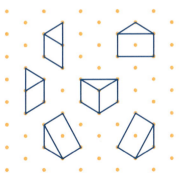

She uses some of them to make the letter F.

How many blocks does she need?

Volume of a cylinder

Reminder: The circumference of a circle is given by $C = 2\pi r$ and the area of a circle by $A = \pi r^2$

A cylinder can be thought of as a circular prism so its volume can be found using

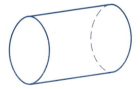

$$\text{volume} = \text{area of cross-section} \times \text{length}$$

$$= \text{area of circular end} \times \text{length}$$

From this we can find a formula for the volume.

We usually think of a cylinder as standing upright so that its length is represented by h (for height).

If the radius of the end circle is r, then the area of the cross-section is πr^2

$\therefore$ $\qquad$ volume $= \pi r^2 \times h$
$\qquad\qquad\qquad = \pi r^2 h$

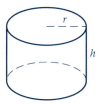

Exercise 15c

Find the volume inside a cylindrical mug of internal diameter 8 cm and height 6 cm. Use the π button on your calculator

Tip As the diameter is 8 cm, the radius is 4 cm.

The volume is given by the formula

$V = \pi r^2 h$

$\qquad = \pi \times 4 \times 4 \times 6 = 301.59\ldots$

Therefore volume of mug is 302 cm^2 (correct to 3 s.f.)

Tip An alternative method would be:

Area of cross section $= \pi r^2$
$\qquad\qquad\qquad\qquad = \pi \times 4 \times 4\,\text{cm}^2 = 50.265\ldots\,\text{cm}^2$

Volume $=$ area of cross section length
$\qquad\quad = (50.265\ldots \times 6)\,\text{cm}^3$
$\qquad\quad = 301.59\ldots\,\text{cm}^3$
$\qquad\quad = 302\,\text{cm}^3$ (correct to 3 s.f.)

Use the value of π on your calculator and give all your answers correct to 3 s.f.

Find the volumes of the following cylinders:

1 Radius 2 cm, height 10 cm

2 Radius 3 cm, height 4 cm

3 Radius 5 cm, height 4 cm

4 Radius 3 cm, height 2.1 cm

5 Diameter 2 cm, height 1 cm

6 Radius 1 cm, height 4.8 cm

7 Diameter 4 cm, height 3 cm

8 Diameter 6 cm, height 1.8 cm

9 Radius 12 cm, height 10 cm

10 Radius 7 cm, height 9 cm

11 Radius 3.2 cm, height 10 cm

12 Radius 6 cm, height 3.6 cm

13 Diameter 10 cm, height 4.2 cm

14 Radius 7.2 cm, height 4 cm

15 Diameter 64 cm, height 22 cm

16 Diameter 2.4 cm, height 6.2 cm

17 Radius 4.8 mm, height 13 mm

18 Diameter 16.2 cm, height 4 cm

19 Radius 76 cm, height 88 cm

20 Diameter 0.02 m, height 0.14 m

Compound shapes

Exercise 15d

Find the volumes of the following solids. Use the value of π on your calculator and give your answers correct to 3 s.f.

Tip Find the area of the cross-section first.

Draw diagrams of the cross-sections but do *not* draw pictures of the solids.

1

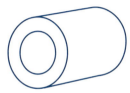

A tube of length 20 cm. The inner radius is 3 cm and the outer radius is 5 cm.

2

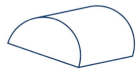

A half-cylinder of length 16 cm and radius 4 cm.

3

A solid of length 6.2 cm, whose cross-section consists of a square of side 2 cm surmounted by a semicircle.

4

A disc of radius 9 cm and thickness 0.8 cm.

5

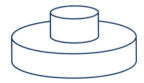

A solid made of two cylinders each of height 5 cm. The radius of the smaller one is 2 cm and of the larger one is 6 cm.

6

A solid made of two half-cylinders each of length 11 cm. The radius of the larger one is 10 cm and the radius of the smaller one is 5 cm.

Puzzle

A set of three encyclopaedias is placed in the normal way on a shelf.

A bookworm takes $\frac{1}{4}$ day to eat through a cover and $\frac{1}{2}$ day to eat through the pages of one book.

How long will it take this bookworm to eat its way from the first page of Volume 1 to the last page of Volume 3?

Library search time again.

This is a Klein Bottle.

Find out what is odd about it.

IN THIS CHAPTER...

you have seen that:

- the volume of a cuboid is found by multiplying its length by its width by its height, i.e. $V = l \times w \times h$

- any solid with a uniform cross-section is called a prism and its volume is equal to the area of the cross-section multiplied by its length

- in particular the volume of a cylinder is given by $V = \pi r^2 h$

16

SQUARES AND SQUARE ROOTS

AT THE END OF THIS CHAPTER...

you should be able to:

1   Find the square of a given number.

2   Find a rough estimate of the square of a number.

3   Find the square of a number using a calculator.

4   Complete a table of given numbers and their squares.

5   Use values in (4) to draw the graph of $y = x^2$.

6   Use the graph of $y = x^2$ to find the squares of numbers.

7   Find, by inspection, the square root of a number.

8   Find rough estimates of square roots.

9   Use a calculator to find square roots of numbers to a given degree of accuracy.

MATHS IS OUT THERE

Did you know that the Pythagoreans associated the number 5 with marriage. This is because they called 2 the first even female number and they called 3 the first odd male number and, as we all know, $2 + 3 = 5$!

BEFORE YOU START

you need to know:

✓ how to multiply decimals

✓ how to draw and scale x and y axes and plot points

✓ how to find the area of a square

KEY WORDS

significant figures, square, square root, the symbol ($\sqrt{}$)

Squares

We obtain the *square* of a number when we multiply the number by itself.

Exercise 16a

Find the squares of the numbers in questions **1** to **15**.

1	3	**6**	50	**11**	0.3
2	5	**7**	300	**12**	2000
3	9	**8**	0.02	**13**	0.004
4	30	**9**	500	**14**	1
5	0.4	**10**	10	**15**	0.03

> **Tip** The square of 3 means 3×3.

> **Tip** Take care with the decimal point.

Write 32 correct to 1 s.f. and use this to give a rough estimate of the square of 32.

$$32 \approx 30$$

$$32^2 \approx 30 \times 30 = 900$$

In questions **16** to **27**, give each number correct to 1 s.f. then use this to give a rough estimate of the square of the number.

16	28	**20**	7.9	**24**	0.0312
17	99	**21**	37.2	**25**	87
18	4.2	**22**	1212	**26**	0.081
19	0.27	**23**	73	**27**	249

Finding squares

Using a calculator

Enter the number to be squared and press the 'square' button, which is usually labelled x^2. If there is no 'square' button, then multiply the number by itself.

Check that the answer you obtain agrees with your rough estimate. Give your answer correct to 4 significant figures unless you are told otherwise.

Exercise 16b

Find the squares of:

1	7.8	**3**	79.2
2	38	**4**	0.41

> **Tip** Find a rough estimate first.

5 0.16	**12** 2.94	**19** 241	**26** 0.072
6 0.032	**13** 1.02	**20** 0.824	**27** 14.2
7 48.2	**14** 13.6	**21** 0.879	**28** 142
8 11.3	**15** 17	**22** 0.0362	**29** 0.142
9 51.3	**16** 1.11	**23** 72.4	**30** 9.73
10 9.8	**17** 7.21	**24** 3.78	**31** 13.9
11 12.1	**18** 11.6	**25** 0.245	**32** 0.0727

33 **a** Copy and complete the following table:

x	0	0.5	1	1.5	2	2.5	3	3.5	4
x^2	0			2.25	4				

b Draw axes for x from 0 to 4 using 2 cm to 1 unit and for y from 0 to 16 using 1 cm to 1 unit. Use the table to draw the graph of $y = x^2$.

c From the graph, find the values of y when $x = 2.2$, 1.8, 3.1 and 2.7.

d Use a calculator to find 2.2^2, 1.8^2, 3.1^2 and 2.7^2. How do these answers compare with your answers to part **c**?

e Repeat parts **c** and **d** with other values of your own choice.

34 **a** Copy and complete the following table:

x	2	4	6	8	10	12	14	15
x^2	4		36		100			225

b Draw axes for x from 0 to 15 using 1 cm $\equiv$ 1 unit and for y from 0 to 240 using 1 cm $\equiv$ 10 units. Use your table to draw the graph of $y = x^2$.

c From the graph, find the values of y when $x = 5.5$, 8.4, 12.8 and 13.6.

d Use a calculator to find 5.5^2, 8.4^2, 12.8^2 and 13.6^2. How do these answers compare with your answers to part **c**?

 Investigation

Is there a shortcut for finding squares of some two-digit numbers?

$$25^2 = (2 \times 3)100 + 25 = 625$$
$$35^2 = (3 \times 4)100 + 25 = 1225$$

Do you see a pattern?

Use the above method to find the following:

$$55^2, 65^2, 75^2, 85^2.$$

Areas of squares

Exercise 16c Find the area of a square of side 7.2 m.

Area $= (7.2 \times 7.2)\,\text{m}^2$

$\approx (7 \times 7)\,\text{m}^2 = 49\,\text{m}^2$

Area $= 51.8\,\text{m}^2$ correct to 3 s.f.

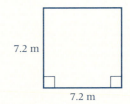

7.2 m

7.2 m

Find the areas of the squares whose sides are given in questions **1** to **9**. Give your answers correct to 3 s.f.

1	2.4 cm	**4**	1.06 m	**7**	0.062 m
2	9.6 m	**5**	17.2 cm	**8**	324 km
3	32.4 cm	**6**	52 mm	**9**	0.31 cm

Tip The area of a square is found by multiplying the length of a side by itself, i.e. by squaring the length of a side.

Square roots

The square root of a number is the number which, when multiplied by itself, gives the original number,

e.g. because $4^2 = 16$, the square root of 16 is 4.

The square root could also be -4 because $(-4) \times (-4) = 16$ but we will deal only with positive square roots in this chapter.

The symbol for the positive square root is $\sqrt{\ }$ so $\sqrt{16} = 4$

Exercise 16d

Find the square roots in questions **1** to **18** without using a calculator.

1	$\sqrt{9}$	**8**	$\sqrt{64}$
2	$\sqrt{25}$	**9**	$\sqrt{1}$
3	$\sqrt{4}$	**10**	$\sqrt{8100}$
4	$\sqrt{81}$	**11**	$\sqrt{0.81}$
5	$\sqrt{100}$	**12**	$\sqrt{0.64}$
6	$\sqrt{36}$	**13**	$\sqrt{4900}$
7	$\sqrt{49}$	**14**	$\sqrt{490\,000}$

Tip Check your answer by squaring it.

15	$\sqrt{0.04}$
16	$\sqrt{400}$
17	$\sqrt{2500}$
18	$\sqrt{10\,000}$

Use the answers to Exercise 16a, questions **1** to **15**, to find the following square roots.

19	$\sqrt{0.09}$	**21**	$\sqrt{0.0004}$	**23**	$\sqrt{4\,000\,000}$
20	$\sqrt{0.16}$	**22**	$\sqrt{250\,000}$	**24**	$\sqrt{0.000\,016}$

Rough estimates of square roots

So far, we have been able to find exact square roots of the numbers we have been given. Most numbers, however, do not have exact square roots; $\sqrt{23}$, for example, lies between 4 and 5 because $4 \times 4 = 16$, and $5 \times 5 = 25$.

$\sqrt{23}$, if given as a decimal, will start with 4.

i.e. $\sqrt{23} = 4.---$

Exercise 16e   Find the first significant figure of the square root of 30.

$$\sqrt{30} = 5.---$$

($\sqrt{30}$ lies between 5 and 6 because $5^2 = 25$ and $6^2 = 36$)

Find the first significant figure of the square roots of the following numbers:

1 17	**4** 40	**7** 85	**10** 0.05	**13** 14.2
2 10	**5** 3	**8** 15	**11** 0.20	**14** 0.50
3 38	**6** 10.2	**9** 4.6	**12** 90	**15** 5.7

Notice that $\sqrt{3} = 1.---$ while $\sqrt{30} = 5.---$

and that $\sqrt{300} = 1-.---$ while $\sqrt{3000} = 5-.---$

Every *pair of figures* added to the original number adds *one* figure to the approximate square root. We can pair off the figures from the decimal point, i.e. $\sqrt{3|00|00}$. Looking at the figure or figures in front of the first dividing line we can find the first significant figure of the square root.

Then $\sqrt{3|00|00.} = 1---.---$

≈ 100 (*Check:* $100 \times 100 = 10\,000$)

and $\sqrt{30|00|00.} = 5---.---$

≈ 500 (*Check:* $500 \times 500 = 250\,000$)

Exercise 16f   Find a rough value for the square root of 5280.

$7^2 = 49$ and $8^2 = 64$ so $\sqrt{52}$ is between 7 and 8

$$\sqrt{52|80.} = 7-.--$$

≈ 70

(*Check:* $70 \times 70 = 4900$)

By finding the first significant figure of the square root, give a rough value for the square root of each of the following numbers:

1	1400	**5**	720	**9**	4160	**13**	756	**17**	729.4
2	62 300	**6**	14 000	**10**	14 860	**14**	75 600	**18**	15.26
3	623	**7**	3260	**11**	396 000	**15**	7 560 000	**19**	3.698
4	7200	**8**	41 600	**12**	396	**16**	4128	**20**	39.46

Finding square roots

Using a calculator

Enter the number, say 5280, then press the square root button which is usually labelled $\sqrt{x}$. You will usually get a number that fills the display; give your answer correct to 4 significant figures. On some calculators you can press the $\sqrt{\ }$ button first.

$$\sqrt{5280} = 72.66$$

Check that it agrees with your rough estimate.

Exercise 16g

Find the square roots of the following numbers correct to 3 s.f. Give a rough estimate first in each case.

1	38.4	**5**	32	**9**	650	**13**	24	**17**	728
2	19.8	**6**	9.8	**10**	65	**14**	19	**18**	7280
3	428	**7**	67	**11**	11.2	**15**	10 300	**19**	61
4	4230	**8**	5.7	**12**	58	**16**	412 000	**20**	7 280 000

21 Find the square roots of the numbers in Exercise 16f.

Investigation

Using the digits 3 and 6 it is possible to make two two-digit numbers, namely 36 and 63.

The difference between the squares of these two numbers is

$$63^2 - 36^2 = 2673$$
$$= 99 \times 27$$
$$= 99 \times 9 \times 3$$
$$= 99 \times \left(\begin{array}{c}\text{the sum of the}\\\text{original digits}\end{array}\right) \times \left(\begin{array}{c}\text{the difference}\\\text{between the original digits}\end{array}\right)$$

Investigate whether or not this is true for other pairs of digits.

Rough estimates of square roots of numbers less than 1

$$0.2 \times 0.2 = 0.04 \qquad \text{so} \qquad \sqrt{0.04} = 0.2$$

and $\qquad \sqrt{0.05} = 0.2- - - \qquad$ also $\quad \sqrt{0.0004} = 0.02$

so $\qquad \sqrt{0.0005} = 0.02- - - \qquad$ but $\quad \sqrt{0.004}$ is neither 0.2 nor 0.02

It is easiest to find a rough estimate of the square root by again pairing off from the decimal point, but this time going to the right instead of to the left: $\sqrt{0.\overline{00}\overline{40}}$, adding a zero to complete the pair.

Now $\sqrt{40} = 6.- - -$ so we see that $\sqrt{0.004} = 0.06- - -$

(*Check:* $\quad 0.06 \times 0.06 = 0.0036 \approx 0.004$)

Using a calculator or tables we find

$$\sqrt{0.004} = 0.0632 \text{ correct to 3 s.f.}$$

Note that each pair of zeros after the decimal point gives one zero after the decimal point in the answer.

Exercise 16h Find the square roots of 0.007 32 and 0.000 732 correct to 3 s.f.

$$\sqrt{0.\overline{00}\overline{73}\,2} = 0.08- - -$$

$$\sqrt{0.007\,32} = 0.0856 \qquad \text{correct to 3 s.f.}$$

$$\sqrt{0.\overline{00}\overline{07}\,32} = 0.02- - -$$

$$\sqrt{0.000\,732} = 0.0271 \qquad \text{correct to 3 s.f.}$$

Find a rough estimate (as far as the first significant figure) and then use your calculator to find the square root of each of the following numbers correct to 3 s.f.

1 0.042	**8** 0.278	**15** 0.0432
2 0.42	**9** 0.0278	**16** 0.009 61
3 0.014	**10** 0.002 78	**17** 0.832
4 0.56	**11** 0.3	**18** 0.32
5 0.000 14	**12** 0.173	**19** 0.052
6 0.5	**13** 0.2	**20** 0.75
7 0.6014	**14** 0.69	**21** 0.000 073

Exercise 16i

Find the side of the square whose area is 50 m².

Length of the side = $\sqrt{50}$ m

$$= 7.\text{---} \text{ m}$$

Length of the side = 7.07 m correct to 3 s.f.

50 m²

Find the sides of the squares whose areas are given below. Give your answers correct to 3 s.f.

1 85 cm²

2 120 cm²

3 500 m²

4 32 m²

5 0.06 m²

6 15.1 cm²

7 749 mm²

8 84 300 km²

9 0.0085 km²

10 59 cm²

11 241 m²

12 61 cm²

Puzzle

An explorer leaves his tent and walks 1 km south, he then walks 2 km due east and finally 1 km north back to his tent. Where is his tent?

Did you know that the Pythagoreans believed that everything could be explained in terms of whole numbers? When they discovered that $\sqrt{2}$ could not be written as a ratio of whole numbers (i.e. a fraction), they tried to keep it secret.

IN THIS CHAPTER...

you have seen that:

- the square of a number bigger than 1 is bigger than the original number whereas the square of a number less than one is smaller than the original number

- you can find the first significant figure of a square root by pairing the numbers out from the decimal point and estimating the square root of the first pair

- the square root of a number bigger than 1 is less than the number

- the square root of a number smaller than one is bigger than the number

17 TRAVEL GRAPHS

AT THE END OF THIS CHAPTER...

you should be able to:

1 Read from a distance–time graph, the distance or time of motion of a moving object.

2 Draw distance–time graphs using suitable scales.

3 Calculate the distance travelled in a given time, by an object moving at a constant speed.

4 Calculate the time taken to travel a given distance at a constant speed.

5 Calculate the average speed of a body, given the distance travelled and time taken.

6 Calculate the average speed for a body covering different distances at different speeds.

7 Read information from a distance–time graph.

MATHS IS OUT THERE

Q: What goes from New York to San Francisco without moving?

A: A railway line

BEFORE YOU START

you need to know:
✓ how to work with fractions and decimals
✓ how to read and draw graphs
✓ metric and Imperial units of distance
✓ units of time, including the 12-hour and 24-hour clock, and how to convert between them

KEY WORDS   average speed, constant speed, distance, knot, speed

Finding distance from a graph

When we went on holiday we travelled by car to our holiday resort at a steady speed of 30 kilometres per hour (km/h), i.e. in each hour we covered a distance of 30 km.

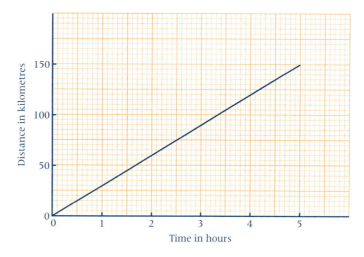

This graph shows our journey. It plots distance against time and shows that

in 1 hour we travelled 30 km	in 4 hours we travelled 120 km
in 2 hours we travelled 60 km	in 5 hours we travelled 150 km
in 3 hours we travelled 90 km	

Exercise 17a

The graphs that follow show four different journeys. For each journey find:

a the distance travelled

b the time taken

c the distance travelled in 1 hour

> **Tip** Make sure you understand what the subdivisions on the scales represent.

1

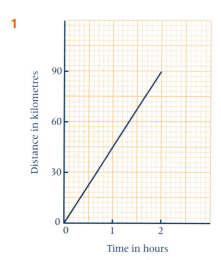

2

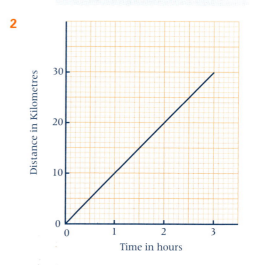

3

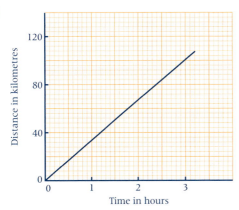

4

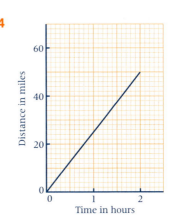

Drawing travel graphs

If Peter walks at 6 km/h, we can draw a graph to show this, using 2 cm to represent 6 km on the distance axis and 2 cm to represent 1 hour on the time axis.

Plot the point which shows that in 1 hour he has travelled 6 km. Join the origin to this point and produce the straight line to give the graph shown. From this graph we can see that in 2 hours Peter travels 12 km and in 5 hours he travels 30 km.

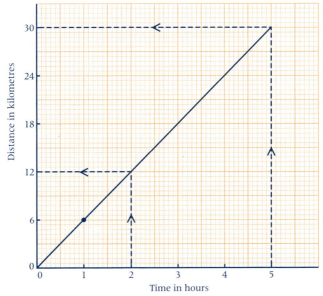

Alternatively we could say that

if he walks 6 km in 1 hour

he will walk $6 \times 2 \, \text{km} = 12 \, \text{km}$ in 2 hours

and he will walk $6 \times 5 \, \text{km} = 30 \, \text{km}$ in 5 hours

The distance walked is found by multiplying the speed by the time,

i.e. $$\text{Distance} = \text{speed} \times \text{time}$$

Exercise 17b   Draw a travel graph to show a journey of 150 km in 3 hours. Plot distance along the vertical axis and time along the horizontal axis.

Let 4 cm represent 1 hour and 2 cm represent 50 km.

Draw a line from 0 to the point above 3 on the time axis and along from 150 on the distance axes.

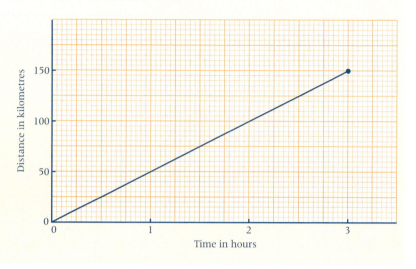

Draw travel graphs to show the following journeys. Plot distance along the vertical axis and time along the horizontal axis. Use the scales given in brackets.

1 60 km in 2 hours
(4 cm ≡ 1 hour, 1 cm ≡ 10 km)

2 180 km in 3 hours
(4 cm ≡ 1 hour, 2 cm ≡ 50 km)

3 300 km in 6 hours
(1 cm ≡ 1 hour, 1 cm ≡ 50 km)

4 80 miles in 2 hours
(6 cm ≡ 1 hour, 1 cm ≡ 10 miles)

5 140 miles in 4 hours
(2 cm ≡ 1 hour, 1 cm ≡ 25 miles)

6 100 km in $2\frac{1}{2}$ hours
(2 cm ≡ 1 hour, 2 cm ≡ 25 km)

7 105 km in $3\frac{1}{2}$ hours
(2 cm ≡ 1 hour, 4 cm ≡ 50 km)

8 75 miles in $1\frac{1}{4}$ hours
(8 cm ≡ 1 hour, 2 cm ≡ 25 miles)

9 40 m in 5 sec
(2 cm ≡ 1 sec, 2 cm ≡ 10 m)

10 240 m in 12 sec
(1 cm ≡ 1 sec, 2 cm ≡ 50 m)

11 Alan walks at 5 km/h. Draw a graph to show him walking for 3 hours. Take 4 cm to represent 5 km and 4 cm to represent 1 hour. Use your graph to find how far he walks in   **a** $1\frac{1}{2}$ hours   **b** $2\frac{1}{4}$ hours.

12 Julie can jog at 10 km/h. Draw a graph to show her jogging for 2 hours. Take 1 cm to represent 2 km and 8 cm to represent 1 hour. Use your graph to find how far she jogs in   **a** $\frac{3}{4}$ hour   **b** $1\frac{1}{4}$ hours.

13 Jo drives at 35 m.p.h. Draw a graph to show her driving for 4 hours. Take 1 cm to represent 10 miles and 4 cm to represent 1 hour. Use your graph to find how far she drives in   **a** 3 hours   **b** $1\frac{1}{4}$ hours.

<u>14</u> John walks at 4 m.p.h. Draw a graph to show him walking for 3 hours. Take 1 cm to represent 1 m.p.h. and 4 cm to represent 1 hour. Use your graph to find how far he walks in   **a** $\frac{1}{2}$ hour   **b** $3\frac{1}{2}$ hours.

The remaining questions should be solved by calculation.

<u>15</u> An express train travels at 200 km/h. How far will it travel in

   a 4 hours   **b** $5\frac{1}{2}$ hours ?

<u>16</u> Ken cycles at 24 km/h. How far will he travel in

   a 2 hours   **b** $3\frac{1}{2}$ hours   **c** $2\frac{1}{4}$ hours ?

<u>17</u> An aeroplane flies at 300 m.p.h. How far will it travel in

   a 4 hours   **b** $5\frac{1}{2}$ hours ?

<u>18</u> A bus travels at 60 km/h. How far will it travel in

   a $1\frac{1}{2}$ hours   **b** $2\frac{1}{4}$ hours ?

<u>19</u> Susan can cycle at 12 m.p.h. How far will she ride in

   a $\frac{3}{4}$ hour   **b** $1\frac{1}{4}$ hours ?

<u>20</u> An athlete can run at 10.5 m/s. How far will he travel in

   a 5 sec   **b** 8.5 sec ?

<u>21</u> A boy cycles at 12 m.p.h. How far will he travel in

   a 2 hours 40 min   **b** 3 hours 10 min ?

<u>22</u> Majid can walk at 8 km/h. How far will he walk in

   a 30 min   **b** 20 min   **c** 1 hour 15 min ?

<u>23</u> A racing car travels at 111 m.p.h. How far will it travel in

   a 20 min   **b** 1 hour 40 min ?

<u>24</u> A bullet travels at 100 m/s. How far will it travel in

   a 5 sec   **b** $8\frac{1}{2}$ sec ?

<u>25</u> A Boeing 747 travels at 540 m.p.h. How far does it travel in

   a 3 hours 15 min   **b** 7 hours 45 min ?

<u>26</u> A racing car travels around a 2 km circuit at 120 km/h. How many laps will it complete in

   a 30 min   **b** 1 hour 12 min ?

Calculating the time taken

Georgina walks at 6 km/h so we can find how long it will take her to walk

a 24 km   **b** 15 km.

a   If she takes 1 hour to walk 6 km, she will take $\frac{24}{6}$ hours, i.e. 4 h, to walk 24 km.

b   If she takes 1 hour to walk 6 km, she will take $\frac{15}{6}$ hours, i.e. $2\frac{1}{2}$ hours, to walk 15 km.

i.e.
$$\text{time} = \frac{\text{distance}}{\text{speed}}$$

Exercise 17c

1   How long will Zena, walking at 5 km/h, take to walk

    a   10 km   **b**   15 km ?

2   How long will a car travelling at 80 km/h, take to travel

    a   400 km   **b**   260 km ?

3   How long will it take David, running at 10 m.p.h. to run

    a   5 miles   **b** $12\frac{1}{2}$ miles ?

4   How long will it take an aeroplane flying at 450 m.p.h. to fly

    a   1125 miles   **b**   2400 miles ?

5   A cowboy rides at 14 km/h. How long will it take him to ride

    a   21 km   **b**   70 km ?

6   A rally driver drives at 50 m.p.h. How long does it take him to cover

    a   75 miles   **b**   225 miles ?

7   An athlete runs at 8 m/s. How long does it take him to cover

    a   200 m   **b**   1600 m ?

8   A dog runs at 20 km/h. How long will it take him to travel

    a   8 km   **b**   18 km ?

9   A liner cruises at 28 nautical miles per hour. How long will it take to travel

    a   6048 nautical miles   **b**   3528 nautical miles ?

10   A car travels at 56 m.p.h. How long does it take to travel

    a   70 miles   **b**   154 miles ?

11 A cyclist cycles at 12 m.p.h. How long will it take him to cycle

    a 30 miles   **b** 64 miles ?

12 How long will it take a car travelling at 64 km/h to travel

    a 48 km   **b** 208 km ?

Average speed

Russell Compton left home at 8 a.m. to travel the 50 km to his place of work. He arrived at 9 a.m. Although he had travelled at many different speeds during his journey he covered the 50 km in exactly 1 hour. We say that his *average speed* for the journey was 50 kilometres per hour, or 50 km/h. If he had travelled at the same speed all the time, he would have travelled at 50 km/h.

Judy Smith travelled the 135 miles from her home to Georgetown in 3 hours. If she had travelled at the same speed all the time, she would have travelled at $\frac{135}{3}$ m.p.h., i.e. 45 m.p.h. We say that her average speed for the journey was 45 m.p.h.

In each case:
$$\text{average speed} = \frac{\text{total distance travelled}}{\text{total time taken}}$$

This formula can also be written:
$$\text{distance travelled} = \text{average speed} \times \text{time taken}$$

and
$$\text{time taken} = \frac{\text{distance travelled}}{\text{average speed}}$$

Suppose that a car travels 35 km in 30 min, and we wish to find its speed in kilometres per hour. To do this we must express the time taken in hours instead of minutes,

i.e. $$\text{time taken} = 30 \text{ min} = \tfrac{1}{2} \text{ hour}$$

Then $$\text{average speed} = \frac{35}{\frac{1}{2}} \text{ km/h} = 35 \times \frac{2}{1} \text{ km/h}$$
$$= 70 \text{ km/h}$$

Great care must be taken with units. If we want a speed in kilometres per hour, we need the distance in kilometres and the time in hours. If we want a speed in metres per second, we need the distance in metres and the time in seconds.

Exercise *17d*

Find the average speed for each of the following journeys:

1 80 km in 1 hour	**5** 80 m in 4 sec	**9** 245 miles in 7 hours	
2 120 km in 2 hours	**6** 135 m in 3 sec	**10** 104 miles in 13 hours	
3 60 miles in 1 hour	**7** 150 km in 3 hours	**11** 252 m in 7 sec	
4 480 miles in 4 hours	**8** 520 km in 8 hours	**12** 255 m in 15 sec	

Find the average speed in km/h for a journey of 39 km which takes 45 min.

To find a speed in km/h you need the distance in kilometres and the time in hours.

First, convert the time taken to hours:

$$45 \, min = \frac{45}{60} \, hour = \frac{3}{4} \, hour$$

Then

$$average \ speed = \frac{distance \ travelled}{time \ taken}$$

$$= \frac{39 \, km}{\frac{3}{4} \, hour}$$

$$= 39 \times \frac{4}{3} \, km/h$$

$$= 52 \, km/h$$

Find the average speed in km/h for a journey of:

13 40 km in 30 min

14 60 km in 40 min

15 48 km in 45 min

16 66 km in 33 min

Find the average speed in km/h for a journey of:

17 4000 m in 20 min

18 6000 m in 45 min

19 40 m in 8 sec

20 175 m in 35 sec

> **Tip** Make sure that the time is in hours and the distance is in kilometres.

Find the average speed in m.p.h. for a journey of:

21 27 miles in 30 min

22 18 miles in 20 min

23 25 miles in 25 min

24 28 miles in 16 min

The following table shows the distances in kilometres between various towns in the West Indies.

	St John's	Roseau	Castries	Basseterre	Kingstown	St Georges	Port of Spain
Roseau	174						
Castries	382	211					
Basseterre	100	478	621				
Kingstown	446	272	74	557			
St Georges	549	570	554	1040	118		
Port of Spain	723	659	534	1218	528	176	
Georgetown	1234	1224	1099	1694	1093	741	565

Use this table to find the average speeds for journeys between:

25 St John's, leaving at 1025 h, and Kingstown, arriving at 1625 h

26 St Georges, leaving at 0330 h, and Castries, arriving at 0730 h

27 Basseterre, leaving at 1914 h, and St Georges, arriving at 2044 h

28 Port of Spain, leaving at 0620 h, and St Johns, arriving at 0750 h

29 Roseau, leaving at 1537 h, and St Georges, arriving at 1907 h

30 Castries, leaving at 1204 h, and Georgetown, arriving at 1624 h

31 Roseau, leaving at 1014 h, and Port of Spain, arriving at 1638 h

Problems frequently occur where different parts of a journey are travelled at different speeds in different times but we wish to find the average speed for the whole journey.

Consider for example a motorist who travels the first 50 miles of a journey at an average speed of 25 m.p.h. and the next 90 miles at an average speed of 30 m.p.h.

One way to find his average speed for the whole journey is to complete the following table by using the relationship:

$$\text{time in hours} = \frac{\text{distance in miles}}{\text{speed in m.p.h.}}$$

	Speed in m.p.h.	Distance in miles	Time in hours
First part of journey	25	50	2
Second part of journey	30	90	3
Whole journey		**140**	**5**

We can add the distances to give the total length of the journey, and add the times to give the total time taken for the journey.

$$\text{average speed for whole journey} = \frac{\text{total distance}}{\text{total time}}$$

$$= \frac{140 \text{ miles}}{5 \text{ hours}}$$

$$= 28 \text{ m.p.h.}$$

Note: Never add or subtract average speeds.

We could also solve this problem, without using a table, as follows:

$$\text{time to travel 50 miles at 25 m.p.h.} = \frac{\text{distance}}{\text{speed}}$$

$$= \frac{50 \text{ miles}}{25 \text{ m.p.h.}}$$

$$= 2 \text{ hours}$$

$$\text{time to travel 90 miles at 30 m.p.h.} = \frac{\text{distance}}{\text{speed}}$$

$$= \frac{90 \text{ miles}}{30 \text{ m.p.h.}}$$

$$= 3 \text{ hours}$$

∴ total distance of 140 miles is travelled in 5 hours

i.e.
$$\text{average speed for whole journey} = \frac{\text{total distance}}{\text{total time}}$$

$$= \frac{140 \text{ miles}}{5 \text{ hours}}$$

$$= 28 \text{ m.p.h.}$$

Exercise 17e

1 I walk for 24 km at 8 km/h, and then jog for 12 km at 12 km/h. Find my average speed for the whole journey.

> **Tip** To find the average speed you need the *total distance* travelled and the *total time* taken.

2 A cyclist rides for 23 miles at an average speed of $11\frac{1}{2}$ m.p.h. before his cycle breaks down, forcing him to push his cycle the remaining distance of 2 miles at an average speed of 4 m.p.h. Find his average speed for the whole journey.

3 An athlete runs 6 miles at 8 m.p.h., then walks 1 mile at 4 m.p.h. Find his average speed for the total distance.

4 A woman walks 3 miles at an average speed of $4\frac{1}{2}$ m.p.h. and then runs 4 miles at 12 m.p.h. Find her average speed for the whole journey.

5 A motorist travels the first 30 km of a journey at an average speed of 120 km/h, the next 60 km at 60 km/h, and the final 60 km at 80 km/h. Find the average speed for the whole journey.

6 Phil Sharp walks the 1 km from his home to the bus stop in 15 min, and catches a bus immediately which takes him the 9 km to the airport at an average speed of 36 km/h. He arrives at the airport in time to catch the plane which takes him the 240 km to Antigua at an average speed of 320 km/h. Calculate his average speed for the whole journey from home to Antigua.

7 A liner steaming at 24 knots takes 18 days to travel between two ports. By how much must it increase its speed to reduce the length of the voyage by 2 days? (A knot is a speed of 1 nautical mile per hour.)

Getting information from travel graphs

Exercise 17f

The graph opposite shows the journey of a coach that calls at three service stations A, B and C on a motorway. B is 60 km north of A and C is 20 km north of B. Use the graph to answer the following questions:

a At what time does the coach leave A?

b At what time does the coach arrive at C?

c At what time does the coach pass B?

d How long does the coach take to travel from A to C?

e What is the average speed of the coach for the whole journey?

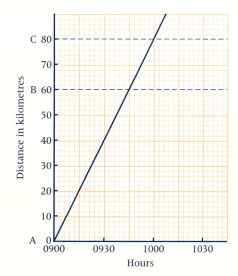

a The coach leaves A at 0900.

b It arrives at C at 1000. (Go from C on the distance axis across to the graph then down to the time axis).

c It passes through B at 0945.

d Time taken to travel from A to C is 1000 − 0900, i.e. 1 hour.

e Distance from A to C = 80 km (reading from the vertical axis)

Time taken to travel from A to C = 1 hour.

$$\text{average speed} = \frac{\text{distance travelled}}{\text{time taken}} = \frac{80\,\text{km}}{1\,\text{hour}} = 80\,\text{km/h}$$

1 The graph overleaf shows the journey of a car through three towns, Axeter, Bexley and Canton, which lie on a straight road. Axeter is 100 km south of Bexley and Canton is 60 km north of it. Use the graph to answer the following questions:

> **Tip** Make sure that you understand what the subdivisions on the scales represent.

a At what time does the car

    i leave Axeter   **ii** pass through Bexley   **iii** arrive at Canton?

b How long does the car take to travel from Axeter to Canton?

c How long does the car take to travel

    i the first 80 km of the journey?   **ii** the last 80 km of the journey?

d What is the average speed of the car for the whole journey?

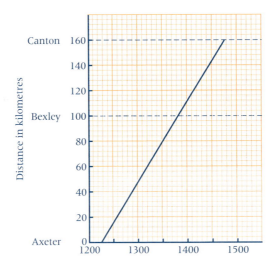

2   A car leaves Kingston at noon on its journey to Port Antonio via Morant Bay. The graph below shows its journey.

a   How far it is from

   i   Kingston to Morant Bay

   ii Morant Bay to Port Antonio?

b   How long does the car take to travel from Kingston to Port Antonio?

c   What is the car's average speed for the whole journey?

d   How far does the car travel between 1.30 p.m. and 2.30 p.m.?

e   How far is the car from

   i   Kingston

   ii Morant Bay, after travelling for $1\frac{1}{2}$ hours?

> **Tip**   You need the difference between these two values on the distance axis.

> **Tip**   Go up from 1.30 p.m. on the time axis to the graph then across to the distance axis. Do the same for 2.30 p.m. Then find the difference between these readings on the distance axis.

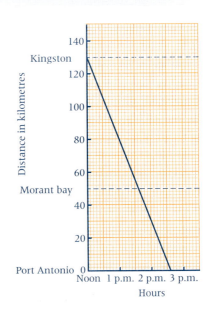

<u>3</u> Father used the family car to transport the children from their home to a summer camp and then returned home. The graph shows the journey.

a   How far is it from home to the camp?

b   How long did it take the family to get to the camp?

c   What was the average speed of the car on the journey to the camp?

d   How long did the car take for the return journey?

e   What was the average speed for the return journey?

f   What was the car's average speed for the round trip?

Tip The 'down hill' section of the graph represents the return journey.

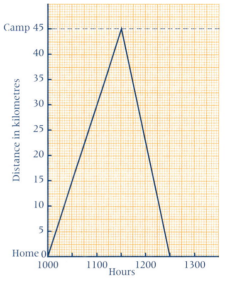

<u>4</u> The graph shows the journey of a car through three towns A, B and C.

a   Where was the car at

  i   0900 h   ii   0930 h?

b   What was the average speed of the car between

  i   A and B   ii   B and C?

c   For how long does the car stop at B?

d   How long did the journey take?

e   What was the average speed of the car for the whole journey? Give your answer correct to 1 s.f.

Tip The car arrives at B at the point where the graph stops going uphill and leaves B at the point where the graph starts going uphill again.

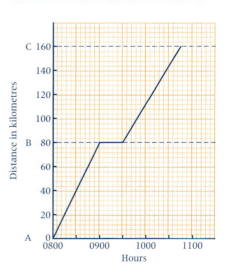

Exercise 17g

The graph shows Mrs Webb's journey on a bicycle to go shopping in the nearest town. Use it to answer the following questions:

a How far is town from home?

b How long did she take to get to town?

c How long did she spend in town?

d At what time did she leave for home?

e What was her average speed on the outward journey?

a Town is where Mrs Webb stops moving away from home, i.e. where the graph stops going uphill. The graph shows that it is 6 km from home to town.

b Mrs Webb left home at 1320 h and arrived in town at 1350 h. The journey therefore took 30 min.

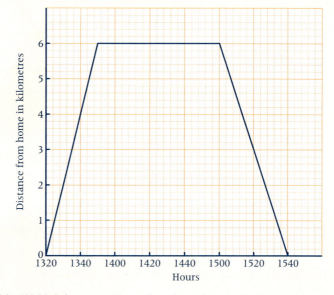

c Mrs Webb left town at the point where the graph starts going downhill. She arrived in town at 1350 h and left at 1500 h. She therefore spent 1 hour 10 min there.

d Mrs Webb left for home at 1500.

e On the outward journey:

$$\text{Average speed} = \frac{\text{distance travelled}}{\text{time taken}}$$

$$= \frac{6\,\text{km}}{30\,\text{min}}$$

$$= \frac{6\,\text{km}}{\frac{1}{2}\,\text{hour}} \quad (\text{time must be in hours})$$

$$= 6 \times \frac{2}{1}\,\text{km/h}$$

$$= 12\,\text{km/h}$$

1 The graph shows the journey of a plane from St Vincent to Martinique and back again. Use the graph to answer the following questions:

a How far is St Vincent from Martinique?

b How long did the outward journey take?

c What was the average speed for the outward journey?

d How long did the plane remain in Martinique?

e At what time did the plane leave Martinique, and how long did the return journey take?

f What was the average speed on the return journey?

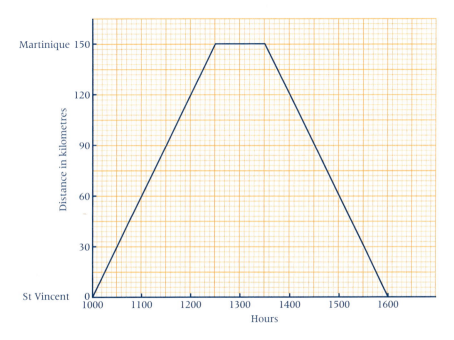

2 Overleaf is the travel graph for two motorists travelling between Kingston and Montego Bay which are 110 miles apart. The first leaves Montego Bay at 0900 h for Kingston, having a short break en route. The second leaves Kingston at 1015 h and travels non-stop to Montego Bay. Use your graph to find

a the average speed of each motorist for the complete journey,

b when and where they pass,

c their distance apart at 1200 h.

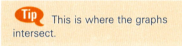
Tip This is where the graphs intersect.

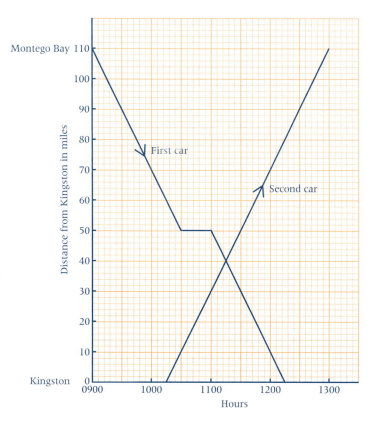

3 The graph below represents the journey of a motorist from Kingston to Mandeville and back again. Use this graph to find

a the distance between the two cities,

b the time the motorist spent in Mandeville,

c his average speed on the outward journey,

d the average speed on the homeward journey (including the stop).

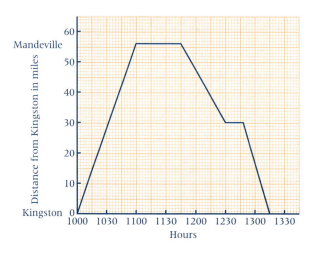

4 The graph below shows Judith's journeys between home and school.

    a At what time did she leave home

        i in the morning   **ii** in the afternoon ?

    b How long was she in school during the day ?

    c How long was she away from school for her mid-day break ?

    d What was the average speed for each of these journeys ?

    e Find the total time for which she was away from home.

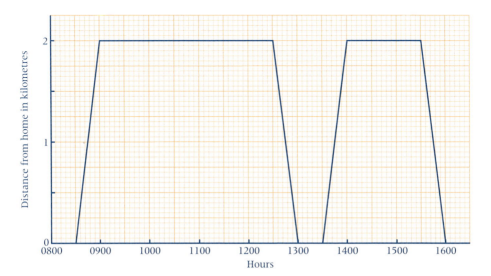

5 The graph on the right shows the journeys of two cars between two towns, A and B, which are 180 km apart. Use the graph to find

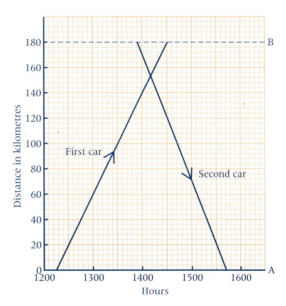

    a the average speed of the first motorist and his time of arrival at B

    b the average speed of the second motorist and the time at which she leaves B

    c when and where the two motorists pass

    d their distance apart at 1427 h.

<u>6</u> The graph represents the bicycle journeys of three school friends, Audrey, Betty and Chris, from the village in which they live to Spanish Town, the nearest main town, which is 30 km away.

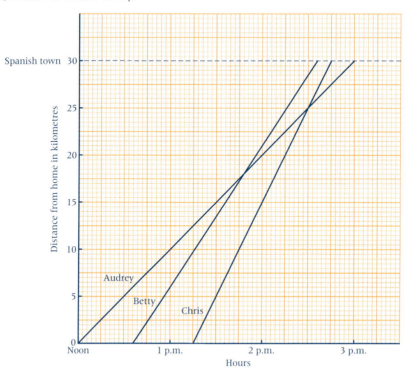

Use the graph to find:

a their order of arrival at Spanish Town

b Audrey's average speed for the journey

c Betty's average speed for the journey

d Chris's average speed for the journey

e where and when Chris passes Audrey

f how far each is from town at 2 p.m.

<u>7</u> Jane leaves home at 1 p.m. to walk at a steady 4 m.p.h. towards Grand Bay, which is 16 miles away, to meet her boyfriend Tim. Tim leaves Grand Bay at 2.18 p.m. and jogs at a steady 6 m.p.h. to meet her. Draw a graph for each of these journeys taking $4\,cm \equiv 1\,hour$ on the time axis and $1\,cm \equiv 1\,mile$ on the distance axis. From your graph find:

   a when and where they meet   **b** their distance apart at 3 p.m.

<u>8</u> A and B are motorway service areas 110 miles apart. A car leaves A at 2.16 p.m. and travels at a steady 63 m.p.h. towards B while a motorcycle leaves B at 2.08 p.m. and travels towards A at a steady 45 m.p.h. Draw a graph for the journeys taking $6\,cm \equiv 1\,hour$ and $1\,cm \equiv 5\,miles$. From your graph find:

   a when and where they pass

   b where the motorcycle is when the car starts

   c where the motorcycle is when the car arrives at B.

Mixed exercises

Exercise **17h**

1 The graph shows John's walk from home to his grandparents' home.

   a How far away do they live?

   b How long did the journey take him?

   c What was his average walking speed?

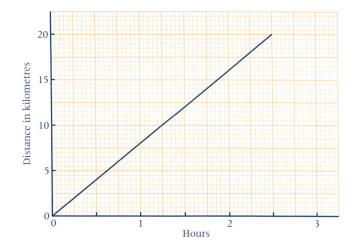

2 Jenny runs at 20 km/h. Draw a graph to show her running for $2\frac{1}{2}$ hours. Use your graph to find:

   a how far she has travelled in $1\frac{3}{4}$ hours

   b how long she takes to run the first 25 km.

3 A ship travels at 18 nautical miles per hour. How long will it take to travel:

   a 252 nautical miles   **b** 1026 nautical miles?

4 Find the average speed in km/h of a journey of 48 km in 36 min.

5 I left Antigua airport at 1147 h to travel the 315 miles to Barbados. If I arrived at 1232 h, what was the plane's average speed?

6 I walk $\frac{2}{3}$ mile in 10 min and then run $\frac{1}{3}$ mile in 2 min. What is my average speed for the whole journey?

7 The graph overleaf shows Paul's journey in a sponsored walk from A to B. On the way his sister, who is travelling by car in the opposite direction from B to A, passes him.

   a How far does Paul walk?

   b How long does he take?

   c How much of this time does he spend resting?

   d What is his average speed for the whole journey?

   e What is his sister's average speed?

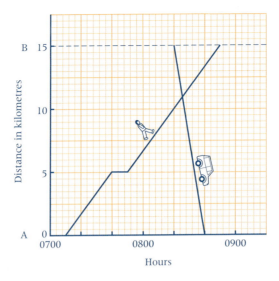

Exercise *17i*

1 The graph shows the journey of a scheduled flight from my island to Castries.

   a How far is my home from Castries?

   b How long did the journey take?

   c What happened during the journey that was not intended?

   d What was the average speed of the flight for the first part of the journey?

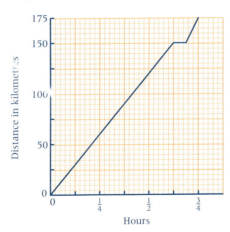

2 Draw a travel graph to show a journey of 440 km in 4 hours.

3 A horse runs at 15 m/sec. How far will it run in

   a 1 min   **b** $1\frac{3}{4}$ min?   Express its running speed in km/h.

4 How long will a coach travelling at 72 km/h take to travel   **a** 216 km   **b** 126 km?

5 Which speed is the faster, and by how much: 50 m/sec or 200 km/h?

6 Find the average speed (km/h) for an 1800 m journey in 9 min.

7 A motorist wants to make a 300 mile journey in $5\frac{1}{2}$ h. He travels the first 60 miles at an average speed of 45 m.p.h., and the next 200 miles at an average speed of 60 m.p.h. What must be his average speed for the remaining part of the journey if he is to arrive on time?

Puzzle

A train, 400 m long and travelling at 120 km/h, enters a tunnel that is 5.6 km long. For what time is any part of the train in the tunnel?

Did you know that

$3025 = (30 + 25)^2$ and that $2025 = (20 + 25)^2$?

Does this work for 4025 and 1025?

IN THIS CHAPTER...

you have seen that:

- a journey at constant speed can be represented by a straight line on a graph

- when you read values from a graph you need to make sure that you understand the meaning of the subdivisions on the scales on the axes

- the formula 'distance = speed × time' can be used to find one quantity when the other two are known

- when you are working out speeds, you must make sure that the units are consistent, e.g. to find a speed in kilometres per hour, the distance must be in kilometres and the time must be in hours

- the average speed for a journey is equal to the total distance travelled divided by the total time taken

REVIEW TEST 2: CHAPTERS 10 TO 17

In questions **1** to **9**, choose the letter for the correct answer.

1 P′ is the image of P under reflection in a mirror line *l*.
At what angle does PP′ cut *l*?

   A 30 degrees   **B** 45 degrees   **C** 60 degrees   **D** 90 degrees

2 The image of $(3, -1)$ under the translation $\begin{pmatrix} 4 \\ 2 \end{pmatrix}$ is

   A $(7, 1)$ **B** $(1, 3)$ **C** $(2, 6)$ **D** $(-1, -3)$

3 A minibus travels between two towns at an average speed of 40 km per hour.
How far will it travel in $2\frac{1}{2}$ hours?

   A 100 km   **B** $80\frac{1}{2}$ km   **C** 80 km   **D** $42\frac{1}{2}$ km

4 A cyclist rides a distance of 9 km in 15 minutes. What is his average speed in
kilometres per hour?

   A 36   **B** 80   **C** 144   **D** 180

5 The circumference of a circle is 88 cm. What is its radius? (Take $\pi = \frac{22}{7}$)

   A 11 cm   **B** 14 cm   **C** 22 cm   **D** 44 cm

6 The side LM of a triangle measures 5 cm. The triangle is enlarged by a scale
factor 2. What is the measure of the image of LM?

   A 3 cm   **B** 7 cm   **C** 10 cm   **D** 25 cm

7 What is the volume of a cylinder of radius $3\frac{1}{2}$ cm and length 4 cm?
(Take $\pi = \frac{22}{7}$)

   A 14 cm³   **B** 44 cm³   **C** 88 cm³   **D** 154 cm³

8 In the diagram LM =

   A 14 cm
   B 15 cm
   C 16 cm
   D 20 cm

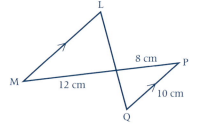

9 The image of $(2, 2)$ under a clockwise rotation of 90 degrees about the origin is

   A $(-2, 2)$ **B** $(2, -2)$ **C** $(-2, 0)$ **D** $(2, 0)$

10 Draw a triangle PQR with vertices at P($2, 5$), Q($3, 8$) and R($7, 1$). Draw the
images of PQR after reflections in

   a the *x*-axis   **b** the *y*-axis   **c** the line $y = x$

Name the coordinates of the images in each case.

11 State whether or not triangles ABC and PQR are similar. If they are, find the length of PR.

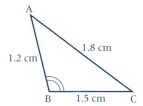 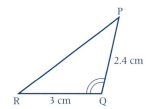

12 This shape consists of a semicircle on a rectangle. Calculate the area of the figure, correct to 2 decimal places. (Take $\pi = 3.142$)

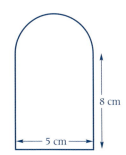

13 A prism of length 20 cm has its cross-section in the shape of a right-angled triangle with sides 6 cm, 8 cm and 10 cm. Calculate the volume of the prism.

14 a Find, correct to 3 s.f., **i** 43.2^2 **ii** $\sqrt{0.00044}$

b The vertices of a triangle XYZ are at X(-3, 1), Y(0, 5) and Z(5, -2). If XYZ is rotated a quarter of a turn clockwise about the origin, find the coordinates of the images of X, Y and Z.

15 A vehicle is travelling at a constant speed of 20 cm/s. Copy and complete the following table showing the time t and distance s.

t	0	1	2	3	4	6	8	10
s	0	20	40			120		

Using a scale of 1 cm to represent 2 seconds on the t-axis and 1 cm to represent 20 cm on the s-axis, draw a graph showing the time and distance travelled.

From the graph find

a the distance travelled in 7 seconds   **b** the time taken to travel 90 cm.

16 A man travelling to a town 4 kilometres away travels the first 2 kilometres in 30 minutes. He then rests for 20 minutes and afterwards continues his journey, reaching the town 30 minutes later.

Show this information on a distance-time graph. Use the graph to find the average speed for the last 30 minutes.

AT THE END OF THIS CHAPTER...

you should be able to:

1   State the ratio for the tangent of an angle in a right-angled triangle.

2   Use a calculator to find the tangent of a given angle.

3   Calculate the length of the adjacent (opposite) side to a given angle, in a right-angled triangle given the opposite (adjacent) side.

4   Solve problems using the tangent ratio in a right-angled triangle.

5   Use a calculator to find an angle, given the value of its tangent.

6   Find an angle given two sides of a right-angled triangle.

MATHS IS OUT THERE

According to legend, Euclid, who was the father of geometry, posed t̲ puzzle.

'A mule and a donkey were carrying a load of sacks. When the donkey groaned the mule looked at him and said; 'Why are you complaining? If you gave me one sack, I would have twice as many as you; and if I gave you one of my sacks, we would have equal loads.'

How many sacks was each carrying?

BEFORE YOU START

you need to know:
✓ how to work to a given place value or significant figure
✓ how to multiply and divide with decimals
✓ how to convert a fraction to a decimal
✓ how to solve a linear equation containing fractions
✓ the angle sum of a triangle

Investigating relationships

Sam wants to know the height of this tree.

He can find this by measuring how far he is from the tree and then measuring the angle of elevation of the top of the tree. Showing this information in a diagram gives this right-angled triangle.

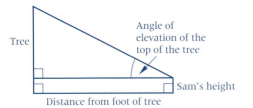

The diagram can be drawn to scale, and the height of the tree measured from the scale drawing.

The disadvantage of this method is that it takes time and precision to get a reasonably accurate result. In this chapter we find how to calculate such a length by a method which, with the help of a scientific calculator, is fast and whose accuracy depends only on the accuracy of the initial measurements.

In this chapter we are going to look at the relationship between the sizes of the angles and the lengths of the sides in right-angled triangles.

Exercise 18a

1 a Draw the given triangle accurately using a protractor and a ruler.

 b Measure $\hat{A}$.

 c Find $\dfrac{BC}{AB}$ as a decimal.

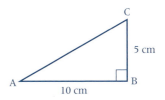

Repeat question **1** for the triangles in questions **2** to **5**.

2

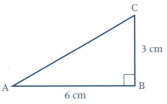

3

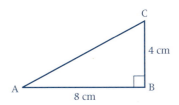

4

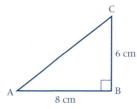

5

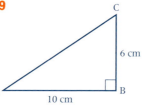

6 Are the triangles in questions **1** to **5** similar?

Repeat question **1** for the triangles in questions **7** to **12**.

7

9

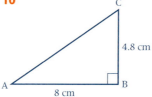

11

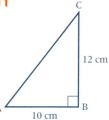

8

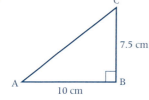

10

12

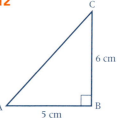

13

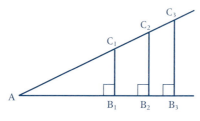

Similar triangles can be drawn so that they overlap, as in this diagram. Copy the diagram above on squared paper. Choose your own measurements but make sure that the lengths of the horizontal lines are whole numbers of centimetres. Measure $\hat{A}$.

Find $\dfrac{B_1C_1}{AB_1}$, $\dfrac{B_2C_2}{AB_2}$ and $\dfrac{B_3C_3}{AB_3}$ as decimals.

14 Copy and complete the table using the information from questions **1** to **11**.

Angle A	$\dfrac{BC}{AB}$
$26\frac{1}{2}^\circ$	0.5

Tangent of an angle

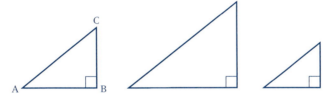

If we consider the set of all triangles that are similar to △ABC then, for every triangle in the set,

the angle corresponding to $\hat{A}$ is the same

the ratio corresponding to $\dfrac{BC}{AB}$ is the same

where BC is the side *opposite* to $\hat{A}$

and   AB is the *adjacent* (or neighbouring) side to $\hat{A}$

From the last exercise you can see that in a right-angled triangle the ratio $\dfrac{\text{opposite side}}{\text{adjacent side}}$ is always the same for a given angle whatever the size of that triangle.

The ratio $\dfrac{\text{opposite side}}{\text{adjacent side}}$ is called the *tangent* of the angle.

$$\text{tangent of the angle} = \frac{\text{opposite side}}{\text{adjacent side}}$$

Or, briefly,

$$\tan(\text{angle}) = \frac{\text{opp}}{\text{adj}}$$

The information about this ratio is used so often that we need a more complete and more accurate table than the one made in the last exercise. The information is stored in scientific calculators.

Finding tangents of angles

To find the tangent of 33°, press $\boxed{\text{tan}}\ \boxed{3}\ \boxed{3}\ \boxed{=}$. You will obtain a number that fills the display. Write down one more figure than the accuracy required.

e.g. $\tan 33° = 0.64940\ldots$

                    $= 0.6494$ correct to 4 significant figures.

If you do not get the correct answer, one reason could be that your calculator is not in 'degree mode'. For all trigonometric work at this stage, angles are measured in degrees, so make sure that your calculator is in the correct mode. Calculators also vary in the order in which buttons have to be pressed; if the order given above does not work try $\boxed{3}\ \boxed{3}\ \boxed{\text{tan}}$.

Find the tangents of the following angles correct to 3 s.f.:

1	20°	**4**	53°	**7**	19°	**10**	45°	**13**	4°	**16**	89°
2	28°	**5**	59°	**8**	12°	**11**	61°	**14**	37°	**17**	52°
3	72°	**6**	9°	**9**	21°	**12**	70°	**15**	44°	**18**	35°

19 Find the tangents of the angles listed in question **14** in Exercise 18a. How do the answers you now have compare with the decimals you worked out?

If they are different, give a reason for this.

Decimals of degrees

Sometimes we need the tangent of an angle that is not a whole number of degrees.

Find the tangent of 34.2°

tan 34.2° = 0.680 correct to 3 s.f.

Find the tangents of the following angles, correct to 3 s.f.:

1	15.5°	**4**	60.1°	**7**	30.6°	**10**	3.8°	**13**	42.4°	**16**	58.8°
2	29.6°	**5**	70.7°	**8**	15.9°	**11**	49.0°	**14**	71.2°	**17**	65.3°
3	11.4°	**6**	46.5°	**9**	10.2°	**12**	32.7°	**15**	49.5°	**18**	63.2°

The names of the sides of a right-angled triangle

Before we can use the tangent for finding sides and angles we need to know which is the side opposite to the given angle and which is the adjacent side.

a The longest side, that is the side opposite the right angle, is called the *hypotenuse*.

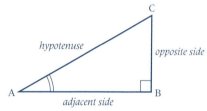

b The side next to the angle (not the hypotenuse) is called the *adjacent side*.

c The third side is the *opposite side*. It is opposite the particular angle we are concerned with.

Sometimes the triangle is in a position different from the one we have been using.

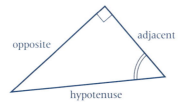

Exercise 18d

Sketch the following triangles. The angle we are concerned with is marked with a double arc like this ◿. Label the sides 'hypotenuse', 'adjacent' and 'opposite'. If necessary, turn the page round so that you can see which side is which.

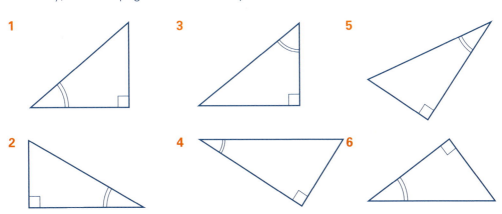

Finding a side of a triangle

We can now use the tangent of an angle to find the length of the opposite side in a right-angled triangle provided that we know an angle and the length of the adjacent side.

Exercise 18e

In this exercise use a calculator. Give your answers correct to 3 s.f.

In △ABC, $\widehat{B} = 90°$, $\widehat{A} = 32°$ and AB = 4 cm.

Find the length of BC.

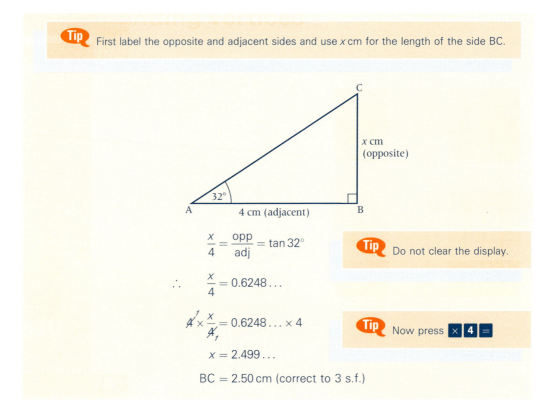

Tip First label the opposite and adjacent sides and use x cm for the length of the side BC.

$$\frac{x}{4} = \frac{\text{opp}}{\text{adj}} = \tan 32°$$

Tip Do not clear the display.

$$\therefore \quad \frac{x}{4} = 0.6248\ldots$$

$$\cancel{4} \times \frac{x}{\cancel{4}} = 0.6248\ldots \times 4$$

Tip Now press × 4 =

$$x = 2.499\ldots$$

$$BC = 2.50 \text{ cm (correct to 3 s.f.)}$$

Find the length of BC in questions **1** to **6**.

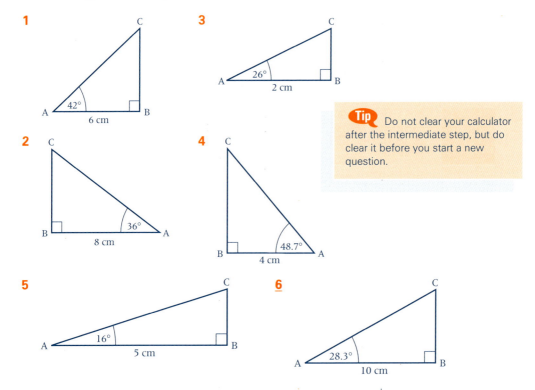

1

3

Tip Do not clear your calculator after the intermediate step, but do clear it before you start a new question.

2

4

5

6

In questions **7** to **10** different letters are used for the vertices of the triangles. In each case find the side required.

7 Find PQ

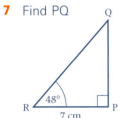

9 Find AC

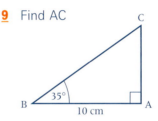

8 Find YZ

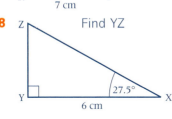

10 Find AC

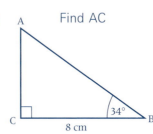

Find BC in questions **11** to **13**. Turn the page round if necessary to identify the opposite and adjacent sides.

11

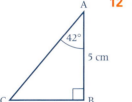

12

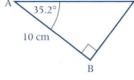

13

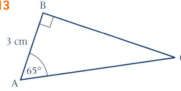

14 In △ABC, $\hat{B} = 90°$, AB = 6 cm and $\hat{A} = 41°$. Find BC.

15 In △PQR, $\hat{Q} = 90°$, PQ = 10 m and $\hat{P} = 16.7°$. Find QR.

16 In △DEF, $\hat{F} = 90°$, DF = 12 cm and $\hat{D} = 56°$. Find EF.

Finding a side adjacent to the given angle

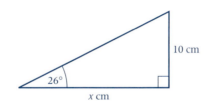

Sometimes the side whose length we are asked to find is adjacent to the given angle instead of opposite to it. Using $\frac{10}{x}$ instead of $\frac{x}{10}$ can lead to an awkward equation so we work out the size of the angle opposite x and use it instead. In this case this other angle is 64° and we label the sides 'opposite' and 'adjacent' to this angle.

Using 64°,

$$\frac{x}{10} = \frac{\text{opposite}}{\text{adjacent}} = \tan 64°$$

so $\quad \dfrac{x}{10} = 2.05 \quad \text{giving} \quad x = 20.5.$

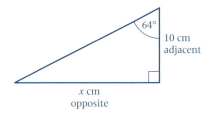

Exercise **18f**

Use a calculator. Give your answers correct to 3 s.f.

In △PQR, $\widehat{P} = 90°$, $\widehat{Q} = 51°$ and PR = 4 cm.

Find the length of PQ.

> **Tip** First find the other angle, i.e. $\widehat{R}$.

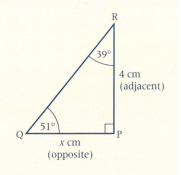

$\widehat{R} = 90° - 51° = 39°$

$$\frac{x}{4} = \frac{\text{opp}}{\text{adj}} = \tan 39°$$

$$\frac{x}{4} = 0.8097\ldots$$

> **Tip** Do not clear the display.

$$\cancel{4}^{1} \times \frac{x}{\cancel{4}_{1}} = 0.8097\ldots \times 4$$

$$x = 3.239\ldots$$

> **Tip** Leave the number on the display then press × 4 =

PQ = 3.24 cm (correct to 3 s.f.)

1 Find ZY

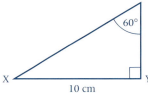

3 Find XZ

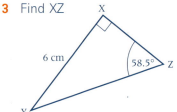

2 Find QP

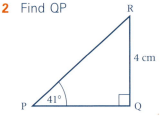

4 Find FD

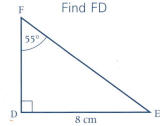

5 In △PQR, $\hat{Q} = 90°$, $\hat{R} = 31°$ and PQ = 6 cm. Find RQ.

6 In △XYZ, $\hat{Z} = 90°$, $\hat{Y} = 38°$ and ZX = 11 cm. Find YZ.

7 In △DEF, $\hat{D} = 90°$, $\hat{E} = 34.8°$ and DF = 24 cm. Find DE.

8 In △ABC, $\hat{C} = 90°$, $\hat{A} = 42.4°$ and CB = 3.2 cm. Find AC.

9 In △LMN, $\hat{L} = 90°$, $\hat{N} = 15°$ and LM = 4.8 cm. Find LN.

10 In △STU, $\hat{U} = 90°$, $\hat{S} = 42.2°$ and TU = 114 cm. Find SU.

Problems

*Exercise **18g***

Give your answers correct to 3 s.f.

Rita is at a point A, 20 metres from the base of a tree that is standing on level ground. From A the angle of elevation of the top, C, of the tree is 23°. What is the height of the tree?

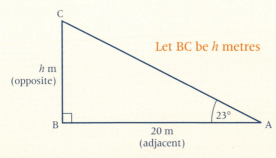

Let BC be h metres

$$\frac{h}{20} = \frac{\text{opp}}{\text{adj}} = \tan 23°$$

$$\frac{h}{20} = 0.4244\ldots$$

$$20 \times \frac{h}{20} = 0.4244\ldots \times 20$$

$$h = 8.489\ldots$$

The height of the tree is 8.49 m correct to 3 s.f.

1 In a triangle ABC, $\hat{A}$ is 35°, $\hat{B}$ is 90° and the length of BC is 10 cm. Find the length of AB.

2 Triangle PQR has a right angle at Q, the length of side PQ is 15 cm and $\hat{P}$ is 50°. Find the length of QR.

3 In triangle PQR, $\hat{P}$ is 90° and $\hat{Q}$ is 34.2°. The length of PQ is 12 cm. Find the length of PR.

4 Triangle XYZ has side XY of length 11 cm, $\hat{Y}$ is a right angle and $\hat{X}$ is 42.5°. Find the length of YZ.

5 A pole stands on level ground. A is a point on the ground 10 m from the foot of the pole. The angle of elevation of the top, C, from A is 27°. What is the height of the pole?

6 ABCD is a rectangle. AB = 42 m and BÂC = 59°. Find the length of BC.

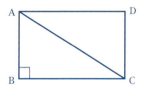

7 In rectangle ABCD, the angle between the diagonal AC and the side AB is 22°, AB = 8 cm. Find the length of BC.

8 In △PQR, PQ = QR. From symmetry, S is the midpoint of PR. P̂ = 72°. PR = 20 cm. Find the height QS of the triangle.

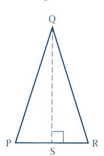

9

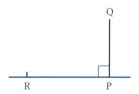

A point R is 14 m from the foot of a flagpole PQ. The angle of elevation of the top of the pole from R is 22°. Find the height of the pole.

10

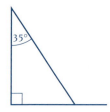

A ladder leans against a vertical wall so that it makes an angle of 35° with the wall. The top of the ladder is 2 m up the wall. How far out from the wall is the foot of the ladder?

11

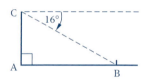

A boat B is 60 m out to sea from the foot A of a vertical cliff AC. From C the angle of depression of B is 16°.

a Find B̂.

b Find the height of the cliff.

 Investigation

Use what you have learnt in this chapter to find the height of a tall tree or a chimney on a building near you.

You will need a long measuring tape and a theodolite to measure angles.

You can make a rough theodolite from a protractor, a length of string, some blu-tack and a small weight such as a pebble.

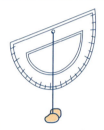

Finding an angle given its tangent

If we are given the value of the tangent of an angle then we can use a calculator to find that angle.

Using a calculator

To find the angle whose tangent is 0.732, enter 0.732 and then press the inverse button followed by the 'tan' button. (If this does not work, consult your instruction book.) The number you see filling the display is the size of the angle in degrees. Give the angle correct to 3 significant figures.

If $\tan \hat{A} = 0.732$

then $\hat{A} = 36.2°$

Exercise *18h*

Find the angles whose tangents are given below.

> **Tip** Know the syntax for your calculator: do you press 2.2 $\tan^{-1}$ or $\tan^{-1}$ 2.2?

1 2.2	**7** 0.16		
2 0.36	**8** 0.62		
3 0.41	**9** 0.6752	**13** 0.697	**17** 1.113
4 4.1	**10** 0.992 93	**14** 0.811	**18** 1.7
5 1.4	**11** 2.666 666	**15** 1.14	**19** 1.01
6 0.31	**12** 0.333 33	**16** 3.59	**20** 1.21

Tangents in the form of fractions

If we are given the value of a tangent in fraction form, then we need to change it to a decimal before we can find the angle.

Exercise *18i*

Find the angle whose tangent is $\frac{2}{3}$

$\tan A = \frac{2}{3} = 0.6666\ldots$

$A = 33.69\ldots° = 33.7°$

(correct to 1 d.p.)

> **Tip** Angle A can be found in one step on the calculator, press
>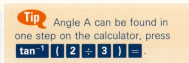

Find the angles whose tangents are given below.

1 $\frac{3}{5}$ 8 $\frac{3}{8}$

Tip First express $\frac{3}{5}$ as a decimal.

2 $\frac{4}{5}$ 9 $2\frac{1}{4}$

3 $\frac{1}{2}$ 10 $1\frac{1}{2}$ 15 $\frac{1}{6}$ 20 $\frac{2}{9}$

4 $\frac{2}{5}$ 11 $\frac{3}{25}$ 16 $\frac{5}{6}$ 21 $\frac{5}{7}$

5 $\frac{7}{10}$ 12 $2\frac{2}{5}$ 17 $\frac{7}{6}$ 22 $\frac{7}{3}$

6 $\frac{3}{20}$ 13 $\frac{1}{3}$ 18 $\frac{5}{3}$ 23 $\frac{4}{9}$

7 $\frac{5}{4}$ 14 $\frac{1}{7}$ 19 $\frac{3}{7}$ 24 $\frac{4}{3}$

Finding an angle given two sides of a triangle

We can now find an angle in a right-angled triangle if we are given the opposite and adjacent sides.

Exercise 18j   In △ABC, $\hat{B} = 90°$, AB = 8 cm and BC = 7 cm. Find $\hat{A}$.

On a copy of the diagram, mark the angle to be found and then label the opposite and adjacent sides to this angle.

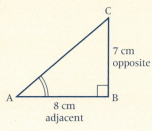

$$\tan \hat{A} = \frac{\text{opp}}{\text{adj}} = \frac{7}{8}$$

$$= 0.875$$

$$\hat{A} = 41.18\ldots°$$

$$= 41.2° \text{ correct to 1 d.p.}$$

Find $\hat{A}$ in questions **1** to **8**.

1
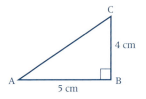

3
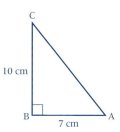

5
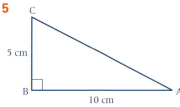

2
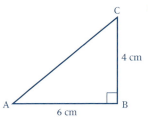

4
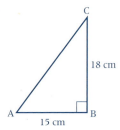

6
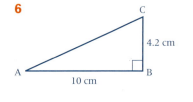

7

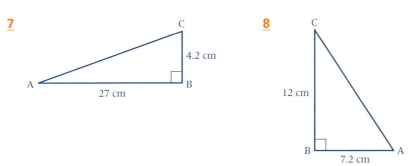

8

In questions **9** to **12**, different letters are used.

9 Find $\hat{P}$

11 Find $\hat{N}$

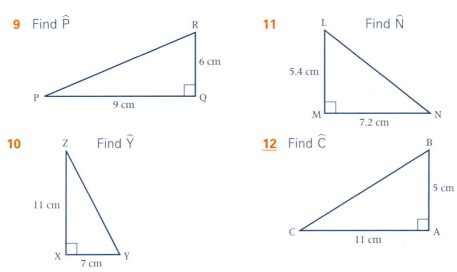

10 Find $\hat{Y}$

12 Find $\hat{C}$

Find $\hat{A}$ in questions **13** and **18**. Turn the page round if necessary before labelling the sides.

13

15

17

14

16

18

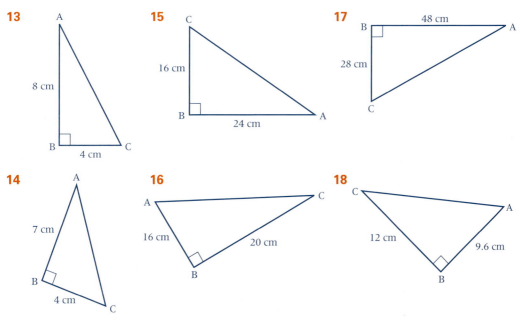

16

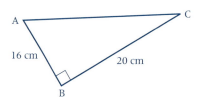

18

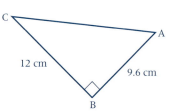

 19 In $\triangle ABC$, $\hat{B} = 90°$, $AB = 12\,cm$, $BC = 11\,cm$. Find $\hat{A}$.

> **Tip** Draw a diagram.

20 In $\triangle PQR$, $\hat{P} = 90°$, $PQ = 3.2\,m$, $PR = 2.8\,m$. Find $\hat{Q}$.

21 In $\triangle DEF$, $\hat{D} = 90°$, $DE = 108\,m$, $DF = 72\,m$. Find $\hat{F}$.

22 In $\triangle XYZ$, $\hat{Z} = 90°$, $YZ = 12\,cm$, $XZ = 11\,cm$. Find $\hat{X}$.

Investigation

John has a problem. For safety reasons he has to find the inclination of the ladder to the vertical.

All he has with him is a scientific calculator and a straight stick. He does not have a ruler or a measuring tape or a protractor and he does not know how long the ladder is.

Investigate how John can calculate the angle that the ladder makes with the wall and how accurate an answer he can expect.

Problems

Exercise 18k

A man walks due north for 5 km from A to B, then 4 km due east to C. What is the bearing of C from A?

We start by drawing a diagram showing all the information given. The bearing of C from A is angle A in the triangle, so we mark this angle.

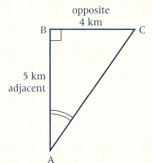

$$\tan \hat{A} = \frac{opp}{adj} = \frac{4}{5}$$

$$= 0.8$$

$$\hat{A} = 38.65\ldots° = 38.7°$$
(correct to 1 d.p.)

The bearing of C from A is $038.7°$.

Notice that we add '0' to make a three-figure bearing.

1

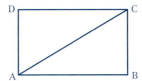

ABCD is a rectangle. AB = 60 m and BC = 36 m. Find the angle between the diagonal and the side AB.

2

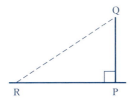

A flagpole PQ is 10 m high. R is a point on the ground 20 m from the foot of the pole. Find the angle of elevation of the top of the pole from R (i.e. $\hat{R}$).

3

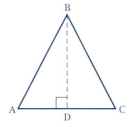

In △ABC, AB = BC. AC = 12 cm. D is the midpoint of AC. The height BD of the triangle is 10 cm. Find $\hat{C}$ and the other angles of the triangle.

4

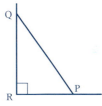

A ladder leans against a vertical wall. Its top, Q, is 3 m above the ground and its foot, P, is 2 m from the foot of the wall. Find the angle of slope of the ladder (that is, $\hat{P}$).

5 The bearing of town A from town B is 032.4°. A is 16 km north of B. How far east of B is it?

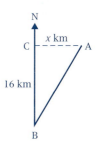

6 In a square, ABCD, of side 8 cm, A is joined to the midpoint E of BC. Find $E\hat{A}B$, $C\hat{A}B$ and $C\hat{A}E$. Notice that AE does *not* bisect $C\hat{A}B$.

7 A ladder leans against a vertical wall. It makes an angle of 72° with the horizontal ground and its foot is 1 m from the foot of the wall. How high up the wall does the ladder reach?

8 Sketch axes for *x* and *y* from 0 to 5. A is the point (1, 0) and B is (5, 2). What angle does the line AB make with the *x*-axis?

9 In a rhombus the two diagonals are of lengths 6.2 cm and 8 cm. Find the angles of the rhombus.

10

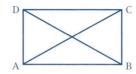

In rectangle ABCD, AB = 24 cm and BC = 11 cm. Find CÂB and hence find the obtuse angle between the diagonals.

11 In △ABC, AB = BC, CA = 10 cm and Ĉ = 72°. Find the height BD of the triangle.

Did you know that the first 'theodolite' was an instrument for measuring horizontal angles?

A vertical circle or arc, enabling angles of elevation and depression to be measured, was added much later.

IN THIS CHAPTER...

you have seen that:

- in a right-angled triangle the tangent of an angle is found by dividing the length of the side opposite the angle by the length of the side adjacent to the angle; this value can be given as a decimal or a fraction

- you use the tangent of an angle to find the length of a side in a right-angled triangle if an angle and a side, other than the hypotenuse, are given

- you can also use the tangent of an angle to find an angle in a right-angled triangle when you are given the opposite and adjacent sides to that angle

- when you use a calculator for more than one step, you do not need to clear the display after the first step

- you can also use the bracket keys on your calculator to help reduce the number of steps in your calculation

19

SINE AND COSINE OF AN ANGLE

AT THE END OF THIS CHAPTER...

you should be able to:

1   Use a calculator to find the sine of a given acute angle.

2   Use a calculator to find acute angles whose sines are given.

3   Use the sine ratio to find a side or an angle of a right-angled triangle.

4   Use a calculator to find the cosine of a given acute angle.

5   Use a calculator to find acute angles whose cosines are given.

6   Use the cosine ratio to find a side or an angle of a right-angled triangle.

7   Use the appropriate trig ratio – sine, cosine or tangent – to solve problems on right-angled triangles.

Did you know that astronomers use trigonometry for calculating distances to stars.

BEFORE YOU START

you need to know:
✓ how to work to a given place value or significant figure
✓ how to multiply and divide with decimals
✓ how to convert a fraction to a decimal
✓ how to solve a linear equation containing fractions
✓ the angle sum of a triangle
✓ the meaning of the tangent of an angle and the terms adjacent side, opposite side and hypotenuse

KEY WORDS

acute angle, adjacent side, chord, cosine of an angle, hypotenuse, isosceles, opposite side, ratio, sine of an angle, tangent of an angle

Trigonometry: sine of an angle

The tangent of an angle was useful when the opposite and adjacent sides of a right-angled triangle were involved.

Sometimes we are interested, instead, in the opposite side and the hypotenuse. These two sides form a different ratio which is called the sine of the angle (or sin for short).

In this diagram

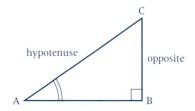

$$\sin \hat{A} = \frac{\text{opp}}{\text{hyp}} = \frac{CB}{AC}$$

All right-angled triangles containing an angle of 40°, for example, are similar so $\dfrac{\text{side opposite } 40°}{\text{hypotenuse}}$ always has the same value.

In all right-angled triangles,

$$\sin \hat{A} = \frac{\text{opp}}{\text{hyp}}$$

The values of this ratio for all acute angles are stored in most calculators.

Exercise 19a

Find the sines of the following angles correct to 3 s.f.:

1 26°	**3** 25.4°	**5** 78.9°	**7** 16.8°	**9** 62.4°
2 84°	**4** 37.1°	**6** 72°	**8** 4.2°	**10** 71.1°

Find the angles whose sines are given below. Give answers correct to 3 s.f.

11 0.834

12 0.413

13 0.639

14 0.704

15 0.937

16 0.07

17 0.647

18 0.357

19 0.428

20 0.261

Tip You will need the **sin⁻¹** button on your calculator. Check how to use it in your manual.

Using the sine ratio to find a side or an angle

Exercise 19b Find the length of BC.

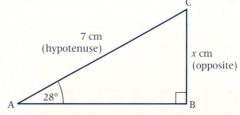

Tip On a copy of the diagram, label the sides in relation to the angle.

$$\sin 28° = \frac{\text{opp}}{\text{hyp}} = \frac{x}{7}$$

Tip We write the equation the other way round as it is easier to handle when the term containing x is on the l.h.s.

$$\frac{x}{7} = 0.4694\ldots$$

Do not clear the display $x = 7 \times 0.4694\ldots = 3.286\ldots$

press × 7   BC = 3.29 cm (correct to 3 s.f.)

1 Find BC.

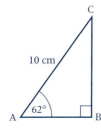

3 Find AC.

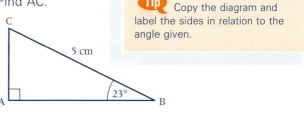

Tip Copy the diagram and label the sides in relation to the angle given.

2 Find BC.

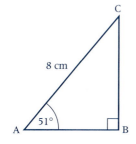

4 Find BC.

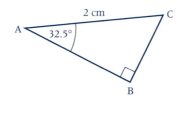

5 Find LM.

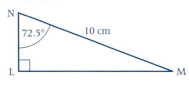

6 Find PQ.

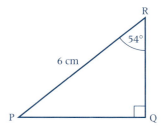

In △ABC, $\widehat{B} = 90°$, AC = 4 cm and BC = 3 cm. Find $\widehat{A}$.

(Label the sides first.)

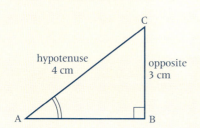

$$\sin \widehat{A} = \frac{\text{opp}}{\text{hyp}} = \frac{3}{4}$$

$$= 0.75$$

$$\widehat{A} = 48.59\ldots°$$

$$\widehat{A} = 48.6° \text{ to 3 s.f.}$$

7 Find $\widehat{A}$.

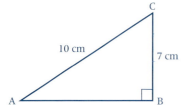

10 Find $\widehat{M}$.

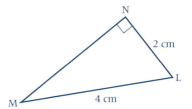

8 Find $\widehat{A}$.

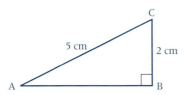

11 Find $\widehat{A}$.

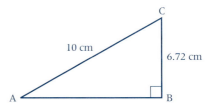

9 Find $\widehat{P}$.

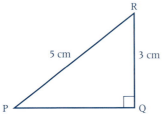

12 Find $\widehat{Q}$.

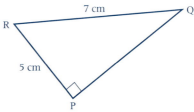

13 In △ABC, $\widehat{B} = 90°$, $\widehat{C} = 36°$ and AC = 3.5 cm. Find AB.

14 In △PQR, $\hat{R} = 90°$, PQ = 7 cm and $\hat{P} = 71.6°$. Find QR.

15 In △ABC, $\hat{B} = 90°$, AB = 3.2 cm and AC = 4 cm. Find $\hat{C}$ and $\hat{A}$.

16 In △PQR, $\hat{Q} = 90°$, PQ = 2.6 cm and PR = 5.5 cm. Find $\hat{R}$.

Puzzle

Solve this cryptogram
$$\begin{array}{r} S\,A\,V\,E \\ +\,M\,O\,R\,E \\ \hline M\,O\,N\,E\,Y \end{array}$$

Cosine of an angle

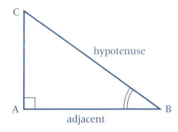

If we are given the adjacent side and the hypotenuse, then we can use a third ratio, $\frac{\text{adjacent side}}{\text{hypotenuse}}$. This is called the *cosine* of the angle (cos for short).

$$\cos \hat{B} = \frac{\text{adj}}{\text{hyp}} = \frac{AB}{BC}$$

Cosines of acute angles are stored in most calculators.

Exercise 19c

Find the cosines of the following angles to 3 s.f.:

1 59°	**3** 4°	**5** 60.1°	**7** 82°	**9** 17.5°
2 48°	**4** 44.9°	**6** 67°	**8** 13.8°	**10** 25.3°

In questions **11** to **20**, cos $\hat{A}$ is given. Find $\hat{A}$ to 3 s.f.

11 0.435   **16** 0.24

12 0.714   **17** 0.938

13 0.7   **18** 0.739

14 0.943   **19** 0.628

15 0.820   **20** 0.843

> **Tip** Use the **cos⁻¹** button on your calculator.

Using the cosine ratio to find a side or an angle

Exercise 19d

In △ABC, $\widehat{B} = 90°$ and AC = 9 cm.

Find AB.

(Label the sides first.)

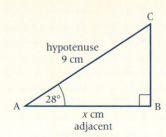

$$\frac{x}{9} = \frac{adj}{hyp} = \cos 28°$$

$$\frac{x}{9} = 0.8829\ldots$$

$$\cancel{9} \times \frac{x}{\cancel{9}} = 0.8829\ldots \times 9$$

$$x = 7.946\ldots$$

$$AB = 7.95\,cm \quad (\text{correct to 3 s.f.})$$

In the following triangles find the required lengths to 3 s.f.

Tip Remember not to clear the display on your calculator after the intermediate step.

1 Find AB.

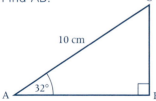

2 Find AB.

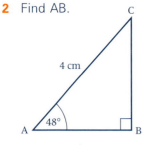

3 Find PQ.

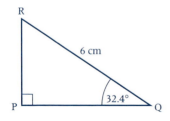

4 Find XZ.

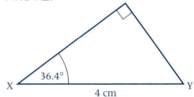

5 Find PR.

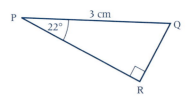

6 Find BC.

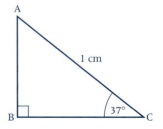

In △ABC, $\hat{B} = 90°$, AB = 4 cm and AC = 6 cm. Find $\hat{A}$.

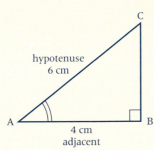

$$\cos \hat{A} = \frac{adj}{hyp} = \frac{4}{6}$$

$$= 0.6666\ldots$$

$$\hat{A} = 48.18\ldots°$$

$$= 48.2° \text{ to 1 d.p.}$$

Give angles correct to 1 d.p.

7 Find $\hat{A}$.

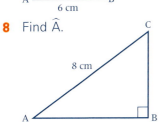

8 Find $\hat{A}$.

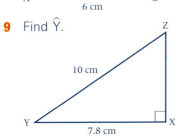

9 Find $\hat{Y}$.

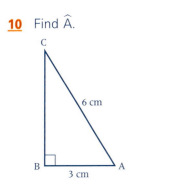

10 Find $\hat{A}$.

11 Find $\hat{Q}$.

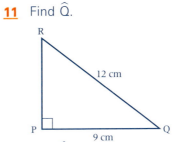

12 Find $\hat{C}$.

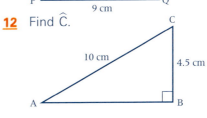

13 Find $\hat{P}$.

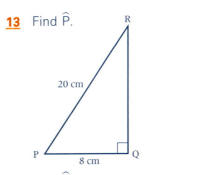

14 Find $\hat{X}$.

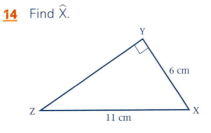

Use of all three ratios

To remember which ratio is called by which name, some people use the word SOHCAHTOA.

$$Sin \ \widehat{A} = \frac{Opposite}{Hypotenuse} \quad SOH$$

$$Cos \ \widehat{A} = \frac{Adjacent}{Hypotenuse} \quad CAH$$

$$Tan \ \widehat{A} = \frac{Opposite}{Adjacent} \quad TOA$$

Exercise 19e State whether sine, cosine or tangent should be used for the calculation of the marked angle.

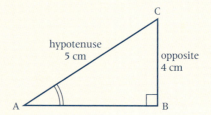

The opposite side and the hypotenuse are given so we should use sin $\widehat{A}$.

In questions **1** to **6** label the sides whose lengths are known, 'hypotenuse', 'opposite' or 'adjacent'. Then state whether sine, cosine or tangent should be used to calculate the marked angle.

1

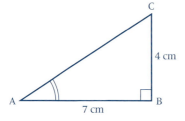

3

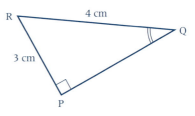

2

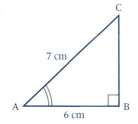

4

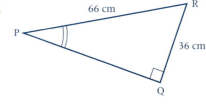

5

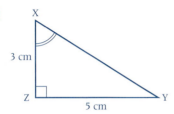

6

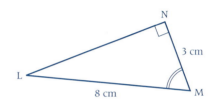

State whether sine, cosine or tangent should be used to calculate to find x.

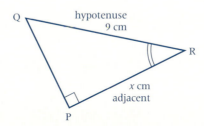

We are given the hypotenuse and wish to find the adjacent side so we should use $\cos \hat{R}$.

In questions **7** to **10**, using 'opposite', 'adjacent' or 'hypotenuse', label the side whose length is given and the side whose length is to be found. Then state whether sine, cosine or tangent should be used for the calculation to find x.

7

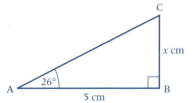

9

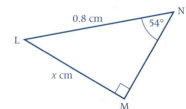

8

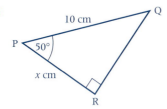

10

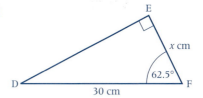

11 Calculate the marked angles in questions **1** to **6** and the lengths given as x cm in questions **7** to **10**. Give answers correct to 3 s.f.

In questions **12** to **15**, find the marked angles.

12

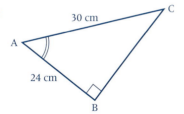

13

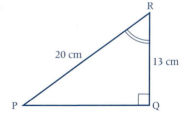

14

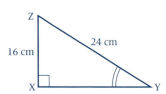

15

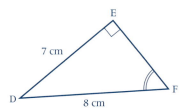

In questions **16** to **19**, find the length of the side marked x cm.

16

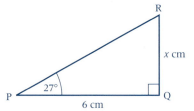

18

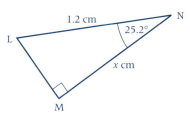

17

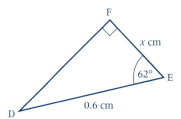

19

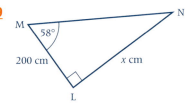

Exercise 19f

Notice that you can find the size of the third angle of the triangle by using the fact that the three angles add up to 180°.

In $\triangle ABC$, $\hat{B} = 90°$, $AC = 15$ cm and $BC = 11$ cm. Find $\hat{A}$, then $\hat{C}$.

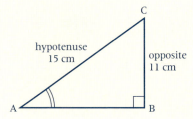

(We are given the hypotenuse and the opposite side so we should use sin $\hat{A}$.)

$$\sin \hat{A} = \frac{\text{opp}}{\text{hyp}} = \frac{11}{15} = 0.7333$$

$$\hat{A} = 47.16\ldots° = 47.2° \text{ to 1 d.p.}$$

$$\hat{C} = 90° - 47.2° \quad \text{(angles of a triangle add up to 180°)}$$

$$= 42.8°$$

1 Find $\hat{A}$, then $\hat{C}$.

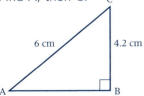

2 Find AC.

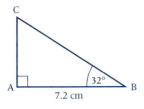

3 Find X, then $\hat{Z}$

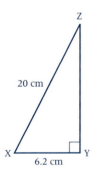

4 Find LM.

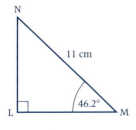

5 Find AB.

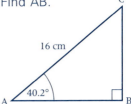

6 Find PQ.

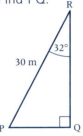

7 Find XZ.

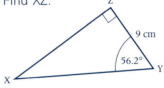

8 Find $\hat{A}$.

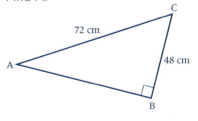

In questions **9** to **12** ABC is a triangle in which $\hat{B} = 90°$. Find the length or angle marked with a cross.

	AB	BC	CA	$\hat{A}$	$\hat{C}$
9	7 cm		10 cm	×	
10		×	5 cm	32.68	
11		×	8 cm		468
12	16 cm	22 cm		×	×

Problems

In an isosceles triangle PQR, PQ = QR = 5 cm and PR = 6 cm.

Find the angles of the triangle.

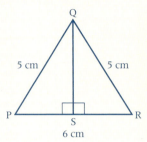

Tip By dividing the triangle down the middle with the line QS, we create two identical right-angled triangles. We can then draw one of these triangles to work with.

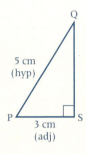

$$\cos \widehat{P} = \frac{adj}{hyp} = \frac{PS}{PQ} = \frac{3}{5}$$

Tip PS = $\frac{1}{2}$PR = 3 cm

$\widehat{P} = 53.1°$ (correct to 1 d.p.)

$\widehat{R} = 53.1°$ (isosceles △; base angle equal)

$P\widehat{Q}R = 73.8°$ (angles of a △ add up to 180°)

Tip Identify one right-angled triangle to use then draw it.

1 In rectangle ABCD, AC = 4 cm and BC = 3 cm.

Find C$\widehat{A}$B.

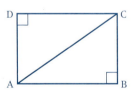

2

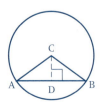

C is the centre of a circle of radius 10 cm. C$\widehat{A}$B = 31°.
Find the distance of the chord AB from the centre, i.e. find DC.

3

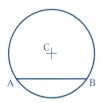

C is the centre of a circle of radius 4 cm. Chord AB is of length 5 cm. Find C$\widehat{A}$B.

4 A ladder 2 m long leans against a wall. Its top is 1.6 m above the foot of the wall. Find the angle that the ladder makes with the ground.

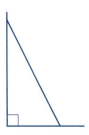

5 A ladder 4 m long stands on horizontal ground and leans against a vertical wall. It makes an angle of 25° with the wall. How far is the foot of the ladder from the foot of the wall?

6 In △ABC, AB = AC = 8 cm and $\hat{B}$ = 68.6°.

Find the height of the triangle.

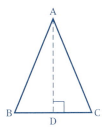

7

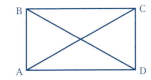

ABCD is a rectangle. AB = 4.2 cm, and AC = 6.3 cm.

Find $\hat{CAB}$ and the acute angle between the diagonals.

8

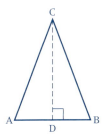

In △ABC, AC = CB = 12 cm and AB = 10 cm. Find $\hat{A}$ and the other angles of △ABC.

If a road gradient is 1 in 5, you rise 1 unit as you walk 5 units up the slope. The angle of the slope is the angle of inclination of the road.

If the gradient of a road is given as 10%, then because $10\% = \frac{1}{10}$, the gradient is 1 in 10.

The gradient of a road is 1 in 5. Find its angle of slope.

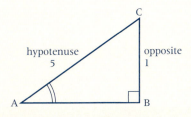

$$\sin \hat{A} = \frac{opp}{hyp} = \frac{1}{5}$$

$$= 0.2$$

$$\hat{A} = 11.5°$$

Therefore the angle of the slope is 11.5°.

9 What is the angle of slope when the gradient of a road is 1 in 8?

10 A road gradient is 1 in 6. What is the angle of slope?

11 Find the angle of slope of a road with a gradient of 10%.

12 Find the angle of slope of a road with a gradient of 5%.

13 From a point P a man walks 5 km on a bearing of 220° to a point Q.

From Q he walks due north to a point R which is due west of P.

Find   **a** the bearing of P from Q   **b** the bearing of P from R

       c the distance of R from P   **d** the distance of R from Q.

14 Kate runs from her home (A) to the Sports' Stadium (B).

From A the stadium is 10 km on a bearing of 137°.

How far is the stadium

a east of her home   **b** south of her home?

15 The village church (C) is 1.7 km from Tom's house (H) on a bearing of 317°.

a Is H east or west of C? Draw a sketch to illustrate your answer.

b Calculate this distance.

c Calculate the distance that C is north or south of H. State clearly whether it is north or south.

16 From a point A a car travels 12 km on a bearing of 057° to a point B. From B it continues its journey for 6 km on a bearing of 330° to a point C.

Calculate how far

a B is east of A   **b** C is west of B

c C is east of A   **d** C is north of A

Exercise 19h

1 Find   **a** $\sin \hat{A}$ where $\hat{A} = 40°$ **b** $\cos \hat{B}$ where $\hat{B} = 50°$.

What do you notice about your answers?

2 Use the diagram to find

a $\sin \hat{A}$ **b** $\cos \hat{C}$.

What is the value of $\hat{A} + \hat{C}$?

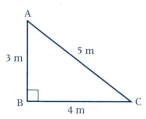

3 If $\sin \hat{A} = 0.3$, write down the value of $\cos \hat{C}$.

4 In △PQR, $\hat{P} = 90°$, and $\cos \hat{Q} = 0.8$. Write down the value of $\sin \hat{R}$.

5 Sketch a triangle ABC in which $\hat{A} = 90°$ and $\hat{B} = 45°$. What is the value of $\hat{C}$? What kind of triangle is △ABC? *Without* using a calculator, write down the value of tan 45°.

Mixed exercises

Exercise 19i

1 Find the sine of 83°.

2 If $\tan \hat{A} = 1.6341$, find $\hat{A}$.

3 Find the cosine of 28°.

4 Find $\hat{X}$ if $\sin \hat{X} = 0.5$

5 Find AB.

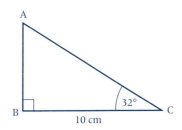

6 Find $\hat{R}$.

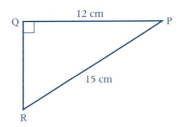

7 Find YZ.

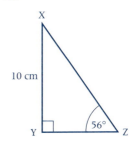

Exercise 19j

1 Find $\cos \hat{A}$ where $\hat{A} = 25°$

2 Find $\hat{C}$ given that $\sin \hat{C} = 0.9311$

3 Find $\tan \hat{Y}$ where $\hat{Y} = 45°$

4 Find $\hat{M}$ given that $\cos \hat{M} = 0.9311$

5 Find MN.

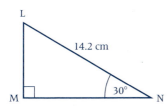

6 Find $\hat{D}$

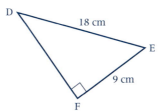

7 Find AC.

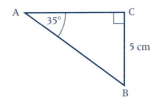

Investigation

Surveyors use a process called 'triangulation' to find distances and heights.

Firstly two 'stations', A and B, are established on the same level and the distance between them is measured. The position of any other point C, such as a point on the opposite side of a river, is then found by measuring the angles between AB and the lines of sight from A to C and from B to C.

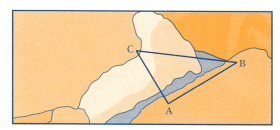

This gives a triangle in which one side is known and the angles at each end of the side. The lengths of the two other sides can then be calculated from this information.

Investigate how this can be done.

Find out what you can about Bertrand Russell (1872–1970) and Gabriel Cramer (1704–52).

IN THIS CHAPTER...

you have seen that:

- in a right-angled triangle,

$$\tan \widehat{A} = \frac{\text{opp}}{\text{adj}}, \quad \sin \widehat{A} = \frac{\text{opp}}{\text{hyp}},$$

$$\cos \widehat{A} = \frac{\text{adj}}{\text{hyp}}$$

- if you draw and label a triangle with the angle and sides in relation to the angle, this will help you to decide which of tan, sin or cos to use

- to use trigonometry in a problem, identify a right-angled triangle you can use then draw this triangle.

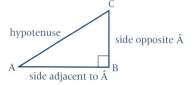

side opposite Â

hypotenuse

A   side adjacent to Â   B

20 PRACTICAL APPLICATIONS OF GRAPHS

Did you know that when you change Caribbean dollars into another currency, you will be given an exchange rate called 'sell' and when you change that currency back into Caribbean dollars you will be given a different exchange rate called 'buy'. This means that if you change Caribbean $ into US $ and the change them straight back to Caribbean $, you will get fewer dollars than you started with. The banks make money this way – and they generally also charge commission!

Graphs involving straight lines

If you were to go to England for a holiday, you would probably have a little difficulty in knowing the cost of things in dollars and cents. If we know the rate of exchange, we can use a simple straight line graph to convert a given number of pounds into dollars or a given number of dollars into pounds.

Given that $1 converts to £0.33, we can draw a graph to convert values from, say, $0–$90 into pounds.

Take 5 cm ≡ $100 and 1 cm ≡ £10.

(≡ means 'is equivalent to'.)

Because $100 ≡ £33 and $200 ≡ £66

we can now plot these points and join them with a straight line.

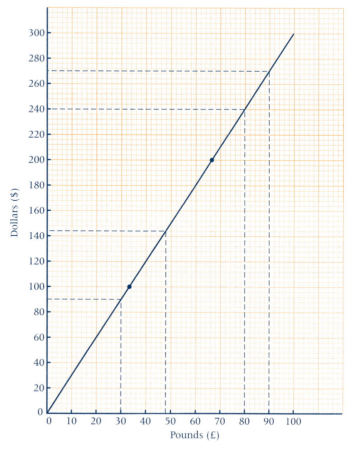

From the graph: $144 ≡ £48 (going from $144 across the graph then down to the £ axis)

$240 ≡ £80

£30 ≡ $90

£90 ≡ $270

Exercise 20a

1 The table gives temperatures in degrees Fahrenheit (°F) and the equivalent values in degrees Celsius (°C).

Temperature in °F	57	126	158	194
Temperature in °C	14	52	70	90

Plot these points on a graph for Celsius values from 0 to 100 and Fahrenheit values from 0 to 220. Let 2 cm represent 20 units on each axis.

Use your graph to convert:

a   97 °F into °C

b   172 °F into °C

c   25 °C into °F

d   80 °C into °F

2   The table shows the conversion from US dollars to £s for various amounts of money.

US dollars	50	100	200
£s	35	70	140

Plot these points on a graph and draw a straight line to pass through them. Let 4 cm represent 50 units on both axes.

Use your graph to convert:

a   160 dollars into £s

b   96 dollars into £s

c   £122 into dollars

d   £76 into dollars

3   The table shows the conversion of various sums of money from US dollars to EC dollars.

US dollars (US $)	100	270	350
EC dollars (EC $)	270	729	945

Plot these points on a graph and draw a straight line to pass through them. Take 2 cm to represent 50 units on the US $-axis and 100 units on the EC $-axis.

Use your graph to convert:

a   160 US $ into EC $

b   330 US $ into EC $

c   440 EC $ into US $

d   980 EC $ into US $

> **Tip**   When you choose scales for axes, make them easy to read. On graph paper, you have 5 subdivisions between each centimetre, so choose a multiple of 5, e.g. in question **4** you could choose 2 cm ≡ 5 km and in question **5** you could choose 2 cm ≡ 500 km.

4   The table shows the distance a girl walks in a given time.

Time walking in hours	0	1	$2\frac{1}{2}$	4	5
Distance walked in km	0	6	15	24	30

Draw a graph of these results. What do you conclude about the speed at which she walks?

How far has she walked in   a 2 hours   b $3\frac{1}{2}$ hours?

How long does she take to walk   c 10 km   d 21 km?

5 The table shows the distance an aircraft has travelled at various times on a particular journey.

Time after departure in hours	0	1	$3\frac{1}{2}$	6
Distance travelled from take-off in km	0	550	1925	3300

Draw a graph of these results. What can you conclude about the speed of the aircraft?

How far does it fly in   **a** $1\frac{1}{2}$ hours   **b** $4\frac{1}{2}$ hours?

How long does it take to fly   **c** 1000 km   **d** 2500 km?

6 Marks in an examination range from 0 to 65. Draw a graph that enables you to express the marks in percentages from 0 to 100. Note that a mark of 0 is 0% while a mark of 65 is 100%.

Use your graph

a   to express marks of 35 and 50 as percentages

b   to find the original mark for percentages of 50% and 80%.

7 Deductions from the wages of a group of employees amount to $35 for every $100 earned. Draw a graph to show the deductions made from gross pay in the range $0–$400 per week.

How much is deducted from an employee whose gross weekly pay is   **a** $125   **b** $240   **c** $335?

How much is earned each week by an employee whose weekly deductions amount to   **d** $40   **e** $88?

8 The table shows the fuel consumption figures for a car in both miles per gallon (X) and in kilometres per litre (Y).

m.p.g. (X)	30	45	60
km/litre (Y)	10.5	15.75	21

Plot these points on a graph taking 2 cm $\equiv$ 10 units on the X-axis and 4 cm $\equiv$ 5 units on the Y-axis. Your scale should cover 0–70 for X and 0–25 for Y.

Use your graph to find:

a   12 km/litre in m.p.g.   **b**   64 m.p.g. in km/litre

c   22.5 km/litre in m.p.g.   **d**   23 m.p.g. in km/litre

9 The table gives various speeds in kilometres per hour with the equivalent values in metres per second.

Speed in km/h (S)	0	80	120	200
Speed in m/s (V)	0	22.2	33.3	55.5

Plot these values on a graph taking 4 cm $\equiv$ 50 units on the S-axis and 4 cm $\equiv$ 10 units on the V-axis.

Use your graph to convert:

a 140 km/h into m/s

b 46 m/s into km/h

c 18 m/s into km/h

d 175 km/h into m/s

10 A number of rectangles, measuring *l* cm by *b* cm, all have a perimeter of 24 cm. Copy and complete the following table:

l	1	2	3	4	6	8
b			9			4

Draw a graph of these results using your own scale. Use your graph to find *l* if *b* is

a 2.5 cm

b 6.2 cm

and to find *b* if *l* is

c 5.5 cm

d 2.8 cm

Puzzle

Trains leave London for Edinburgh every hour on the hour. Trains leave Edinburgh for London every hour on the half-hour. The journey takes five hours each way.

Carrie takes a train from London to Edinburgh. How many trains from Edinburgh bound for London pass her train? Do not count any trains that may be in the stations at either end of the journey.

Graphs involving curves

When two quantities that are related are plotted one against the other, we often find that the points do not lie on a straight line. They may, however, lie on a smooth curve.

Consider the table below, which gives John's height on his birthday over a period of 8 years.

Age in years	11	12	13	14	15	16	17	18	19
Height in cm	138	140	144	150	158	165	170	172	173

These points can be plotted on a graph and joined to give a smooth curve through the points, as shown.

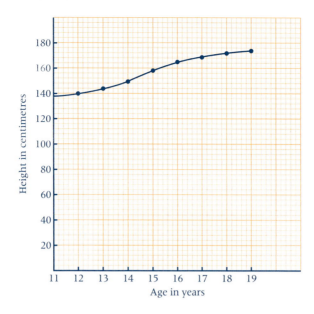

The graph enables us to estimate that:

a He was 162 cm tall when he was $15\frac{1}{2}$ years old,

b He was 146 cm tall when he was 13 years 5 months,

c When he was 17 years 6 months, he was 171 cm tall.

We can also deduce that:

i The fastest increase in height was between his fourteenth and fifteenth birthdays – the curve is steepest between these two birthdays,

ii He grew very little between his eighteenth and nineteenth birthdays – the curve is quite flat in this region.

We could obtain more accurate results if we took 100 cm as the lowest height on the vertical axis and used a larger scale.

Exercise 20b

1 The weights of lead spheres of various diameters are shown in the table.

Diameter in mm (D)	4	5.2	6.4	7.2	7.9	8.8
Weight in grams (W)	380	840	1560	2230	2940	4070

Plot this information on a graph and draw a smooth curve through the points. Use 2 cm ≡ 1 unit on the D-axis and 2 cm ≡ 500 units on the W-axis.

Use your graph to estimate

a the weight of a lead sphere of diameter 6 mm

b the diameter of a lead sphere of weight 2 kg

2 Recorded speeds of a motor car at various times after starting from rest are shown in the table.

Time in seconds	0	5	10	15	20	25	30	35	40
Speed in km/h	0	62	112	148	172	187	196	199	200

Taking $2\,cm \equiv 5\,sec$ and $1\,cm \equiv 10\,km/h$, plot these results and draw a smooth curve to pass through these points.

Use your graph to estimate

a the time which passes before the car reaches

   i 100 km/h   **ii** 150 km/h

b its speed after

   i 13 sec   **ii** 27 sec

3 The weight of a puppy at different ages is given in the table.

Age in days (A)	10	20	40	60	80	100	120	140
Weight in grams (W)	50	100	225	425	750	875	950	988

Draw a graph to represent this data, taking $1\,cm \equiv 10\,days$ on the A-axis and $1\,cm \equiv 50\,g$ on the W-axis.

Hence estimate

a the weight of the puppy after **i** 50 days **ii** 114 days

b the age of the puppy when it weighs **i** 500 g **ii** 1000 g

c the weight it puts on between day 25 and day 55

d its birth weight

4 The speed of a particle (v metres per second) at various times (t seconds) after starting is given in the table.

t	0	1	2	3	4	5	6	7
v	0	35	60	76.5	83	83	76	57

Plot this information on a graph using $2\,cm \equiv 1\,unit$ on the t-axis and $2\,cm \equiv 10\,units$ on the v-axis.

Use your graph to find:

a the greatest speed of the particle and the time at which it occurs

b its speed after **i** 3.5 sec **ii** 6.8 sec

c when its speed is 65 m/sec

5 In the United Kingdom the cost of fuel (£C) per nautical mile for a ship travelling at various speeds (v knots) is given in the table.

V	12	14	16	18	20	22	24	26	28
C	18.15	17.16	16.67	16.5	16.5	16.67	16.94	17.36	17.82

Draw a graph to show how cost changes with speed.
Use 1 cm ≡ 1 knot and 10 cm ≡ £1. (Take £16 as the lowest value for C.)

Use your graph to estimate:

a the most economical speed for the ship and the corresponding cost per nautical mile

b the speeds when the cost per nautical mile is £17

c the cost when the speed is **i** 13 knots   **ii** 24.4 knots

6 Cubes made from a certain metal with edges of the given lengths have weights as given in the table.

Length of edge in cm (L)	1	2	3	4	5	6
Weight of cube in grams (W)	9	72	243	576	1125	1944

Plot this information on a graph, joining the points with a smooth curve.
Take 2 cm ≡ 1 unit on the L-axis and 1 cm ≡ 100 g on the W-axis.

From your graph find:

a the weight in grams of a cube with edge

    i 3.5 cm   **ii** 5.3 cm

b the length of the edge of a cube with weight

    i 500 g   **ii** 1500 g

7 The temperatures, taken at 2-hourly intervals, at my home on a certain day were as given in the table.

Draw a graph to show this data taking 1 cm ≡ 1 hour and 1 cm ≡ 1 °C.

Use your graph to estimate:

a the temperature at 11 a.m. and at 11 p.m.

b the times at which the temperature was 29 °C.

Time	Temperature in °C
midnight	26.6
2 a.m.	26.0
4 a.m.	25.8
6 a.m.	26.0
8 a.m.	27.4
10 a.m.	28.2
noon	29.4
2 p.m.	30.2
4 p.m.	30.0
6 p.m.	29.6
8 p.m.	28.8
10 p.m.	28.0
midnight	27.6

8 The time of sunset at Tobago on different dates, each two weeks apart, is given in the table.

	Sept		Oct		Nov		Dec	
Date (D)	5	29	12	26	10	24	7	21
Time (T)	1820	1815	1811	1800	1750	1742	1736	1730

Using 1 cm ≡ 1 week on the D-axis and 2 cm ≡ 10 minutes on the T-axis, plot these points on a graph and join them with a smooth curve. Take 1720 as the lowest value for T.

From your graph estimate:

a   the time of sunset on 17 Nov

b   the date on which the sun sets at 1734

9 A rectangle measuring l cm by b cm has an area of 24 cm². The table gives different values of l with the corresponding values of b.

l	1	2	3	4	6	8	12	16
b	24		8		4		2	1.5

Complete the table and draw a graph to show this information, joining the points with a smooth curve. Take 1 cm ≡ 1 unit on the l-axis and 1 cm ≡ 2 units on the b-axis.

Use your graph to estimate the value of

a l when b is   **i** 14 cm   **ii** 2.4 cm

b b when l is   **i** 18 cm   **ii** 2.8 cm

Q: Which travels faster, heat or cold?

A: Heat, because you can catch cold.

IN THIS CHAPTER...

you have seen that:

● you can draw a graph and use it to convert a quantity from one unit to another.

21

COORDINATES AND THE STRAIGHT LINE

MATHS IS OUT THERE

Q: Why can't a pin stand on its point?

A: Euclid said that 'a point has no magnitude' so there's nothing for it to stand on.

The equation of a straight line

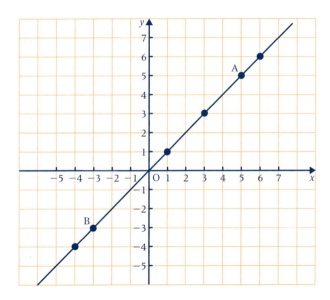

If we plot the points with coordinates $(-4, -4)$, $(1, 1)$, $(3, 3)$ and $(6, 6)$, we can see that a straight line can be drawn through these points that also passes through the origin.

For each point the y-coordinate is the same as the x-coordinate.

This is also true for any other point on this line,

e.g. the coordinates of A are $(5, 5)$ and of B are $(-3, -3)$.

Hence y-coordinate $= x$-coordinate

or simply $y = x$

This is called the equation of the line.

We can also think of a line as a set of points, i.e. this line is the set of points, or *ordered number pairs*, such that $\{(x, y)\}$ satisfies the relation $y = x$.

It follows that if another point on the line has an x-coordinate of -5, then its y-coordinate is -5 and if a further point has a y-coordinate of 4, its x-coordinate is 4.

In a similar way we can plot the points with coordinates $(-2, -4)$, $(1, 2)$, $(2, 4)$ and $(3, 6)$.

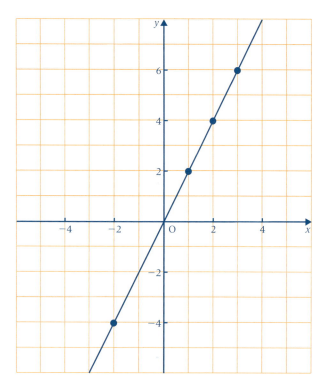

These points also lie on a straight line passing through the origin.

In each case the y-coordinate is twice the x-coordinate.

The equation of this line is therefore $y = 2x$ and we often refer, briefly, to 'the line $y = 2x$'.

If another point on this line has an x-coordinate of 4,

its y-coordinate is 2×4, i.e. 8,

and if a further point has a y-coordinate of -5,

its x-coordinate must be $-2\frac{1}{2}$.

Exercise 21a

1 Find the y-coordinates of points on the line $y = x$ that have x-coordinates of

   a 2   **b** 3   **c** 7   **d** 12.

2 Find the y-coordinates of points on the line $y = x$ which have x-coordinates of

   a -1 **b** -6 **c** -8 **d** -20.

3 Find the y-coordinates of points on the line $y = -x$ that have x-coordinates of

    a $3\frac{1}{2}$ **b** $-4\frac{1}{2}$ **c** 6.1 **d** -8.3

4 Find the x-coordinates of points on the line $y = -x$ that have y-coordinates of

    a 7 **b** -2 **c** $5\frac{1}{2}$ **d** -4.2.

5 Find the y-coordinates of points on the line $y = 2x$ that have x-coordinates of

    a 5 **b** -4 **c** $3\frac{1}{2}$ **d** -2.6.

6 Find the x-coordinates of points on the line $y = -3x$ that have y-coordinates of

    a 3 **b** -9 **c** 6 **d** -4.

7 Find the x-coordinates of points on the line $y = \frac{1}{2}x$ that have y-coordinates of

    a 6 **b** -12 **c** $\frac{1}{2}$ **d** -8.2.

8 Find the x-coordinates of points on the line $y = -4x$ that have y-coordinates of

    a 8 **b** -16 **c** 6 **d** -3.

9 If the points $(-1, a)$ $(b, 15)$ and $(c, -20)$ lie on the straight line with equation $y = 5x$, find the values of a, b and c.

> **Tip** $A(-1, a)$ lies on $y = 5x$.
> Replace y by a and x by -1. Then solve the equation to find a.

10 If the points $(3, a)$, $(-12, b)$ and $(c, -12)$ lie on the straight line with equation $y = -\frac{2}{3}x$, find the values of a, b and c.

11 Using 1 cm to 1 unit on each axis, plot the points $(-2, -6)$, $(1, 3)$, $(3, 9)$ and $(4, 12)$. What is the equation of the straight line that passes through these points?

12 Using 1 cm to 1 unit on each axis, plot the points $(-3, 6)$ $(-2, 4)$, $(1, -2)$ and $(3, -6)$. What is the equation of the straight line that passes through these points?

13 Using the same scale on each axis, plot the points $(-6, 2)$, $(0, 0)$, $(3, -1)$ and $(9, -3)$. What is the equation of the straight line that passes through these points?

14 Using the same scale on each axis, plot the points $(-6, -4)$, $(-3, -2)$, $(6, 4)$ and $(12, 8)$. What is the equation of the straight line that passes through these points?

15 Which of the points $(-2, -4)$, $(2.5, 4)$, $(6, 12)$ and $(7.5, 10)$ lie on the line $y = 2x$?

16 Which of the points $(-5, -15)$, $(-2, 6)$, $(1, -3)$ and $(8, -24)$ lie on the line $y = -3x$?

17 Which of the following points lie

    a above the line $y = \frac{1}{2}x$

    b below the line $y = \frac{1}{2}x$

    $(2, 2)$, $(-2, 1)$, $(3, 0)$, $(-4.2, -2)$, $(-6.4, -3.2)$?

Plotting the graph of a given equation

If we want to draw the graph of $y = 3x$ for values of x from -3 to $+3$, then we need to find the coordinates of some points on the line.

As we know that it is a straight line, two points are enough. However, it is sensible to find three points, the third point acting as a check on our working. It does not matter which three points we find, so we will choose easy values for x, one at each extreme and one near the middle.

If $\qquad x = -3, \quad y = 3 \times (-3) = -9$

If $\qquad x = 0, \quad y = 3 \times 0 = 0$

If $\qquad x = 3, \quad y = 3 \times 3 = 9$

These look neater if we write them in table form:

x	-3	0	3
y	-9	0	9

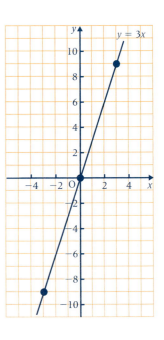

Exercise **21b**

In questions **1** to **6**, draw the graphs of the given equations on the same set of axes. Use the same scale on both axes, taking values of x between -4 and 4, and values of y between -6 and 6. You should take at least three x values and record the corresponding y values in a table. Write the equation of each line somewhere on it.

1 $y = x$ $\qquad\qquad$ **3** $y = \frac{1}{2}x$ $\qquad\qquad$ **5** $y = \frac{1}{3}x$

2 $y = 2x$ $\qquad\qquad$ **4** $y = \frac{1}{4}x$ $\qquad\qquad$ **6** $y = \frac{3}{2}x$

In questions **7** to **12**, draw the graphs of the given equations on the same set of axes.

7 $y = -x$ $\qquad\qquad$ **9** $y = -\frac{1}{2}x$ $\qquad\qquad$ **11** $y = -\frac{1}{3}x$

8 $y = -2x$ $\qquad\qquad$ **10** $y = -\frac{1}{4}x$ $\qquad\qquad$ **12** $y = -\frac{3}{2}x$

We can conclude from these exercises that the graph of an equation of the form $y = mx$ is a straight line that:

- passes through the origin
- gets steeper as m increases
- makes an acute angle with the positive x-axis if m is positive
- makes an obtuse angle with the positive x-axis if m is negative.

Gradient of a straight line

The gradient or slope of a line is defined as the amount the line rises vertically divided by the distance moved horizontally.

i.e.   gradient or slope of AB $= \dfrac{BC}{AC}$

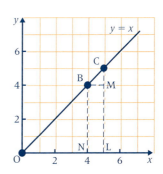

The gradient of any line is defined in a similar way.

Considering any two points on a line, the gradient of the line is given by

$$\dfrac{\text{the increase in } y \text{ value}}{\text{the increase in } x \text{ value}}$$

If we plot the points $O(0, 0)$, $B(4, 4)$ and $C(5, 5)$, all of which lie on the line with equation $y = x$, then:

$$\text{gradient of OC} = \dfrac{CL}{OL} = \dfrac{5}{5} = 1$$

$$\text{gradient of OB} = \dfrac{BN}{ON} = \dfrac{4}{4} = 1$$

$$\text{gradient of BC} = \dfrac{CM}{BM} = \dfrac{5-4}{5-4} = \dfrac{1}{1} = 1$$

These show that, whichever two points are taken, the gradient of the line is 1.

Similarly, if we plot the points $P(-3, 6)$, $Q(-1, 2)$ and $R(4, -8)$, all of which lie on the line with equation $y = -2x$, then:

$$\text{gradient of PR} = \dfrac{\text{increase in } y \text{ value from P to R}}{\text{increase in } x \text{ value from P to R}}$$

$$= \dfrac{y\text{-coordinate of R} - y\text{-coordinate of P}}{x\text{-coordinate of R} - x\text{-coordinate of P}}$$

$$= \dfrac{(-8) - (6)}{(4) - (-3)}$$

$$= \dfrac{-8 - 6}{4 + 3} = \dfrac{-14}{7} = -2$$

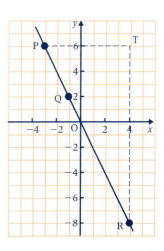

Exercise 21c Draw axes for x and y, for values between -6 and $+6$, taking 1 cm as 1 unit on each axis.

Plot the points A(-4, 4), B(2, -2) and C(5, -5), all of which lie on the line $y = -x$. Find the gradient of

a AB   **b** BC   **c** AC

a Gradient of AB

$$= \frac{(-2)-(4)}{(2)-(-4)} = \frac{-6}{6} = -1$$

b Gradient of BC

$$= \frac{(-5)-(-2)}{(5)-(2)} = \frac{-3}{3} = -1$$

c Gradient of AC

$$= \frac{(-5)-(4)}{(5)-(-4)} = \frac{-9}{9} = -1$$

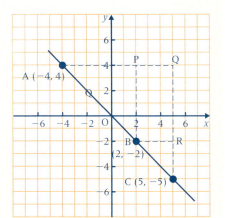

1 Using 2 cm to 1 unit on each axis, draw axes that range from 0 to 6 for x and from 0 to 10 for y. Plot the points A(2, 4), B(3, 6) and C(5, 10), all of which lie on the line $y = 2x$. Find the gradient of

a AB   **b** BC   **c** AC

2 Draw the x-axis from -4 to 4 taking 2 cm as 1 unit, and the y-axis from -16 to 12 taking 0.5 cm as 1 unit. Plot the points X(-3, 12), Y(-1, 4) and Z(4, -16), all of which lie on the line $y = -4x$. Find the gradient of

a XY   **b** YZ   **c** XZ

3 Choosing your own scale and range of values for both x and y, plot the points D(-2, -6), E(0, 0) and F(4, 12), all of which lie on the line $y = 3x$. Find the gradient of

a DE   **b** EF   **c** DF

4 Taking 2 cm as 1 unit for x and 1 cm as 1 unit for y, draw the x-axis from -1.5 to 2.5 and the y-axis from -10 to 6. Plot the points A(-1.5, 6), B(0.5, -2) and C(2.5, -10), all of which lie on the line $y = -4x$. Find the gradient of

a AB   **b** BC   **c** AC

Copy and complete the following table and use it to draw the graph of $y = 1.5x$.

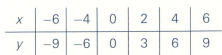

x	−6	−4	0	2	4	6
y						

Choosing your own points, find the gradient of this line using two different sets of points.

x	−6	−4	0	2	4	6
y	−9	−6	0	3	6	9

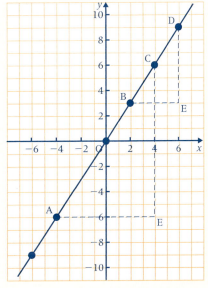

Four points, A, B, C and D, have been chosen.

$$\text{Gradient of line} = \frac{\text{CE}}{\text{AE}} = \frac{6 - (-6)}{4 - (-4)} = \frac{12}{8} = 1.5$$

$$\text{Gradient of line} = \frac{\text{DF}}{\text{BF}} = \frac{9 - 3}{6 - 2} = \frac{6}{4} = 1.5$$

(Finding the gradient using any other two points also gives a value of 1.5.)

5 Copy and complete the following table and use it to draw the graph of $y = 2.5x$.

x	−3	−1	0	2	4
y					

Choose your own pairs of points to find the gradient of this line at least twice.

6 Copy and complete the following table and use it to draw the graph of $y = -0.5x$.

Choose your own pairs of points to find the gradient of this line at least twice.

x	−6	−2	3	4
y				

7 Determine whether the straight lines with the following equations have positive or negative gradients:

a $y = 5x$

b $y = -7x$

c $y = 12x$

d $y = -\frac{1}{4}x$

d $3y = -x$

e $5y = -12x$

These exercises, together with the worked examples, confirm our conclusions on p. 333, namely that

- the larger the value of m the steeper is the slope
- lines with positive values for m make an acute angle with the positive x-axis
- lines with negative values for m make an obtuse angle with the positive x-axis.

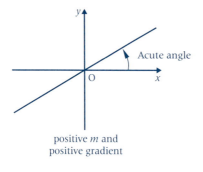

positive m and
positive gradient

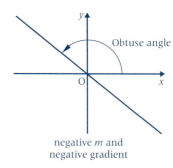

negative m and
negative gradient

Exercise 21d

For each of the following pairs of lines, state which line is the steeper. Show both lines on the same sketch.

1 $y = 5x,\quad y = \frac{1}{5}x$

2 $y = 2x,\quad y = 5x$

3 $y = \frac{1}{2}x,\quad y = \frac{1}{3}x$

4 $y = -2x,\quad y = -3x$

5 $y = 10x,\quad y = 7x$

6 $y = -\frac{1}{2}x,\quad y = -\frac{1}{4}x$

7 $y = -6x,\quad y = -3x$

8 $y = 0.5x,\quad y = 0.75x$

Determine whether each of the following straight lines makes an acute angle or an obtuse angle with the positive x-axis.

9 $y = 4x$

10 $y = -3x$

11 $y = -\frac{1}{2}x$

12 $y = 3.6x$

13 $y = \frac{1}{3}x$

14 $y = 0.7x$

15 $y = 10x$

16 $y = 0.5x$

17 $y = -6x$

18 $y = -\frac{2}{3}x$

19 $y = -\frac{3}{4}x$

20 $y = -0.4x$

21 Estimate the gradient of each of the lines shown in the sketch.

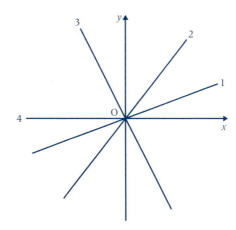

Puzzle

Here is a very ingenious method of guessing the values of three dice thrown by a friend, without seeing them.

Tell him to think of the first die.

Multiply by 2. Add 5. Multiply by 5.

Add the value of the second die.

Multiply by 10. Add the value of the third die.

Now ask the total. From this total subtract 250.

The three digits of your answer will be the values of his three dice.

As an example, if the answer was 705, then 706 − 250 = 456. The three dice were therefore 3, 5 and 6. Try it and see. Why does it work?

Lines that do not pass through the origin

If we plot the points (−3, −1), (1, 3), (3, 5), (4, 6) and (6, 8), and draw the straight line that passes through these points, we can use it to find

a the equation of the line

b its gradient

c the distance from the origin to the point where the line crosses the y-axis.

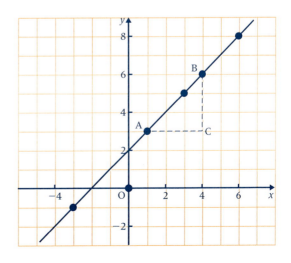

a In each case, the y-coordinate is 2 more than the x-coordinate, i.e. all the points lie on the line with equation $y = x + 2$

b Using the points A and B, the gradient of the line is given by $\dfrac{BC}{AC}$, i.e. $\dfrac{3}{3} = 1$.

c The line crosses the y-axis at the point (0, 2) which is 2 units above the origin. This quantity is called the y-intercept.

Exercise 21e

Draw the graph of $y = -4x + 3$ for values of x between -4 and $+4$. Hence find

a the gradient of the line

b its y-intercept.

x	-3	0	3
y	15	3	-9

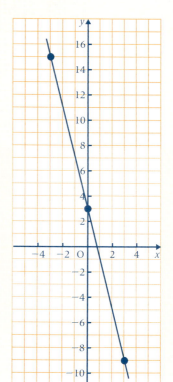

a Moving from the point $(0, 3)$ to $(-3, 15)$ the gradient is

$$\frac{3 - 15}{0 - (-3)} = \frac{-12}{3} = \frac{-4}{1} = -4$$

b The y-intercept is 3

In the following questions, draw the graph of the given equation using the given x values. Hence find the gradient of the line and its intercept on the y-axis. Use 1 cm as 1 unit on each axis with x values ranging from -8 to $+8$ and y values ranging from -10 to $+10$.

Compare the values you get for the gradient and the y-intercept with the numbers in the right-hand side of each equation.

1 $y = 3x + 1$; x values $-3, 1, 3$
Use your graph to find the value of y when x is **a** -2 **b** 2

2 $y = -3x + 4$; x values $-2, 2, 4$
Use your graph to find the value of y when x is **a** -1 **b** 3

3 $y = \frac{1}{2}x + 4$; x values $-8, 0, 6$
Use your graph to find

    a the value of y when x is -2 **b** the value of x when y is 6

4 $y = x - 3$; x values $-4, 2, 8$
Use your graph to find the value of x when y is **a** 4 **b** -5

5 $y = \frac{3}{4}x + 3$; x values $-4, 0, 8$
Use your graph to find the value of x when y is **a** 6 **b** 4.5

Draw the graph of $y = -2x + 3$ for values of x between -4 and $+4$. Hence find

a the gradient of the line   **b** its y-intercept.

Compare the values for the gradient and the y-intercept with the number of xs and the number term on the right-hand side of the equation.

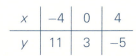

x	-4	0	4
y	11	3	-5

a Gradient of line $= \dfrac{3 - 11}{0 - (-4)} = -\dfrac{8}{4}$

$\qquad\qquad\qquad\quad = -2$

b The y-intercept is 3

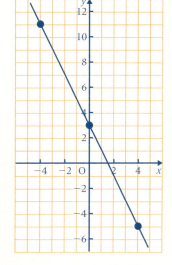

The number of xs on the right-hand side is -2, which is the same as the gradient of the line.

The number term on the right-hand side is 3, which is the same as the y-intercept.

In the following questions, draw a graph for each of the given equations. In each case find the gradient and the y-intercept for the resulting straight line. Take 1 cm as 1 unit on each axis, together with suitable values of x within the range -4 to $+4$. Choose your own range for y when you have completed the table.

Compare the values you get for the gradient and the y-intercept with

a   the number of xs

b   the number term on the right-hand side of the equation.

6 $y = 2x - 2$ **8** $y = 3x - 4$ **10** $y = -\frac{3}{2}x + 3$ **12** $y = -2x - 7$

7 $y = -2x + 4$ **9** $y = \frac{1}{2}x + 3$ **11** $y = 2x + 5$ **13** $y = -3x + 2$

The equation $y = mx + c$

The results of Exercise 21e show that we can 'read' the gradient and the y-intercept of a straight line from its equation.

For example, the line with equation $y = 3x - 4$ has a gradient of 3 and its y-intercept is -4.

In general we can conclude that the equation $y = mx + c$ gives a straight line where m is the gradient of the line and c is the y-intercept.

Write down the gradient, m, and the y-intercept, c, for the straight line with equation $y = 5x - 2$

Comparing the line $y = 5x - 2$ with $y = mx + c$ gives

$$m = 5 \quad \text{and} \quad c = -2$$

Write down the gradient, m, and y-intercept, c, for the straight line with the given equation.

1 $y = 4x + 7$ **3** $y = 3x - 2$ **5** $y = 7x + 6$ **7** $y = \frac{3}{4}x + 7$

2 $y = \frac{1}{2}x - 4$ **4** $y = -4x + 5$ **6** $y = \frac{2}{5}x - 3$ <u>**8**</u> $y = 4 - 3x$

Sketch the straight line with equation $y = 5x - 7$

Comparing $y = 5x - 7$ with $y = mx + c$ shows that we want a line with gradient 5 and y-intercept -7.

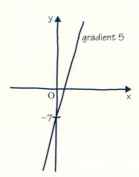

Sketch the straight lines with the given equations.

9 $y = 2x + 5$ **12** $y = -2x - 3$ <u>**15**</u> $y = -5x - 3$

10 $y = 7x - 2$ **13** $y = -\frac{2}{3}x + 8$ <u>**16**</u> $y = 3x + 7$

11 $y = \frac{1}{2}x + 6$ <u>**14**</u> $y = 4x + 2$ <u>**17**</u> $y = \frac{3}{4}x - 2$

Sketch the straight line with equation $y = 2 - 3x$

First rearrange the equation in the form

$y = mx + c$

i.e. $y = -3x + 2$

Comparing $y = -3x + 2$ with $y = mx + c$ shows that we want a line with gradient -3 and y-intercept 2.

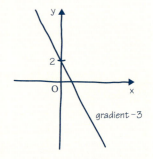

18 $y = 4 - x$ **21** $y = -3 - x$ <u>**24**</u> $y = -5(x - 1)$

19 $y = 3 - 2x$ **22** $y = 2(x + 1)$ <u>**25**</u> $y = 3(4 - x)$

20 $y = 8 - 4x$ <u>**23**</u> $y = 3(x - 2)$ <u>**26**</u> $y = -2(2x + 3)$

Parallel lines

Lines with the same gradient are said to be parallel.

The diagram shows the lines $y = x + 2$ and $y = x - 3$.

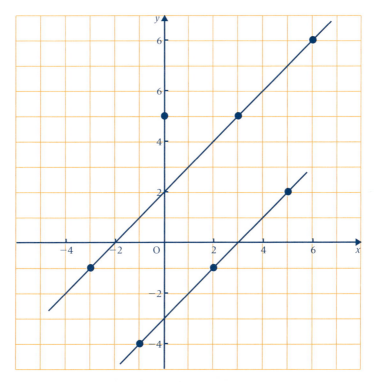

These lines have the same gradient, i.e. they are parallel.

Now consider a third line, parallel to the first two lines and passing through the point A(0, 5).

Its gradient is the same as that of the first lines, i.e. $m = 1$.

It crosses the y-axis at (0, 5) so its y intercept is 5, i.e. $c = 5$.

Therefore the equation of the third line is $y = x + 5$.

Similarly the equation of another parallel line passing through the point (0, −5) is $y = x - 5$.

Exercise 21g

1 Draw the graphs of $y = 3x + 1$ and $y = 3x - 4$ taking x values of −2, 2 and 3.

(Let x range from −5 to +5 and y from −10 to +10.

Take 1 cm to represent 1 unit on each axis.)

What do you notice about these lines?
What do you notice about their m values?

2 Draw the graphs of $y = -2x + 3$ and $y = -2x - 3$ taking x values of -3, 0 and 3.

(Take 1 cm to represent 1 unit on each axis. Let x range from -6 to $+6$ and y from -10 to $+10$.)

What do you notice about these lines?

What do you notice about their m values?

By finding the gradient of each line, determine whether or not the given pairs of equations represent parallel lines.

3 $y = 4x + 2$, $y = 4x - 7$

4 $y = \frac{1}{2}x + 6$, $y = \frac{1}{2}x + 10$

5 $y = x + 4$, $y = 2x + 4$

6 $y = 3x + 5$, $y = x + 7$

7 $y = -x + 4$, $y = -x - 3$

8 $y = -5x + 2$, $y = -5x - 13$

9 $y = \frac{2}{3}x + 3$, $y = \frac{1}{3}x - 4$

10 $y = \frac{1}{2}x - 4$, $y = 0.5x + 2$

Find the gradient of each of the lines $x + y = 4$ and $y = -x + 2$.
Hence determine whether or not the two lines are parallel.

$x + y = 4$ (1)

$y = -x + 2$ (2)

Rearrange (1) in the form $y = mx + c$

Equation (1) gives $y = -x + 4$
the gradient of this line is -1

The gradient of the line $y = -x + 2$ is -1
i.e. the lines have the same gradient and are therefore parallel.

Find the gradient of each of the lines in each question. Hence determine whether or not the two lines are parallel.

11 $y = 2x + 3$, $2y = 4x - 7$

12 $3y = 9x - 2$, $y = 3x + 13$

13 $x + y = 5$, $y = -2x + 3$

14 $3y = 5x + 7$, $6y = 10x - 3$

15 $5y = x + 2$, $3y = x + 2$

16 $x + y = 4$, $y = -x + 6$

Lines parallel to the axes

We began by considering the equation $y = mx$,

i.e. the equation $y = mx + c$ when $c = 0$.

This equation gave a straight line passing through the origin.

Now we will see what happens when $m = 0$.

Think, for example, of the equation $y = 3$.

For every value of x the y-coordinate is 3. This means that the graph of $y = 3$ is a straight line parallel to the x-axis at a distance 3 units above it.

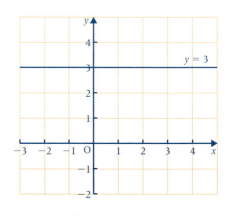

$y = c$ is therefore the equation of a straight line parallel to the x-axis at a distance c away from it. If c is positive, the line is above the x-axis, and if c is negative, the line is below the x-axis.

Similarly $x = b$ is the equation of a straight line parallel to the y-axis at a distance b units from it.

Exercise 21h   Draw, on the same diagram, the straight line graphs of $x = -3$, $x = 5$, $y = -2$ and $y = 4$.

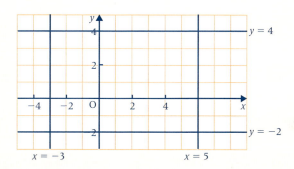

In the following questions, take both x and y in the range -8 to $+10$. Let 1 cm be 1 unit on each axis.

1 Draw the straight line graphs of the following equations in a single diagram:

$x = 2$, $x = -5$, $y = \frac{1}{2}$, $y = -3\frac{1}{2}$

2 Draw the straight line graphs of the following equations in a single diagram:

$y = -5$, $x = -3$, $x = 6$, $y = 5.5$

3 On one diagram, draw graphs to show the following equations:

$x = 5$, $y = -5$, $y = 2x$

Write down the coordinates of the three points where these lines intersect. What kind of triangle do they form?

4 On one diagram, draw the graphs of the straight lines with equations $x = 4$, $y = -\frac{1}{2}x$, $y = 3$

Write down the coordinates of the three points where these lines intersect. What kind of triangle is it?

5 On one diagram, draw the graphs of the straight lines with equations $y = 2x + 4$, $y = -5$, $y = 4 - 2x$

Write down the coordinates of the three points where these lines intersect. What kind of triangle is it?

Simultaneous equations

When we are given an equation we can draw a graph. Any of the equations that occur in this chapter give us a straight line. Two equations give us two straight lines that usually cross one another.

Consider the two equations $x + y = 4$ $\quad y = 1 + x$

Suppose we know that the x-coordinate of the point of intersection is in the range $0 \leqslant x \leqslant 5$:

$x + y = 4$

x	0	4	5
y	4	0	-1

$y = 1 + x$

x	0	2	5
y	1	3	6

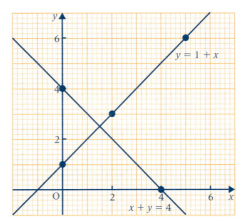

At the point where the two lines cross, the values of x and y are the same for both equations, so they are the solutions of the pair of equations.

From the graph we see that the solution is $x = 1\frac{1}{2}$, $y = 2\frac{1}{2}$.

$x = 1\frac{1}{2}$ and $y = 2\frac{1}{2}$ satisfy both equations. We call a pair of equations like this, simultaneous equations.

Exercise 21i

Solve the following pairs of simultaneous equations graphically. In each case draw axes for x and y and use values in the ranges indicated, taking 2 cm to 1 unit:

1 $x + y = 6$ $\quad 0 \leqslant x \leqslant 6,$ $\quad 0 \leqslant y \leqslant 6,$
$\quad y = 3 + x$

2 $x + y = 5$ $\quad 0 \leqslant x \leqslant 6,$ $\quad 0 \leqslant y \leqslant 6$
$\quad y = 2x + 1$

5 $2x + y = 3$ $\quad 0 \leqslant x \leqslant 3,$ $\quad -3 \leqslant y \leqslant 3$
$\quad x + y = 2\frac{1}{2}$

6 $y = 5 - x$ $\quad 0 \leqslant x \leqslant 5,$ $\quad 0 \leqslant y \leqslant 7$
$\quad y = 2 + x$

7 $3x + 2y = 9$ $\quad 0 \leqslant x \leqslant 4,$ $\quad -2 \leqslant y \leqslant 5$
$\quad 2x - 2y = 3$

3 $y = 4 + x$ $\quad 0 \leqslant x \leqslant 6,$ $\quad 0 \leqslant y \leqslant 6$
$\quad y = 1 + 3x$

4 $x + y = 1$ $\quad -3 \leqslant x \leqslant 2,$ $\quad -2 \leqslant y \leqslant 4$
$\quad y = x + 2$

8 $2x + 3y = 4$ $\quad -2 \leqslant x \leqslant 2,$ $0 \leqslant y \leqslant 4$
$\quad y = x + 2$

9 $x + 3y = 6$ $\quad 0 \leqslant x \leqslant 5,$ $0 \leqslant y \leqslant 5$
$\quad 3x - y = 6$

10 $x = 2y - 3$ $\quad -2 \leqslant x \leqslant 3,$ $\quad 0 \leqslant y \leqslant 4$
$\quad y = 2x + 1$

Special cases

Exercise 21j

Try to solve the following equations graphically. Why do you think the method breaks down?

1 $x + y = 9$ $0 \leqslant x \leqslant 9$
    $x + y = 4$ $0 \leqslant y \leqslant 9$

3 $2x + y = 3$ $0 \leqslant x \leqslant 3$
    $4x + 2y = 7$ $-3 \leqslant y \leqslant 4$

2 $y = 2x + 3$ $0 \leqslant x \leqslant 4$
    $y = 2x - 1$ $-1 \leqslant y \leqslant 11$

4 $y = 2x - 4$ $0 \leqslant x \leqslant 4$
    $2x = y + 4$ $-4 \leqslant y \leqslant 4$

There are other ways of solving two simultaneous equations. This work is done in Book 3.

Mixed exercises

Exercise 21k

1 Find the x-coordinates of the points on the line $y = 3x$ that have y-coordinates of

   a 6   **b** -12 **c** 2.

2 If the points $(6, a)$, $(-\frac{1}{2}, b)$ and $(c, 1)$ lie on the straight line with equation $3y = -2x$, find the values of a, b and c.

3 Determine whether the straight lines with the given equations have positive or negative gradients:

   a $y = 4x$ **b** $y = -2x + 2$ **c** $y = \frac{2}{3}x - 7$

4 Copy and complete the following table and use it to draw the graph of $y = 2x - 3$:

x	-3	0	4
y			

Choose your own points to find the gradient of this line.

5 Determine in each case whether the straight line with the given equation makes an acute angle or an obtuse angle with the positive x-axis.

   a $y = -\frac{2}{3}x$ **b** $y = 5x + 2$

   c $2x + y = 3$ **d** $3y = -4x + 7$

6 Draw on the same axes, using 1 cm as 1 unit in each case, the graphs of $y = 2x - 4$ and $2x + y + 8 = 0$. Write down the coordinates of the point where these lines intersect.

Exercise **21l**

1 Find the y-coordinates of the points on the line $y = 5x$ that have x-coordinates of

a 2   **b** 3   **c** $\frac{1}{2}$.

2 If the points $(-1, a)$, $(b, 15)$ and $(c, -20)$ lie on the straight line with equation $y = 5x$, find the values of a, b and c.

3 Determine whether the straight lines with the given equations have positive or negative gradients:

a $y = 6x$ **b** $y = -3x + 2$ **c** $x + y = 4$

4 Write down the gradients and y-intercepts for the straight lines with the given equations:

a $y = 4x - 7$ **b** $2y = 5x + 2$ **c** $y - 3x = 2$ **d** $3y = -x - 12$

5 Determine whether or not the given pairs of equations represent parallel lines:

a $y = -x + 2$, $x + y = 3$ **b** $2y = 4x + 3$, $y + 2x = 5$

6 Draw, on the same axes, the graphs of $x = -3$, $y = \frac{1}{2}x$ and $y = 4$, for values of x between -4 and $+8$. Write down the coordinates of the three points where these lines intersect.

Exercise **21m**

1 Find the y-coordinates of the points on the line $y = 7x + 4$ that have x-coordinates of

a 1   **b** -2 **c** -5.

2 If the points $(3, a)$, $(-2, b)$ and $(c, -10)$ lie on the straight line with equation $y = 5 - 3x$, find the values of a, b and c.

3 Sketch on the same axes the graphs of the straight lines with equations

a $y = -3x$ **b** $y = 2x + 4$.

4 Draw the graph of $y = 5x - 2$ for values of x between -4 and 4. Use 2 cm as 1 unit on the x-axis and 1 cm as 1 unit on the y-axis. From your graph, or otherwise, find

a the gradient of the line   **b** its y-intercept.

5 Write down the equations of the straight lines that have the given gradients and y-intercepts:

a gradient 2, y-intercept -4

b gradient $\frac{1}{2}$, y-intercept 5

c gradient -4, y-intercept -3

6 Draw, on the same axes, the graphs of $x = 1$, $y = -2x - 2$, $y = 4$ for values of x between -4 and $+4$. Write down the coordinates of the three points where these lines intersect.

7 Draw on the same axes, using 1 cm as 1 unit in each case, the graphs of $2x - y = 4$ and $2x + y = -8$. Add together the two equations. Draw the graph of the new equation on the same set of axes. Do you notice anything special about these three lines?

8 Draw on the same axes the graphs of $x + 2y = 8$, $x + y = 4$ and $2x + 3y = 12$. What do you notice about these three lines?

Did you know that a US gallon is 231 cubic inches, which is equal to the old English wine gallon?

An Imperial gallon is 277.42 cubic inches, which is about 20% more than a US gallon.

IN THIS CHAPTER...

you have seen that:

- you can find the missing coordinate of a point on a line, given the equation of the line and one coordinate, by substituting the given coordinate into the equation and solving it

- you can draw a straight line graph, given its equation by finding the coordinates of three points on the line

- the gradient of a straight line whose equation is $y = mx + c$ is m

- a positive gradient means that the line makes an acute angle with the positive x-axis and a negative gradient means that the line makes an obtuse angle with the positive x-axis

- a straight line whose equation is $y = mx + c$ crosses the y-axis where $y = c$

- if you know the gradient and y-intercept of a straight line you can substitute these values for m and c in $y = mx + c$ and hence give the equation of the line

- two straight lines are parallel if their gradients are equal

- the equation of a straight line parallel to the x-axis is $y = a$ and the equation of a straight line parallel to the y-axis is $x = b$

- two linear simultaneous equations can be solved by drawing the graphs of the equations and finding where they intersect

AT THE END OF THIS CHAPTER...

you should be able to:

1   solve inequalities algebraically

2   represent inequalities on a diagram

3   write down inequalities to describe a region

You know that there are an infinite number of counting numbers: 1 is the 1st, 2 is the 2nd, 3 is the third, 4 is the 4th, and so on.

Did you know that there are exactly the same number of positive even numbers: 2 is the 1st, 4 is the 2nd, 6 is the 3rd, 8 is the 4th, and so on.

BEFORE YOU START

you need to know:
- ✓ the equations of lines parallel to the axes
- ✓ the meaning of the symbols $<$, $\leqslant$, $>$ and $\geqslant$
- ✓ how to draw a number line
- ✓ how to work with positive and negative numbers

KEY WORDS

boundary line, inequality, symbols $<$, $\leqslant$, $>$ and $\geqslant$, range, two-dimensional space, xy–plane

Consider the statement

$$x > 5$$

This is an *inequality* (as opposed to $x = 5$ which is an equality or equation).

This inequality is true when x stands for any number that is greater than 5. Thus there is a range of numbers that x can stand for and we can illustrate this range on a number line.

The circle at the left hand end of the range is 'open', because 5 is not included in the range.

<hr />

Exercise 22a   Use a number line to illustrate the range of values of x for which $x < -1$

(The open circle means that -1 is not included. All values smaller than -1 are to the left of it on the number line.)

Use a number line to illustrate the range of values of x for which each of the following inequalities is true:

1 $x > 7$ **4** $x > 0$ **7** $x < 5$

2 $x < 4$ **5** $x < -2$ **8** $x < 0$

3 $x > -2$ **6** $x > \frac{1}{2}$ **9** $x < 1.5$

10 State which of the inequalities given in questions **1** to **9** are satisfied by a value of x equal to

     a 2   **b** -3 **c** 0   **d** 1.5   **e** 0.0005

11 For each of the questions **1** to **9** give a number that satisfies the inequality and is

     a a whole number   **b** not a whole number

12 Consider the true inequality $3 > 1$

     a Add 2 to each side.   **b** Add -2 to each side.

     c Take 5 from each side.   **d** Take -4 from each side.

     In each case state whether or not the inequality remains true.

13 Repeat question **12** with the inequality $-2 > -3$

14 Repeat question **12** with the inequality $-1 < 4$

15 Try adding and subtracting different numbers on both sides of a true inequality of your own choice.

Solving inequalities

From the last exercise we can see that

> an inequality remains true when the *same* number is added to, or subtracted from, *both* sides.

Now consider the inequality $x - 2 < 3$

Solving this inequality means finding the range of values of x for which it is true.

Adding 2 to each side gives $x < 5$

We have now solved the inequality.

Exercise 22b

Solve the following inequalities and illustrate your solutions on a number line:

1 $x - 4 < 8$

2 $x + 2 < 4$

3 $x - 2 > 3$

4 $x - 3 > -1$

5 $x + 4 < 2$

6 $x - 5 < -2$

7 $x - 3 < -6$

8 $x + 7 < 0$

> **Tip** Add 4 to each side first.

9 $x + 2 < -3$

Solve the inequality $4 - x < 3$

$$4 - x < 3$$

(Aim to get the x term on one side of the inequality and the number term in the other.)

Add x to each side $4 < 3 + x$

Take 3 from each side $1 < x$ or $x > 1$

An inequality remains true if the sides are reversed but you must remember to reverse the inequality sign.

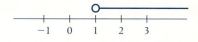

Solve the following inequalities and illustrate your solutions on a number line:

10 $4 - x > 6$

11 $2 < 3 + x$

12 $7 - x > 4$

13 $5 < x + 5$

14 $5 - x < 8$

15 $2 > 5 + x$

16 $3 - x > 2$

17 $6 < x + 8$

18 $2 + x < -3$

19 $2 > x - 3$ **22** $3 - x < 3$ **25** $3 > -x$

20 $4 < 5 - x$ **23** $5 < x - 2$ **26** $4 - x > -9$

21 $1 < -x$ **24** $7 > 2 - x$ **27** $5 - x < -7$

28 Consider the true inequality $12 < 36$

    a Multiply each side by 2   **b** Divide each side by 4

    c Multiply each side by 0.5   **d** Divide each side by 6

    e Multiply each side by -2 **f** Divide each side by -3

In each case state whether or not the inequality remains true.

29 Repeat question **28** with the true inequality $36 > -12$

30 Repeat question **28** with the true inequality $-18 < -6$

31 Repeat question **28** with a true inequality of your own choice.

32 Can you multiply both sides of an inequality by any one number and be confident that the inequality remains true?

An inequality remains true when both sides are multiplied or divided by the same *positive* number.

Multiplication or division of an inequality by a negative number should be avoided, because it destroys the truth of the inequality.

Exercise 22c

Solve the inequality $2x - 4 > 5$ and illustrate the solution on a number line.

$$2x - 4 > 5$$

Add 4 to both sides $2x > 9$

Divide both sides by 2 $x > 4\frac{1}{2}$

Solve the inequalities and illustrate the solutions on a number line:

1 $3x - 2 < 7$ **3** $4x - 1 > 7$ **5** $5 + 2x < 6$ **7** $4x - 5 < 4$

2 $1 + 2x > 3$ **4** $3 + 5x < 8$ **6** $3x + 1 > 5$ **8** $6x + 2 > 11$

Solve the inequality $3 - 2x \leqslant 5$ and illustrate the solution on a number line. ($\leqslant$ means 'less than or equal to')

(As with equations, we collect the letter term on the side with the greater number to start with. In this case we collect on the right.)

$$3 - 2x \leqslant 5$$

Add $2x$ to each side $3 \leqslant 5 + 2x$

Take 5 from each side $\qquad$ $-2 \leqslant 2x$

Divide each side by 2 $\qquad$ $-1 \leqslant x$ i.e. which in reverse is $x \geqslant -1$

(A solid circle is used for the end of the range because -1 *is* included.)

Solve the inequalities and illustrate each solution on a number line:

9 $\ 3 \leqslant 5 - 2x$ $\qquad$ **12** $\ 4 \geqslant 9 - 5x$ $\qquad$ **15** $\ x - 1 > 2 - 2x$ $\qquad$ **18** $\ 2x + 1 \leqslant 7 - 4x$

10 $\ 5 \geqslant 2x - 3$ $\qquad$ **13** $\ 10 < 3 - 7x$ $\qquad$ **16** $\ 2x + 1 \geqslant 5 - x$ $\qquad$ **19** $\ 1 - x > 2x - 2$

11 $\ 4 - 3x \leqslant 10$ $\qquad$ **14** $\ 8 - 3x \geqslant 2$ $\qquad$ **17** $\ 3x + 2 \leqslant 5x + 2$ $\qquad$ **20** $\ 2x - 5 > 3x - 2$

Find, where possible, the range of values of x which satisfy both of the inequalities **a** $x \geqslant 2$ and $x > -1$ $\quad$ **b** $x \leqslant 2$ and $x > -1$ $\quad$ **c** $x \geqslant 2$ and $x < -1$

a

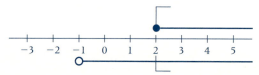

(Illustrating the ranges on a number line, we can see that both inequalities are satisfied for values on the number line where the ranges overlap.)

$\therefore\ x \geqslant 2$ and $x > -1$ are both satisfied for $x \geqslant 2$

b

$x \leqslant 2$ and $x > -1$ are both satisfied for $-1 < x \leqslant 2$

c

There are no values of x for which $x \geqslant 2$ and $x < -1$ are both satisfied.
(The lines do not overlap.)

Find, where possible, the range of values of x for which the two inequalities are both true:

21 a $\ x > 2$ and $x > 3$ $\qquad\qquad\qquad$ **23 a** $\ x \leqslant 4$ and $x > -2$

$\quad$ **b** $\ x \geqslant 2$ and $x \leqslant 3$ $\qquad\qquad\qquad\quad$ **b** $\ x \geqslant 4$ and $x < -2$

$\quad$ **c** $\ x < 2$ and $x > 3$ $\qquad\qquad\qquad\quad$ **c** $\ x \leqslant 4$ and $x < -2$

22 a $\ x \geqslant 0$ and $x \leqslant 1$ $\qquad\qquad\qquad$ **24 a** $\ x < -1$ and $x > -3$

$\quad$ **b** $\ x \leqslant 0$ and $x \leqslant 1$ $\qquad\qquad\qquad\quad$ **b** $\ x < -1$ and $x < -3$

$\quad$ **c** $\ x < 0$ and $x > 1$ $\qquad\qquad\qquad\quad$ **c** $\ x > -1$ and $x < -3$

Solve each of the following pairs of inequalities and then find the range of values of x which satisfy both of them:

25 $x - 4 < 8$ and $x + 3 > 2$

26 $3 + x \leqslant 2$ and $4 - x \leqslant 1$

27 $x - 3 \leqslant 4$ and $x + 5 \geqslant 3$

28 $2x + 1 > 3$ and $3x - 4 < 2$

29 $5x - 6 > 4$ and $3x - 2 < 7$

30 $3 - x > 1$ and $2 + x > 1$

31 $1 - 2x \leqslant 3$ and $3 + 4x < 11$

32 $0 > 1 - 2x$ and $2x - 5 \leqslant 1$

Find the values of x for which $x - 2 < 2x + 1 < 3$

($x - 2 < 2x + 1 < 3$ represents two inequalities,

i.e. $x - 2 < 2x + 1$ and $2x + 1 < 3$, so solve each one separately.)

$x - 2 < 2x + 1$

$\quad -2 < x + 1$

$\quad -3 < x$ i.e. $x > -3$

$2x + 1 < 3$

$\quad 2x < 2$

$\quad x < 1$

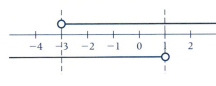

$-3 < x < 1$

Find the range of values of x for which the following inequalities are true:

33 $x + 4 > 2x - 1 > 3$

34 $x - 3 \leqslant 2x \leqslant 4$

35 $3x + 1 < x + 4 < 2$

36 $2 - x < 3x + 2 < 8$

37 $2 - 3x \leqslant 4 - x \leqslant 3$

38 $x - 3 < 2x + 1 < 5$

39 $2x < x - 3 < 4$

40 $4x - 1 < x - 4 < 2$

41 $4 - 3x < 2x - 5 < 1$

42 $x < 3x - 1 < x + 1$

 ## Puzzle

Find two numbers, one of which is twice the other, such that the sum of their squares is equal to the cube of one of the numbers.

Using two-dimensional space

So far we have discussed inequalities in a purely algebraic way. Now we look at them in a more visual way, using graphs.

If we have the inequality $x \geqslant 2$, x can take any value greater than or equal to 2. This can be represented by the following diagram.

On this number line, x can take any value on the heavy part of the line including 2 itself, as indicated by the solid circle at 2.

If $x > 2$ then the diagram is as shown below.

In this case, x cannot take the value 2 and this is shown by the open circle at 2.

It is sometimes more useful to use two-dimensional space with x and y axes, rather than a one-dimensional line. We represent $x \geqslant 2$ by the set of points whose x coordinates are greater than or equal to 2. (y is not mentioned in the inequality so y can take any value.)

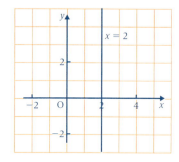

The boundary line represents all the points for which $x = 2$ and the region to the right contains all points with x coordinates greater than 2.

To indicate this, and to make future work easier, we use a continuous line for the boundary when it is included and we shade the region we do *not* want.

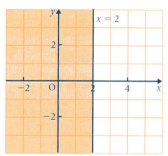

The inequality $x > 2$ tells us that x may not take the value 2. In this case we use a broken line for the boundary.

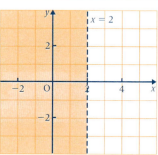

We can draw a similar diagram for $y > -1$

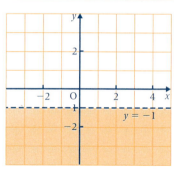

Exercise 22d   Draw diagrams to represent the inequalities

a $x \leqslant 1$ **b** $2 < y$

a $x \leqslant 1$

The boundary line is $x = 1$ (included).

The unshaded region represents $x \leqslant 1$

b $2 < y$

The boundary line is $y = 2$ (not included).

The unshaded region represents $2 < y$

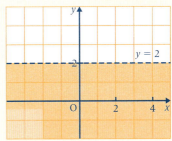

Draw diagrams to represent the following inequalities:

1 $x \geqslant 2$ **3** $x > -1$ **5** $x \geqslant 0$ **7** $x \leqslant -4$

2 $y \leqslant 3$ **4** $y < 4$ **6** $0 > y$ **8** $2 < x$

Draw a diagram to represent $-3 < x < 2$ and state whether or not the points $(1, 1)$ and $(-4, 2)$ lie in the given region.

$-3 < x < 2$ gives two inequalities, $-3 < x$ and $x < 2$ so the boundary lines are $x = -3$ and $x = 2$ (neither included). Shade the regions not wanted.

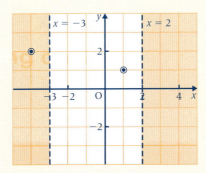

The unshaded region represents $-3 < x < 2$

Plot the points, then you see that $(-4, 2)$ does not lie in the given region.

$(1, 1)$ lies in the given region.

Draw diagrams to represent the following pairs of inequalities:

9 $2 \leqslant x \leqslant 4$

12 $4 < y < 5$

15 $-\frac{1}{2} \leqslant x \leqslant 1\frac{1}{2}$

10 $-3 < x < 1$

13 $0 \leqslant x < 4$

16 $-2 \leqslant y < -1$

11 $-1 \leqslant y \leqslant 2$

14 $-2 < y \leqslant 3$

17 $3 \leqslant x < 5$

18 In each of the questions **9** to **11**, state whether or not the point $(1, 4)$ lies in the unshaded region.

Give the inequality that defines the unshaded region

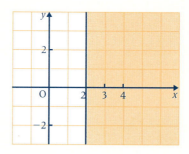

Boundary line $x = 2$ (included)

Inequality is $x \leqslant 2$

Give the inequalities that define the unshaded region

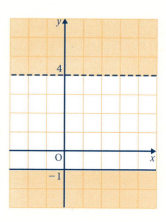

Boundary lines $y = 4$ (not included)

and $y = -1$ (included)

The inequalities are $y < 4$ and $y \geqslant -1$ or $-1 \leqslant y < 4$

Give the inequalities that define the unshaded regions:

19

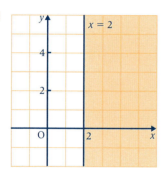

20

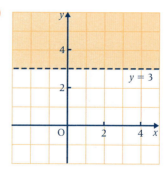

21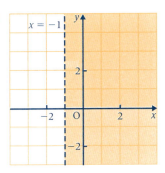

23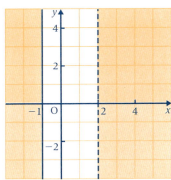

22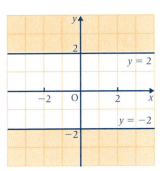

24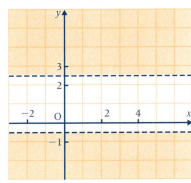

25 In each of the questions **19** to **24** state whether or not the point $(2, -1)$ is in the unshaded region.

Give the inequalities that define the *shaded* regions:

26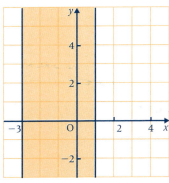

28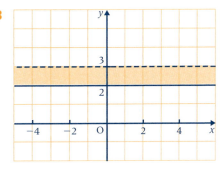

27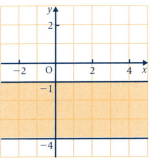

29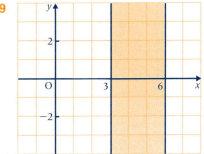

30 In each of the questions **26** to **29** state whether or not the point $(0, 2)$ is in the shaded region.

Exercise 22e Draw a diagram to represent the region defined by the set of inequalities $-1 \leqslant x \leqslant 2$ and $-5 \leqslant y \leqslant 0$

There are four inequalities here: $-1 \leqslant x$, $x \leqslant 2$, $-5 \leqslant y$ and $y \leqslant 0$.

The boundary lines are

$x = -1$: for $-1 \leqslant x$, shade the region on the left of the line

$x = 2$: for $x \leqslant 2$, shade the region on the right of the line

$y = -5$: for $-5 \leqslant y$, shade the region below this line

$y = 0$: for $y \leqslant 0$, shade the region above the line

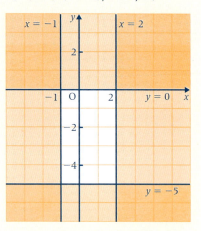

The unshaded region represents the inequalities.

Draw diagrams to represent the regions described by the following sets of inequalities. In each case, draw axes for values of x and y from -5 to 5.

1 $2 \leqslant x \leqslant 4$, $-1 \leqslant y \leqslant 3$

2 $-2 < x < 2$, $-2 < y < 2$

3 $-3 < x \leqslant 2$, $-1 \leqslant y$

4 $0 \leqslant x \leqslant 4$, $0 \leqslant y \leqslant 3$

5 $-4 < x < 0$, $-2 < y < 2$

6 $-1 < x < 1$, $-3 < y < 1$

7 $x \geqslant 0$, $y \geqslant 0$

8 $x \geqslant 1$, $-1 \leqslant y \leqslant 2$

Give the sets of inequalities that describe the unshaded regions:

9

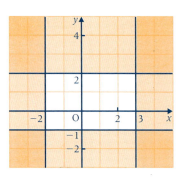

10

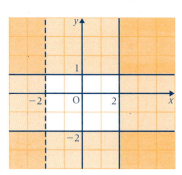

11 Is the point $(2\frac{1}{2}, 0)$ in either of the unshaded regions in questions **9** and **10**?

Give the sets of inequalities that describe the unshaded regions:

12

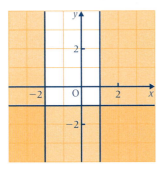

14

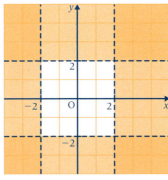

13

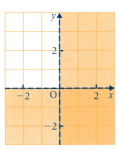

15

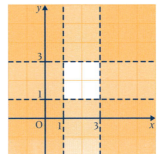

Estimate which distance is the greater
– from A to B or from B to C?

Now check your answer by measuring.

Investigate other optical illusions.

IN THIS CHAPTER...

you have seen that:

- an inequality remains true when the same number is added to, or subtracted from, both sides

- an inequality remains true when both sides are multiplied or divided by the same **positive** number. Do not multiply or divide an inequality by a negative number. It destroys the inequality.

AT THE END OF THIS CHAPTER...

you should be able to:

1   Identify congruent shapes.

2   Identify transformations under which shapes are congruent.

3   State the necessary and sufficient conditions for triangles to be congruent.

4   Use the properties in (3) to solve problems on congruent triangles.

5   Use conditions necessary for congruent triangles to prove properties of parallelograms.

6   Use conditions for congruence to investigate properties of special quadrilaterals.

Did you know that the word geometry comes from the Greek word *geometrein*, which means 'to measure the land'?

BEFORE YOU START

you need to know:

✓ basic facts about angles
✓ the meaning of reflection, rotation, translation and enlargement
✓ how to describe a transformation
✓ the meaning of similar figures
✓ how to draw an accurate copy of a triangle

KEY WORDS

alternate angles, congruent shapes, corresponding angles, corresponding sides, hypotenuse, included angle, interior angles, kite, parallelogram, rectangle, reflection, rhombus, rotation, similar, square, transformations, translation, vertically opposite angles

The basic facts

These facts were introduced earlier. They are revised here because they are needed for the exercises in this chapter.

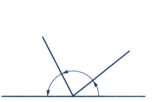

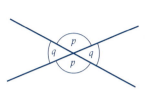

angles on a straight line add up to 180°

angles at a point add up to 360°

vertically opposite angles are equal.

When a transversal cuts a pair of parallel lines, various angles are formed and:

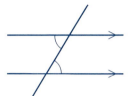

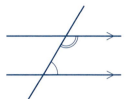

corresponding angles are equal;

alternate angles are equal;

interior angles add up to 180°.

In *any* triangle, whatever its shape or size, the sum of the three angles is 180°, e.g.

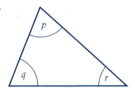

$$p + q + r = 180°$$

The exercise that follows will help you revise these facts. However, you must state which fact you have used as a reason for statements that you make. The facts do not have to be quoted in full but can be shortened using a number of standard abbreviations.

For example, if you state that $x = 60°$ and the reason is that x and 60° are vertically opposite angles then you could write

$$x = 60° \quad (\text{vert. opp. } \angle s)$$

Exercise 23a

Find the size of the angle marked x, giving brief reasons to justify your statements. (Fill in the size of any angles that you find.)

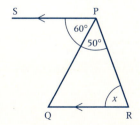

From the diagram SP and QR are parallel

So $\quad P\widehat{Q}R = 60°$ (alt. $\angle s$)

$50° + 60° + x = 180°$ ($\angle$ sum of $\triangle PQR$)

$\therefore \qquad\qquad x = 70°$

STP Caribbean Mathematics 2

In each of the following diagrams find the size of the angle marked x, giving brief reasons for your answer:

> **Tip** Draw a diagram. Mark in all the facts given and any other facts you know. This will help you to find the required angle.

1

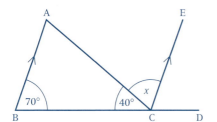

2

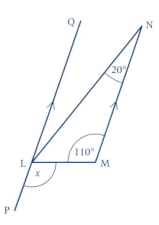

3

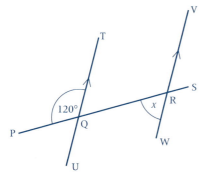

5

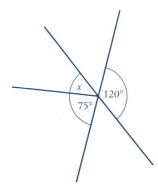

4

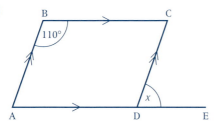

6

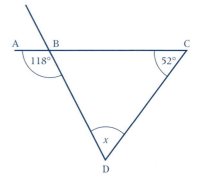

Prove that $\widehat{ABC} = \widehat{CDE}$

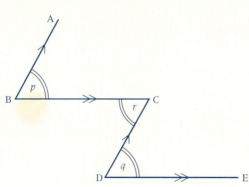

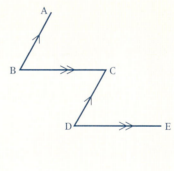

(Mark with a letter the angles you need to refer to. Use a symbol to show equal angles. We have used a double arc.)

From the diagram

$$p = r \quad (\text{alt. } \angle s)$$

$$r = q \quad (\text{alt. } \angle s)$$

$\therefore \qquad\qquad\qquad p = q$

i.e. $\qquad\qquad \widehat{ABC} = \widehat{CDE}$

7

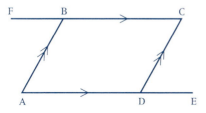

Prove that $\widehat{ABF} = \widehat{CDE}$

9

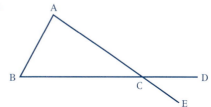

Prove that $\widehat{DCE} + \widehat{CAB} + \widehat{CBA} = 180°$

8

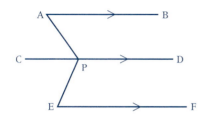

Prove that $\widehat{APE} = \widehat{BAP} + \widehat{FEP}$

10

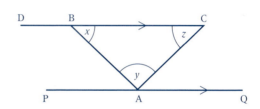

Prove that $x + y + z = 180°$

Congruent shapes

Two shapes are *congruent* if they are exactly the same shape and size, i.e. one shape is an exact copy of the other shape, although not necessarily drawn in the same position.

In each of these diagrams the two figures are congruent:

In each case the second shape is an exact copy of the first shape although it may be turned round or turned over.

In each of these diagrams the two figures are not congruent (they may be similar; i.e. have the same shape but different sizes):

Exercise 23b

In the following questions state whether or not the two shapes are congruent.

If you are not sure, trace one shape and see if it fits exactly over the other shape: Remember 'congruent' means exactly the same shape and size.

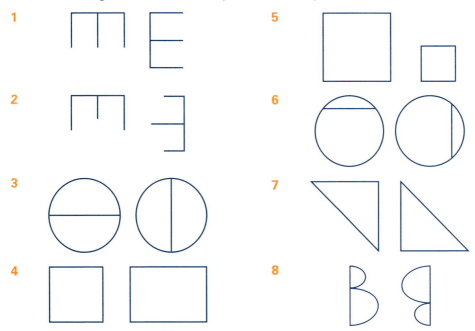

Transformations and congruent figures

The shape and size of a figure are not altered by certain transformations. Reflection produces congruent shapes:

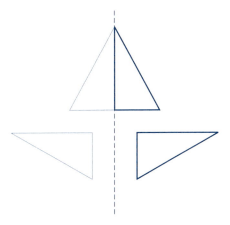

Rotation produces congruent shapes:

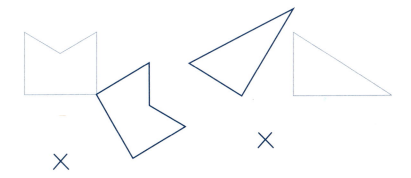

Translation produces congruent shapes:

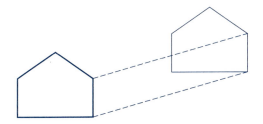

But enlargement does *not* produce congruent shapes, it produces similar shapes:

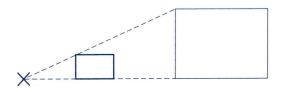

STP Caribbean Mathematics 2

Exercise 23c

Describe the transformation in each of the following cases. The paler blue shape is the image. State whether the object and the image are congruent:

1

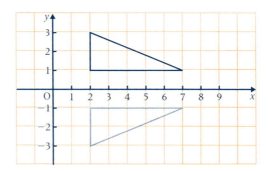

2

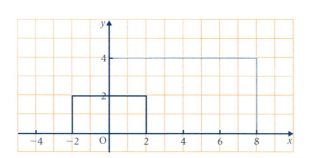

3

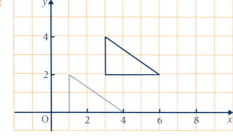

4

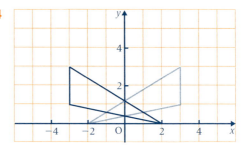

5

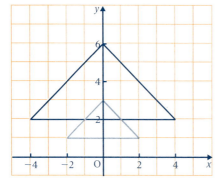

6

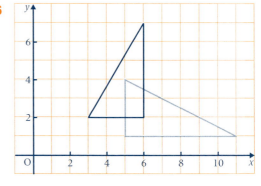

Congruent triangles

It is not always practical or possible to use tracing paper to determine whether two shapes are congruent.

Triangles are simple figures and not very much information is needed to determine whether one triangle is an exact copy of another triangle.

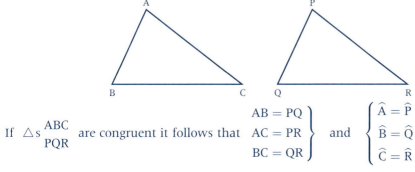

If $\triangle$s $\begin{matrix} ABC \\ PQR \end{matrix}$ are congruent it follows that $\left. \begin{matrix} AB = PQ \\ AC = PR \\ BC = QR \end{matrix} \right\}$ and $\left\{ \begin{matrix} \hat{A} = \hat{P} \\ \hat{B} = \hat{Q} \\ \hat{C} = \hat{R} \end{matrix} \right.$

To make an exact copy of these triangles we do not need to know the lengths of all three sides and the sizes of all three angles: three measurements are usually enough and we now investigate which three measurements are suitable.

Exercise 23d

In each of the following questions make a rough sketch of $\triangle ABC$. Construct a triangle with the same measurements as those given for $\triangle ABC$. Can you construct a different triangle with the given measurements?

1 $\triangle ABC$, in which $AB = 8\,\text{cm}$, $BC = 5\,\text{cm}$, $AC = 6\,\text{cm}$.

> **Tip** You need a sharp pencil for this exercise.

2 $\triangle ABC$, in which $\hat{A} = 40°$, $\hat{B} = 60°$, $\hat{C} = 80°$.

3 $\triangle ABC$, in which $AB = 7\,\text{cm}$, $BC = 12\,\text{cm}$, $AC = 8\,\text{cm}$.

4 $\triangle ABC$, in which $\hat{A} = 20°$, $\hat{B} = 40°$, $\hat{C} = 120°$.

5 What extra information do you need about $\triangle ABC$ in questions 2 and 4 in order to make an exact copy?

Three pairs of sides

From the last exercise you should be convinced that an exact copy of a triangle can be made if the lengths of the three sides are known. Therefore,

> two triangles are congruent if the three sides of one triangle are equal to the three sides of the other triangle.

However if the three angles of one triangle are equal to the three angles of another triangle they may not be congruent. Similar triangles are also called equiangular triangles.

Exercise 23e | Decide whether the following pairs of triangles are congruent. Give brief reasons for your answers.

a

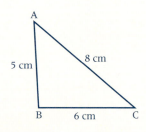 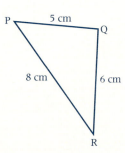

$$AB = PQ$$
$$BC = QR \left.\right\} \quad \therefore \quad \triangle s \, \frac{ABC}{PQR} \text{ are congruent (3 sides)}$$
$$AC = PR$$

(Notice that we write corresponding vertices one under the other.)

b

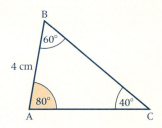 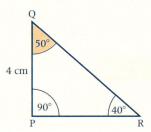

First we need to calculate the angles that are shaded.

In $\triangle ABC$, $\hat{A} = 80°$ (angles of $\triangle$)

In $\triangle PQR$, $\hat{Q} = 50°$ (angles of $\triangle$)

The angles of the triangles are not equal.

$\therefore \triangle ABC$ and $\triangle PQR$ are not congruent.

c

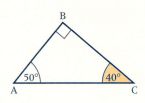 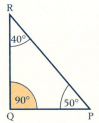

In $\triangle ABC$, $\hat{C} = 40°$ (angles of $\triangle$)

In $\triangle PQR$, $\hat{Q} = 90°$ (angles of $\triangle$)

$\therefore \triangle s \, \frac{ABC}{PQR}$ are similar, but probably not congruent because no dimensions are given.

In questions **1** to **4** state whether or not the two triangles are congruent. Give a brief reason for your answers. All lengths are in centimetres:

1

2
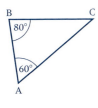 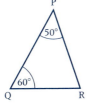

3
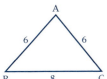 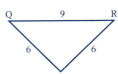

4
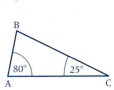 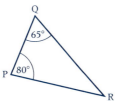

5
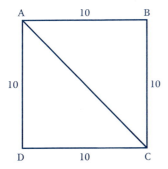

Are △ADC and △ABC congruent?

6
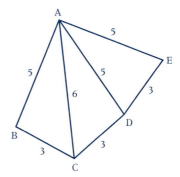

Which triangles are congruent?

7
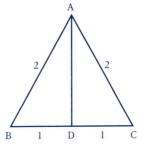

Are △ABD and △ACD congruent?

8
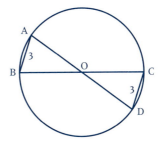

The point O is the centre of the circle and the radius is 5 cm. Are △ABO and △CDO congruent?

Two angles and a side

To make an exact copy of a triangle we need to know the length of at least one side.

Exercise 23f

1 Construct △ABC, in which AB = 6 cm, $\hat{A} = 30°$, $\hat{B} = 60°$

2 Construct △PQR, in which PR = 6 cm, $\hat{P} = 30°$, $\hat{Q} = 60°$

3 Construct △LMN, in which LM = 6 cm, $\hat{L} = 30°$, $\hat{M} = 60°$

4 Construct △XYZ, in which YZ = 6 cm, $\hat{X} = 30°$, $\hat{Y} = 60°$

5 How many of the triangles that you have constructed are congruent?

6 How many different triangles can you construct from the following information: one angle is 40°, another angle is 70° and the length of one side is 8 cm?

> **Tip** You need a sharp pencil.

Now you can see that we are able to make an exact copy of a triangle if we know the sizes of two of its angles and the length of one side provided that we place the side in the same position relative to the angles in both triangles, i.e.

> two triangles are congruent if two angles and one side of one triangle are equal to two angles and the *corresponding* side of the other triangle.

Exercise 23g

Decide whether these triangles are congruent. Give a brief reason for your answer.

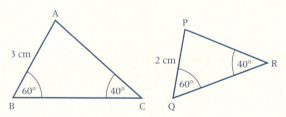

$$\triangle s \; \begin{matrix} ABC \\ PQR \end{matrix} \text{ are similar} \qquad \text{(angles equal)}$$

but not congruent   (AB and PQ are corr. sides and are *not* equal).

In questions **1** to **8** state whether or not the two triangles are congruent. Give brief reasons for your answers. All lengths are in centimetres:

1

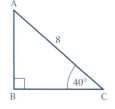

> **Tip** Remember, congruent triangles are identical in all respects.

2

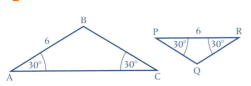

7

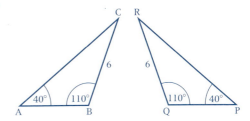

3

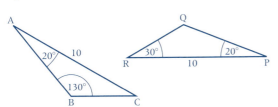

8

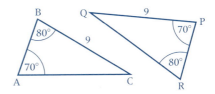

4

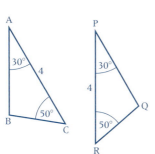

9 Are △ABC and △ADC congruent?

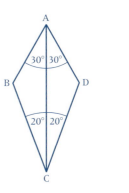

5

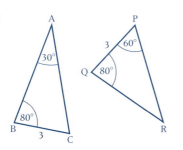

10 Are △ABC and △ADC congruent?

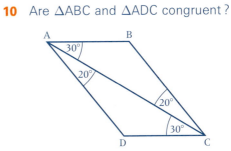

6

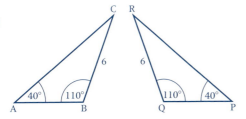

Two sides and an angle

We are now left with one more possible combination of three measurements: if we know the lengths of two sides and the size of one angle in a triangle, does this fix the size and shape of the triangle?

Exercise 23h

Can you make an exact copy of the following triangles from the information given about them? (Try to construct each triangle.)

1   $\triangle$ABC, in which AB $= 8$ cm, BC $= 5$ cm, $\hat{B} = 30°$

2   $\triangle$XYZ, in which XY $= 8$ cm, XZ $= 5$ cm, $\hat{Y} = 30°$

3   $\triangle$PQR, in which $\hat{Q} = 60°$, PQ $= 6$ cm, QR $= 8$ cm

4   $\triangle$LMN, in which LM $= 8$ cm, $\hat{M} = 20°$, LN $= 4$ cm

5   $\triangle$DEF, in which DE $= 5$ cm, $\hat{E} = 90°$, EF $= 6$ cm

Now it is possible to see that we can make an exact copy of a triangle if we know the lengths of two sides and the size of one angle, provided that the angle is between those two sides. Therefore,

two triangles are congruent if two sides and the *included* angle of one triangle are equal to two sides and the *included* angle of the other triangle.

If the angle is not between the two known sides, then we cannot always be sure that we can make an exact copy of the triangle. We will now investigate this case further.

Exercise 23i

Can you make an exact copy of each of the following triangles from the information given about them?

1   $\triangle$ABC, in which AB $= 6$ cm, $\hat{B} = 90°$, AC $= 10$ cm.

2   $\triangle$PQR, in which PQ $= 8$ cm, $\hat{Q} = 40°$, PR $= 6.5$ cm.

3   $\triangle$XYZ, in which XY $= 5$ cm, $\hat{Y} = 90°$, XZ $= 13$ cm.

4   $\triangle$LMN, in which LM $= 5$ cm, $\hat{M} = 60°$, LN $= 4.5$ cm.

5   $\triangle$DEF, in which DE $= 7$ cm, $\hat{E} = 90°$, DF $= 10$ cm.

6   $\triangle$RST, in which RS $= 5$ cm, $\hat{S} = 120°$, RT $= 8$ cm.

Therefore, if we are told that one angle in a triangle is a right angle and we are also given the length of one side and the hypotenuse, then this information fixes the shape and size of the triangle since it is equivalent to knowing the lengths of the three sides.

> Two triangles are congruent if they both have a right angle, and the hypotenuse and a side of one triangle are equal to the hypotenuse and a side of the other triangle.

Exercise 23j

In questions 1 to 8 state whether or not the two triangles are congruent. Give brief reasons for your answers. All lengths are in centimetres:

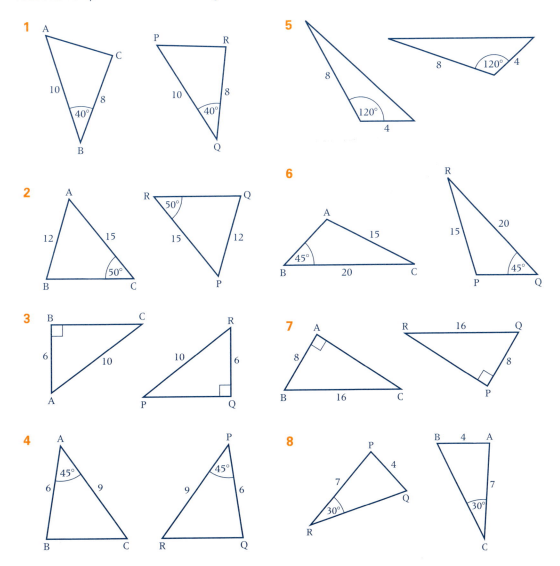

9

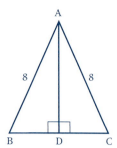

Are △ABD and △ACD congruent?

10

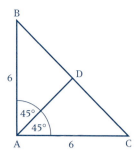

Are △ABD and △ACD congruent?

Summing up, two triangles are congruent if:

either the three sides of one triangle are equal to the three sides of the other triangle (S S S)

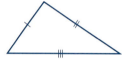 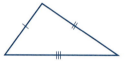

or two angles and a side of one triangle are equal to two angles and the corresponding side of the other triangle (A A S)

or two sides and the included angle of one triangle are equal to two sides and the included angle of the other triangle (S A S)

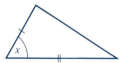 

or two triangles each have a right angle, and the hypotenuse and a side of one triangle are equal to the hypotenuse and a side of the other triangle (R H S).

Exercise 23k

State whether or not each of the following pairs of triangles are congruent. Give brief reasons for your answers. All measurements are in centimetres:

1

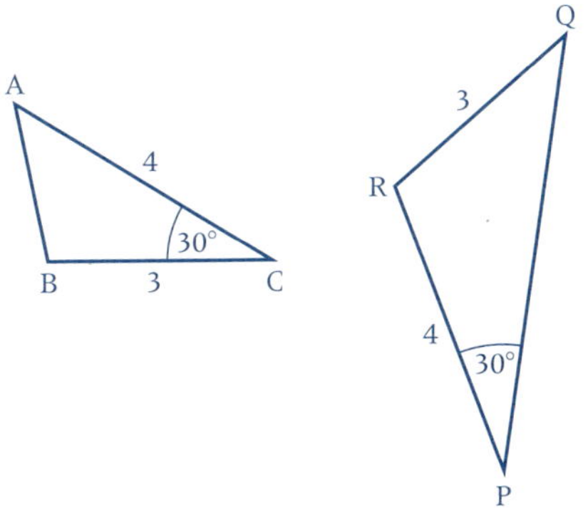

2

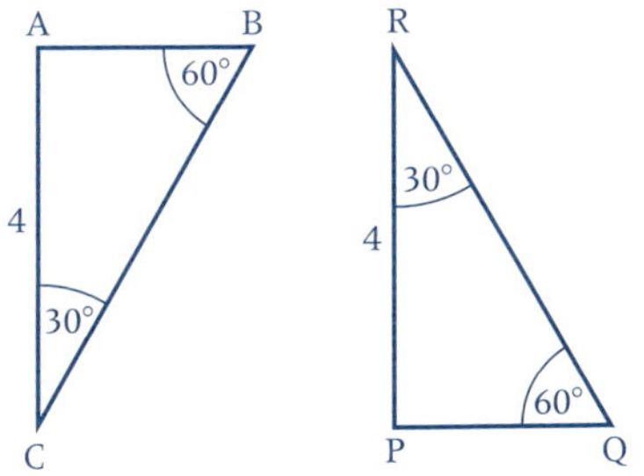

3

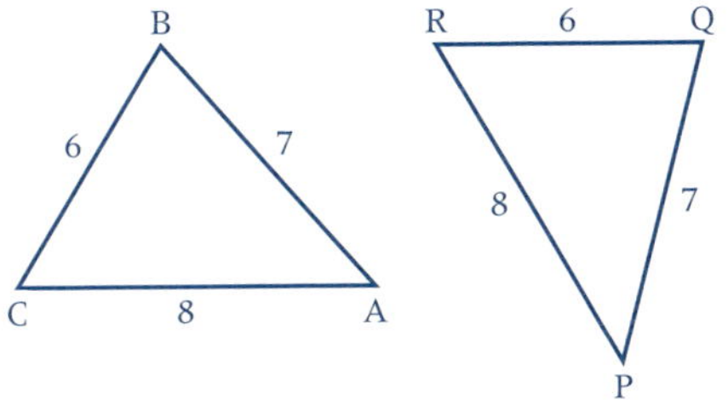

4

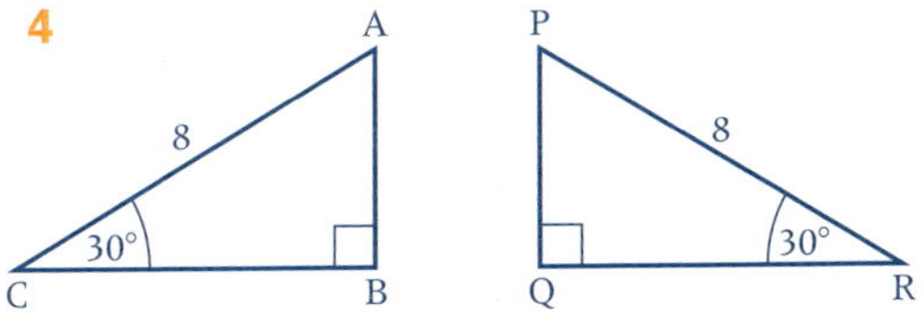

5

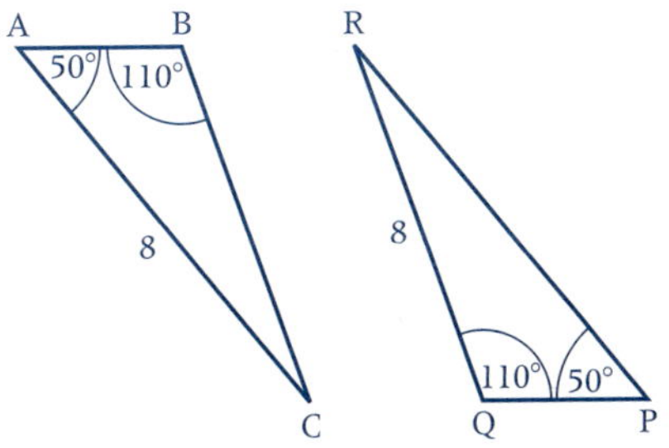

6

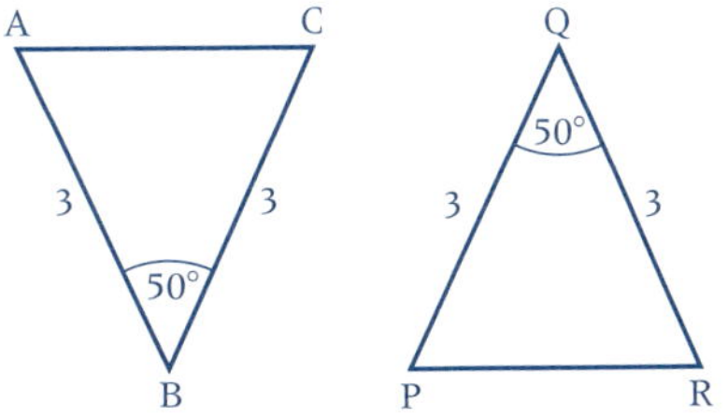

7

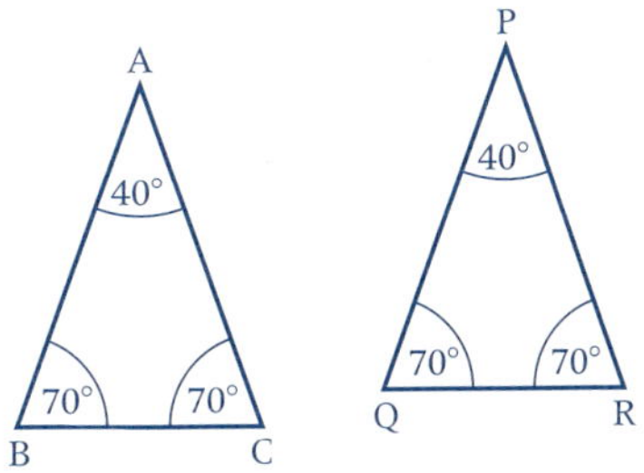

8

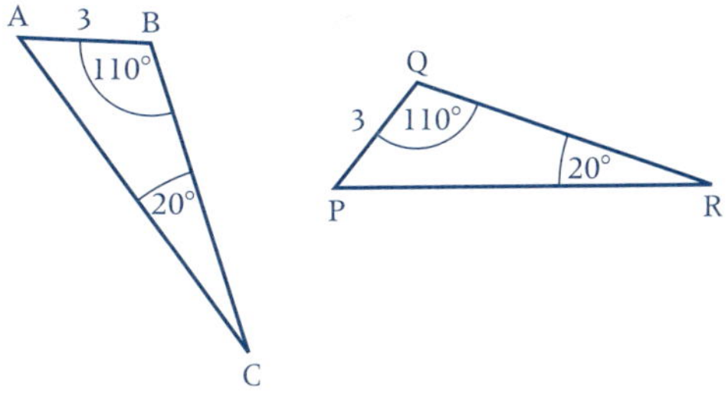

9

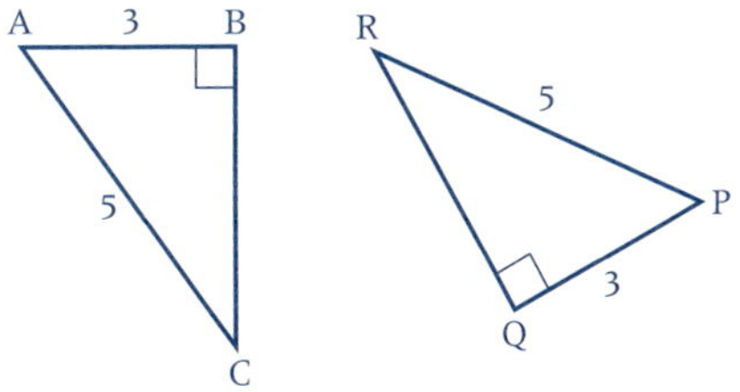

10

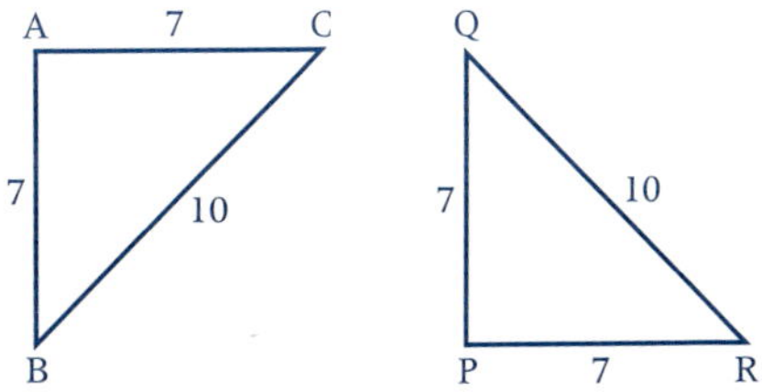

375

11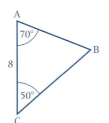

12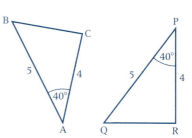

Problems

We do not need to know actual measurements to prove that triangles are congruent. If we can show that a correct combination of sides and angles are the same in both triangles, the triangles must be congruent.

Exercise 23I

In a quadrilateral ABCD, AB = DC and AD = BC. The diagonal BD is drawn. Prove that △ABD and △CDB are congruent.

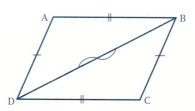

(Mark on your diagram all the information given and any further facts that you discover. The symbol ⌢ on BD indicates that it is common to both triangles.)

In △s ABD, CDB AB = CD (given)

AD = CB (given)

DB is the same for both triangles

∴ △s ABD ⁄ CDB are congruent (SSS).

> **Tip** There are four possible ways of proving that two triangles are congruent. Go through these in turn for each pair of triangles to decide which set of conditions are met.

1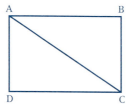

ABCD is a rectangle. Prove that △ABC and △CDA are congruent.

2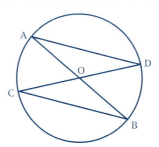

AB and CD are diameters of the circle and O is the centre. Prove that △AOD and △COB are congruent.

3

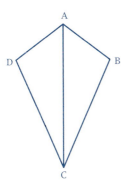

ABCD is a kite in which AD = AB and CD = BC. Prove that △ADC and △ABC are congruent.

4

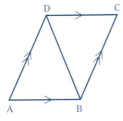

ABCD is a parallelogram. Prove that △ABD and △CDB are congruent.

5

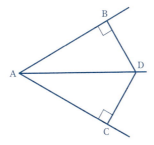

AD bisects BÂC, DB is perpendicular to AB and DC is perpendicular to AC. Prove that △ABD and △ACD are congruent.

6

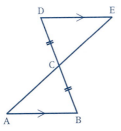

Prove that △ABC and △EDC are congruent.

7

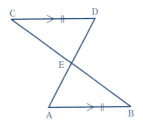

CD and AB are equal and parallel. Prove that △ABE and △DCE are congruent.

8

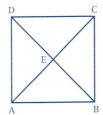

ABCD is a square. Show that △s ABE, BCE, CDE and DAE are all congruent.

9

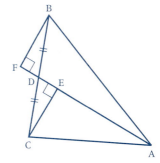

D is the midpoint of BC. CE and BF are perpendicular to AF. Find a pair of congruent triangles.

10

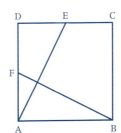

ABCD is a square. E is the midpoint of DC and F is the midpoint of AD. Show that △ADE and △BAF are congruent.

11 ABC is an isosceles triangle in which AB = AC. D is the midpoint of AB and E is the midpoint of AC. Prove that △BDC is congruent with △CEB.

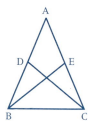

12 ABCD is a rectangle and E is the midpoint of AB. Join DE and CE and show that △ADE is congruent with △BCE.

13 ABCD is a rectangle. E is the midpoint of AB and F is the midpoint of DC. Join DE and BF and show that △ADE is congruent with △CBF.

 Puzzle

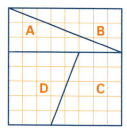 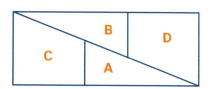

Draw a square of side 8 cm on centimetre graph paper. Divide the square into two congruent triangles and two congruent trapeziums as shown in the diagram on the left.

Rearrange the four shapes to form the rectangle on the right. This rectangle measures 13 cm by 5 cm so has an area of $13 \times 5\,\text{cm}^2 = 65\,\text{cm}^2$.

However, the area of the original square is $8 \times 8\,\text{cm}^2 = 64\,\text{cm}^2$.

How do you explain this apparent contradiction?

Using congruent triangles

Once two triangles have been shown to be congruent it follows that the other corresponding sides and angles are equal. This gives a good way of proving that certain angles are equal or that certain lines are the same length.

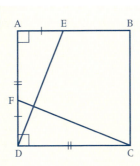

Exercise 23m

ABCD is a square and AE = DF.
Show that DE = CF.

In △s DAE and CDF

$$AE = DF \quad (\text{given})$$
$$DA = CD \quad (\text{sides of a square})$$
$$D\hat{A}E = C\hat{D}F \quad (\text{angles of a square are } 90°)$$

∴ △s $\dfrac{\text{DAE}}{\text{CDF}}$ are congruent (SAS).

(We have written the triangles so that corresponding vertices are lined up. We can then see the remaining corresponding sides and angles.)

∴   DE = CF

1   BD bisects A$\hat{B}$C. BE and BF are equal. Show that triangles BED and BFD are congruent and hence prove that ED = FD.

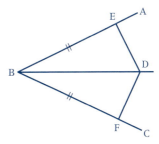

2

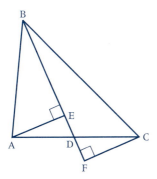

D is the midpoint of AC. AE and CF are both perpendicular to BF. Show that triangles AED and CFD are congruent and hence prove that AE = CF.

3

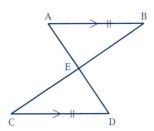

AB and CD are parallel and equal in length. Show that △AEB and △DEC are congruent and hence prove that E is the midpoint of both CB and AD.

379

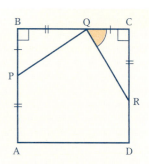

ABCD is a square and P, Q and R are points on AB, BC and CD respectively such that AP = BQ = CR.

Show that $P\hat{Q}R = 90°$

(We will first prove that $\triangle PBQ$ and $\triangle QCR$ are congruent.)

In $\triangle PBQ$ and $\triangle QCR$

$BQ = CR$	(given)
$P\hat{B}Q = Q\hat{C}R = 90°$	(angles of a square)

also $PB = QC \begin{cases} AB = BC \\ AP = BQ \end{cases}$ (sides of a square) (given)

$\therefore \triangle s \begin{matrix} PBQ \\ QCR \end{matrix}$ are congruent.   (SAS)

$\therefore \qquad\qquad B\hat{Q}P = Q\hat{R}C$

In $\triangle QRC \qquad Q\hat{R}C + C\hat{Q}R = 90°$ (angles of triangle)

$\therefore \qquad\qquad B\hat{Q}P + C\hat{Q}R = 90°$ ($B\hat{Q}P = Q\hat{R}C$, proved above)

But $\qquad B\hat{Q}P + P\hat{Q}R + C\hat{Q}R = 180°$ (angles on st. line)

$\therefore \qquad\qquad P\hat{Q}R = 90°$

4

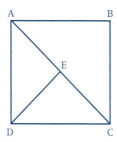

ABCD is a square and E is the midpoint of the diagonal AC. First show that triangles ADE and CDE are congruent and hence prove that DE is perpendicular to AC.

5

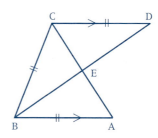

CD is parallel to BA and CD = CB = BA.

Show that $\triangle CDE$ is congruent with $\triangle BAE$ and hence that CA bisects BD.

6 Using the same diagram and the result from question 6, show that $\triangle BEC$ is congruent with $\triangle CED$. Hence prove that CA and BD cut at right angles.

7

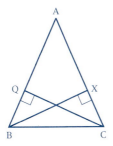

Triangle ABC is isosceles, with AB = AC. BX is perpendicular to AC and CQ is perpendicular to AB. Prove that BX = CQ.

Tip Find a pair of congruent triangles first.

8

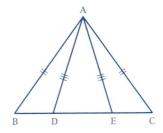

In the diagram, AB = AC and AD = AE. Prove that BD = EC.

Tip Consider triangles ABD and ACE.

9 AB is a straight line. Draw a line AX perpendicular to AB. On the other side of AB, draw a line BY perpendicular to AB so that BY is equal to AX. Prove that $A\hat{X}Y = B\hat{Y}X$.

Properties of parallelograms

In Book 1 we investigated the properties of parallelograms by observation and measurement of a few particular parallelograms. Now we can use congruent triangles to prove that these properties are true for all parallelograms.

A parallelogram is formed when two pairs of parallel lines cross each other.

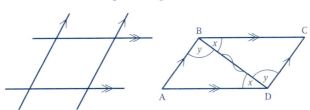

In the parallelogram ABCD, joining BD gives two triangles in which:

the angles marked x are equal (they are alternate angles with respect to parallels AD and BC);
the angles marked y are equal (they are alternate angles with respect to parallels AB and DC);

and BD is the same for both triangles.

∴   △s $\begin{array}{c} BCD \\ DAB \end{array}$ are congruent (A A S).

∴   BC = AD and AB = DC

i.e.   the opposite sides of a parallelogram are the same length.

Also from the congruent triangles

$$\hat{A} = \hat{C}$$

and $\qquad$ $A\hat{B}C = C\hat{D}A$ $\quad(x + y = y + x)$

i.e. $\qquad$ the opposite angles of a parallelogram are equal.

Drawing both diagonals of the parallelogram gives four triangles.

Considering the two triangles BEC, DEA

$$BC = AD \quad (\text{opp. sides of } \|\text{gram})$$

$$E\hat{B}C = E\hat{D}A \quad (\text{alt. } \angle s)$$

$$B\hat{E}C = A\hat{E}D \quad (\text{vert. opp. } \angle s)$$

$\therefore \quad \triangle s \begin{array}{c} BEC \\ DEA \end{array}$ are congruent (A A S).

$\therefore \qquad BE = ED \ \text{ and } \ AE = EC$

i.e. $\qquad$ the diagonals of a parallelogram bisect each other.

The diagrams below summarise these properties.

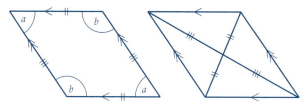

It is equally important to realise that, in general,

$\qquad$ the diagonals are *not* the same length,

$\qquad$ the diagonals do *not* bisect the angles of a parallelogram.

In the exercise that follows, you are asked to investigate the properties of some of the other special quadrilaterals.

Exercise 23n

1 ABCD is a rhombus (a parallelogram in which all four sides are equal in length). Join AC and show that △ABC and △ADC are congruent. What does AC do to the angles of the rhombus at A and C? Does the diagonal BD do the same to the angles at B and D?

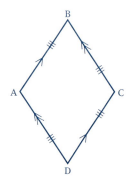

2

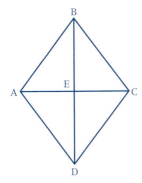

ABCD is a rhombus. Use the results from question 1 to show that △s ABE and BCE are congruent. What can you now say about the angles AEB and BEC?

3

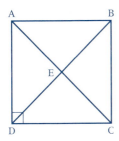

ABCD is a square (a rhombus with right-angled corners). Use the properties of the diagonals of a rhombus to show that △AEB is isosceles. Hence prove that the diagonals of a square are the same length.

Are the two diagonals of *every* rhombus the same length?

4

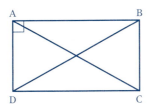

ABCD is a rectangle (a parallelogram with right-angled corners). Prove that △s ADB and DAC are congruent. What can you deduce about the lengths of AC and DB?

5

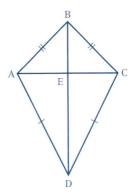

ABCD is a kite in which AB = BC and AD = DC.

Does the diagonal BD bisect the angles at B and D?

Does the diagonal AC bisect the angles at A and C?

Is E the midpoint of either diagonal?

What can you say about the angles at E?

6

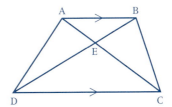

ABCD is a trapezium: it has just one pair of parallel sides. Are there any congruent triangles in this diagram?

7

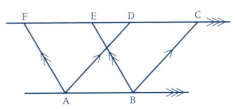

In the diagram, ABCD and ABEF are parallelograms. Show that △s ADF and BCE are congruent.

By considering the shape ABCF and then removing each of the triangles AFD and BEC in turn, what can you say about the areas of the two parallelograms?

383

Puzzle

Penny pushed $3 across the counter at the Post Office. 'Would you give me some 35c stamps, two fewer 30c stamps than 35c stamps and seven 5c stamps please'. How many stamps did she get?

Assume that there was no change.

Using properties of special quadrilaterals

In question **7** you proved a property of two parallelograms.

Two parallelograms with the same base and drawn between the same pair of parallel lines are equal in area.

The diagrams below summarise the other results from Exercise 23n:

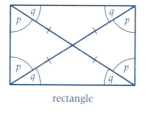

rectangle

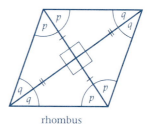

rhombus

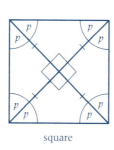

square

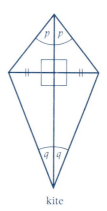

kite

You can now use any of these facts in the following exercise.

Exercise 23p

1

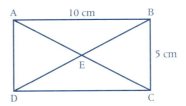

ABCD is a rectangle. The diagonals AC and DB cut at E. How far is E from BC?

2

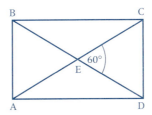

ABCD is a rectangle in which CÊD = 60°. Find EĈD.

3 ABCD is a parallelogram in which AB̂C = 120° and BĈA = 30°. Show that ABCD is also a rhombus.

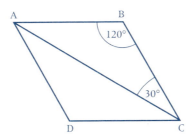

4 In a rectangle ABCD, AB = 6 cm and the diagonal BD is 10 cm. Make a rough sketch of the rectangle and then construct ABCD. Measure AD.

5 In a rhombus ABCD, the diagonal AC is 8 cm and the diagonal BD is 6 cm. Construct the rhombus and measure AB. (Remember first to make a rough sketch.)

6 ABCD is a parallelogram in which the diagonal AC is 10 cm and the diagonal BD is 12 cm. AC and BD cut at E and AÊB = 60°. Make a rough sketch of the parallelogram and then construct ABCD. Measure BC.

7 Construct a rhombus ABCD in which the sides are 5 cm long and the diagonal AC is 8 cm long. Measure the diagonal BD.

8 Construct a square ABCD whose diagonal, AC, is 8 cm long. Measure the side AB.

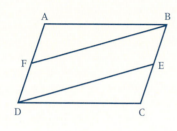

ABCD is a parallelogram. E is the midpoint of BC and F is the midpoint of AD.
Prove that BF = DE.

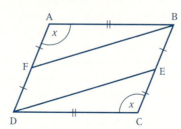

(Remember that in a parallelogram the opposite sides are equal and the opposite angles are equal.)

In △s ABF and CDE

$$AF = EC \quad (\tfrac{1}{2} \text{ opp. sides of parallelogram})$$

$$AB = DC \quad (\text{opp. sides of parallelogram})$$

$$F\hat{A}B = E\hat{C}D \quad (\text{opp. angles of parallelogram})$$

∴ △s $\begin{smallmatrix} ABF \\ CDE \end{smallmatrix}$ are congruent (SAS).

∴ BF = DE

9

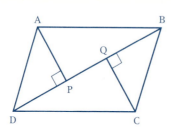

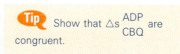

Tip Show that △s $\begin{smallmatrix} ADP \\ CBQ \end{smallmatrix}$ are congruent.

ABCD is a parallelogram. AP is perpendicular to BD and CQ is perpendicular to BD. Prove that AP = CQ.

10 ABCD is a rhombus. P is the midpoint of AD and Q is the midpoint of CD. Prove that BP = BQ.

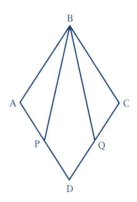

11

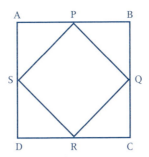

ABCD is a square and P, Q, R and S are the midpoints of AB, BC, CD and DA. Prove that PQRS is a square.

12

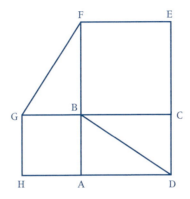

ABCD is a rectangle. ABGH and BCEF are squares. Show that GF = BD.

 # Investigation

a Draw a rectangle measuring 14 cm by 8 cm and cut it into four pieces as shown in the diagram.

Put all four pieces together to form

i a parallelogram

ii a large equilateral triangle

iii a large isosceles triangle

iv a kite

v two congruent triangles.

b Draw another triangle, this time measuring 12 cm by 8 cm. Can you make the same shapes as in part **a**? If your answer is 'yes', do so. If your answer is 'no', explain what is special about the first rectangle so that all the shapes can be made.

c Give the dimensions of another rectangle that will allow you to make all the shapes listed in part **a**. Are there many more?

Did you know that Cathleen Morawetz was the daughter of John Synge, an applied mathematician? She is the first woman to head a mathematics institute – the Courant Institute of Mathematics in New York.

IN THIS CHAPTER...

you have seen that:

- congruent shapes are identical in every way

- similar shapes are the same shape, but different in size

- two triangles are congruent if

    the three sides of one triangle are equal to the three sides of the other triangle,

    or if two angles and one side of one triangle are equal to two angles and the *corresponding* side of the other triangle,

or if two sides and the *included* angle of one triangle are equal to two sides and the *included* angle of the other triangle,

or if both have a right angle, and the hypotenuse and a side of one triangle are equal to the hypotenuse and a side of the other triangle.

These conditions can be abbreviated as SSS, AAS, SAS and RHS

STATISTICS

Statistics wasn't generally taught in schools until the introduction of the electronic calculator.

Imagine trying to do all the calculations you need to do using pencil and paper!

Stem-and-leaf diagrams

This list gives the number of cars in a car park at noon on 20 consecutive Saturdays.

104 99 107 102 115 102 108 98 110 106

108 95 94 118 95 105 114 102 113 97

Grouping these figures in a frequency table loses some of the detail. Another way of showing these numbers is to use a stem-and-leaf diagram. This keeps the detail of individual figures.

The groups form the stem; the numbers are all between 90 and 120, so we use the number of tens as the numbers for the stem, i.e. 9, 10 and 11. Write these down the left-hand side. The leaves are the corresponding units. These are written on the right, next to the appropriate stem, so 104 is represented by 10 in the stem and a 4 in the leaf next to it.

Start working across the first row marking each unit in the correct place. Do not try to put them in order at this stage.

Stem	Leaves
9	9 8 5 4 5 7
10	4 7 2 2 8 6 8 5 2
11	5 0 8 4 3

9 | 5 means 95
10 | 8 means 108, etc.

Check that you have 20 values.

Next redraw the diagram with the numbers in order of size and give a key.

Stem	Leaves
9	4 5 5 7 8 9
10	2 2 2 4 5 6 7 8 8
11	0 3 4 5 8

11 | 4 means 114

Check again that you have 20 values.

Exercise 24a

1 This stem-and-leaf diagram shows the number of pages in a sample of books.

Number of pages

1	01 20 34 41 58 75 92
2	09 10 25 36 71 77 79 80 86
3	10 24 63

1 | 62 means 162

a Write down the number of books in the sample.

b How many books have more than 250 pages?

> **Tip** The key shows that the numbers in the stem are hundreds. More than 250 pages means all the books with over 250 pages.

2 A maths test was marked out of 40.

This stem-and-leaf diagram shows the marks scored by some students in the test.

Marks																1|4 means 14
1	4	4	5													
2	0	4	5	6	7	7	7									
3	0	0	1	2	3	3	4	4	4	4	4	4	5	5	8	9	9

a What do the numbers in the stem represent?

b Calculate the number of students marks shown in the diagram.

c Write down the lowest mark.

d Two other students got marks of 36 and 29.

Add these two marks in the correct place.

3 Moira timed her journey to school on seventeen consecutive days.

These times are shown in this stem-and-leaf diagram.

Time in minutes								2|5 means 25 minutes
1	2	4	5	6	8	8	9	
2	0	0	1	1	5	7	9	
3	0	2	5					

a For the next three days her times were 17 minutes, 19 minutes and 27 minutes.

Add these three values to the stem-and-leaf diagram.

Rewrite your stem-and-leaf diagram so that all the values are in order.

Use your new diagram for these questions.

b Write down the number of times the journey took more than 20 minutes.

c Write down the number of times the journey was less than 25 minutes.

d Write down the number of times that the journey was longer than half an hour.

4 A school library bought some new books. This is a list of their prices.

$4.50 $6.20 $5.40 $4.90 $5.70 $4.39 $6.45 $5.95 $4.75 $5.80

a How many new books were bought?

b The first four prices in the list are shown in this stem-and-leaf diagram.

Price of books	5|70 means $5.70
4	50 90
5	40
6	20

Complete the diagram to show all the prices.

5 These are the recorded playing times, in minutes, of some CDs.

61 37 74 50 50 59 72 41 47 68 59 59 70 72
69 42 59 66 60 61 65 50 55 52 61 45 74 68

a How many CDs were recorded?

b Complete this stem-and-leaf diagram to show these times.
The first time in the list has been entered.

Number of minutes   6|1 means 61

```
3 |
4 |
5 |
6 | 1
7 |
```

c Rewrite the stem and-leaf diagram so that the times are in order.

d How many of these CDs have a playing time of less than 55 minutes?

Mean, mode and median

In Book 1 we saw that, when we have a set of numbers, there are three different measures we can use that attempt to give a 'typical member' that is representative of the set.

Mean

The *mean* (arithmetic average) of a set of n numbers is the sum of the numbers divided by n.

The mean of the set 2, 6, 8, 8, 10, 10, 12, is

$$\frac{2+6+8+8+10+10+12}{7} = \frac{56}{7} = 8$$

The mean value of a set of numbers is the most frequently used form of average, so much so that the word 'average' is often used for 'mean value'.

For example, if you were asked for your average mark in a set of examinations, you would total the marks and divide by the number of examinations, i.e. you would find the mean mark.

Mode

In a set of numbers, the *mode* is the number that occurs most often. For example, for the set 2, 2, 4, 4, 4, 5, 6, 6, the mode is 4 as no other number occurs more than twice.

The mode is easier to find if the numbers are arranged in order of size. If the numbers in a set are all different, there is no mode. For example, the set 1, 2, 3, 5, 8,10, has no mode.

If there are two (or more) numbers which equally occur most often, there are two (or more) modes.

For example, in the set 1, 2, 2, 3, 5, 5, 8, both 2 and 5 are modes.

Median

If we arrange a set of numbers in order of size, the *median* is the number in the middle.

For example, for the seven numbers in the set 2, 4, 5, 7, 7, 8, 9, the median is 7.

When there is an even number of numbers in the set, the median is the mean of the two middle numbers. For example, for the eight numbers in the set 2, 3, 4, 4, 5, 6, 7, 7, the median is the mean of 4 and 5, i.e. 4.5.

For a small set of numbers, say 15, it is easy to find the median and we can see that it is the $\left(\frac{15+1}{2}\right)$th value, i.e. the 8th value. From examples such as this we deduce that, for n numbers arranged in order of size, the median is the $\left(\frac{n+1}{2}\right)$th number.

For example, for 59 numbers, the median is the $\left(\frac{59+1}{2}\right)$th number, i.e. the 30th number.

For 60 numbers, the median is the $\left(\frac{60+1}{2}\right)$th number, i.e. the $(30\frac{1}{2})$th number.

This means the average of the 30th and 31st numbers.

Exercise 24b A page from a novel by George Lamming was chosen at random and the number of letters in each of the first twenty words on that page was recorded:

$$3, 4, 5, 3, 7, 8, 3, 3, 6, 2, 4, 6, 4, 6, 3, 13, 4, 3, 3, 2$$

Find the mean, modal and median number of letters per word.

Arranging the numbers in size order:

$$2, 2, 3, 3, 3, 3, 3, 3, 3, 4, 4, 4, 4, 5, 6, 6, 6, 7, 8, 13$$

10th 11th

The mean is $\frac{92}{20} = 4.6$

The mode is 3

The median is the value of the $\left(\dfrac{20+1}{2}\right)$th number,

i.e. the $(10\frac{1}{2})$th number, which is the average of the 10th and 11th numbers

∴ the median is 4

Find the mean, mode and median of the sets of numbers in questions **1** to **4**. Remember to arrange them in order of size first. Give answers correct to three significant figures where necessary:

1 3, 6, 2, 5, 9, 2, 4

3 1.6, 2.4, 3.9, 1.7, 1.6, 0.2, 1.3, 2.0

2 13, 16, 12, 14, 19, 12, 14, 13

4 1.3, 1.8, 1.7, 1.9, 1.4, 1.5, 1.3, 1.8, 1.2

5 Ten music students took a Grade 3 piano examination.
They obtained the following marks:

$$106, 125, 132, 140, 108, 102, 75, 135, 146, 123.$$

Find the mean and median marks. Which of these two representative measures would be most useful to the teacher who entered the students? (Give *brief* reasons, do not write an essay on the subject.)

6 A small firm employs ten people. The salaries of the employees are as follows:

$30000, $8000, $5000, $5000, $5000, $5000, $5000, $4000, $3000, $1500

Find the mean, mode and median salary.

Which of these three figures is a trade union official unlikely to be interested in, and why?

7 Thirty 15-year-olds were asked how much pocket money they received each week and the following amounts (in cents) were recorded:

0, 0, 0, 50, 50, 50, 100, 100, 100, 100, 100, 100, 100, 150, 150, 200, 200, 200, 200, 200, 200, 200, 200, 200, 200, 250, 250, 250, 500, 1000.

Find the mean, mode and median amount.

If you were presenting your parents with an argument for an increase in pocket money, which of the three representative measures would you use and why?

8 The first eight customers at a supermarket one Saturday spent the following amounts:

$25.10, $3.80, $20.50, $15.70, $38.40, $9.60, $46.20, $10.46.

Find the mean and median amount spent.

9 This stem-and-leaf diagram shows the number of pages in a sample of books.

Number of pages   1 | 34 means 134

1	01 19 34
2	09 10 25 36
3	10

a Write down the number of pages in the longest book.

> **Tip** Find the middle value for the numbers of pages

b Write down the number of pages in the shortest book.

c Find the median number of pages.

> **Tip** First add the values, i.e. 101 + 119 + ...

d Find the mean number of pages per book.

10 A maths test was marked out of 40. This stem-and-leaf diagram shows the marks scored by the students in one group.

Marks | 1 | 4 means 14

1	4 4 5
2	0 5 6 7 7 7
3	0 0 1 2 3 3 4 4 4 4 4 4 5 5 6 8 9 9

a How many marks are shown here?   **b** Write down the highest mark.

c Find the median mark   **d** Find the mean mark.

11 This stem-and-leaf diagram shows the number of tomatoes harvested from each plant in Joe's garden one week.

Number of tomatoes | 1 | 3 means 13

0	2 8 9
1	3 3 5 5 5 5 6 6 6 6 6 8 8 9
2	1 3 3 5 5 5 5 7

Find

a the mode   **b** the mean

c the median number of tomatoes harvested.

12 This stem-and-leaf diagram shows the number of telephone calls received by an hotel each day for a month.

Number of calls | 2 | 6 means 26

2	3 5 5 6 6 6 7
3	2 4 4 6 8 8 8 8 9 9 9
4	0 1 2 4 4 4 5 7 8 8

Find

a the modal number of calls.   **b** the median

c the mean number of calls per day.

Finding the mode from a frequency table

The frequency table shows the number of houses in a village that are occupied by different numbers of people:

Number of people living in one house	0	1	2	3	4	5	6
Frequency	2	10	8	15	25	12	4

The highest frequency is 25 so there are more houses with four people living in them than any other number, i.e. the mode is 4.

Finding the mean from a frequency table

The pupils in class 3G were asked to state the number of children in their own family and the following frequency table was made:

Number of children per family	1	2	3	4	5
Frequency	7	15	5	2	1

Total no. of families $= 30$

This information has not been grouped: all the numbers are here so we can total this set. We have seven families with one child giving seven children, fifteen with two children giving thirty children and so on, giving the total number of children as

$$(7 \times 1) + (15 \times 2) + (5 \times 3) + (2 \times 4) + (1 \times 5) = 65$$

There are 30 numbers in the set, so the mean is

$$\frac{65}{30} = 2.2 \text{ (to 1 d.p.)}$$

i.e. there are, on average, 2.2 children per family.

To avoid unnecessary errors, this kind of calculation needs to be done systematically and it helps if the frequency table is written vertically.

We can then add a column for the number of children in each group and sum the numbers in this column for the total number of children.

Number of children per family x	Frequency f	fx
1	7	7
2	15	30
3	5	15
4	2	8
5	1	5
	No. of families = 30	No. of children = 65

$$\text{mean} = \frac{65}{30} = 2.2 \text{ (to 1 d.p.)}$$

Exercise 24c

1

Number of tickets bought per person for a football match	1	2	3	4	5	6	7
Frequency	250	200	100	50	10	3	1

Find the mean number of tickets bought per person.

2

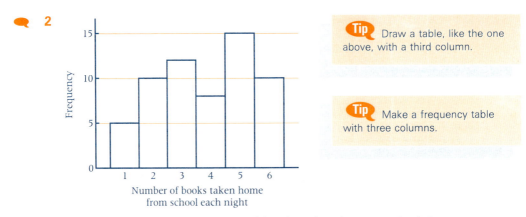

Tip Draw a table, like the one above, with a third column.

Tip Make a frequency table with three columns.

Find the mean for the number of books taken home each night.

3 This table shows the results of counting the number of prickles per leaf on 50 rose trees.

Number of prickles	1	2	3	4	5	6
Frequency	4	2	8	7	20	9

Find
a the mean number of prickles per leaf   b the mode.

4 A six-sided die was thrown 50 times. The table gives the number of times each score was obtained.

Score	1	2	3	4	5	6
Frequency	7	8	10	8	5	12

Find
a the mean score per throw   b the mode.

5 Three coins were tossed together 30 times and the number of heads per throw was recorded.

Number of heads	0	1	2	3
Frequency	3	12	10	5

Find
a the mean number of heads per throw
b the mode.

Finding the median from a frequency table

Exercise 24d

1 A group of students gathered this information about themselves.

Number of children in each family	Frequency
1	8
2	12
3	4
4	2

Find the median number of children per family.

Tip You want to find the middle value.

First find the number of families (add up the frequencies).

You can find where in the order the middle value is by adding 1 to the total, then dividing this by 2.

This tells you which family or families you want.

Is it in the first 8 families? This would give a median of 1 child.

Is it in the next 12 families, i.e. 9 to 20?

2 Once every five minutes, Debbie counted the number of people queuing at a checkout. Her results are shown in this table.

Number of people queuing at a supermarket checkout	Frequency
0	4
1	6
2	5
3	2
4	2

Write down the median number of people queuing.

3 This frequency table shows the distribution of scores when a die is rolled 20 times.

Score	1	2	3	4	5	6
Frequency	3	2	5	3	3	4

Find the median score.

4 In a shooting competition a competitor fired 50 shots at a target and got the following scores

Score	1	2	3	4	5
Frequency	3	4	18	16	9

Find

a the median score **b** the mode **c** the mean.

5 The table shows the distribution of goals scored by the home teams one Saturday.

Score	0	1	2	3	4	5
Frequency	3	8	4	3	5	2

Find

a the median score **b** the mode **c** the mean.

 Puzzle

A man left an estate of $604 500 to be divided among his widow, 4 sons and 5 daughters. He directed that every daughter should have three times as much as a son and that every son twice as much as their mother. What was the widow's share?

Investigation

a Count the number of letters in the surname of each of the teachers in your school.

Enter this information in a frequency table like this.

Number of letters in surname	3	4	5	6	7	8		
Frequency								

You do not have to start your table with 3. Start with the number of letters in the shortest surname. Go up as far as you need to.

b Now find, from these data,

i the mode   **ii** the median   **iii** the mean.

c Repeat this investigation for the students in your class. Compare the mode, median and mean for your class with that for the teachers. Are the values similar or quite different?

IN THIS CHAPTER...

you have seen that:

- stem-and-leaf diagrams preserve individual details, unlike grouped frequency tables, bar charts and pie charts

- when the stem gives the number of tens, the leaves give the units; when the stem gives the number of hundreds, the leaves give the tens and units

- you find the mean by adding all the values then dividing this sum by the number of values

- the mode is the value or values that occur most often, i.e. with the greatest frequency

- the median is the middle value after a set of values has been arranged in order of size; when there are two middle values the median is half-way between them

- the mean, mode and median can all be found from a stem-and-leaf diagram and from an ungrouped frequency table.

25 SETS

Did you know that Venn Diagrams are named after John Venn (1834–1923), an Englishman born in Yorkshire who studied logic at Cambridge University?

Set notation

A *set* is a collection of things having something in common.

Things that belong to a set are called *members* or *elements*. These elements are usually separated by commas and written down between curly brackets or braces.

Instead of writing 'the set of Jamaican reggae stars', we write {Jamaican reggae stars}.

The symbol $\in$ means 'is a member of' so that 'History is a member of the set of school subjects' may be written History $\in$ {school subjects}

Similarly the symbol $\notin$ means 'is not a member of'.

'Elm is not a breed of dog' may be written Elm $\notin$ {breeds of dogs}

Exercise 25a

1   Use the correct set notation to write down the following sets.

   a   the set of teachers in my school   b   the set of books I have read.

2   Write down two members from each of the sets given in question **1**.

   Describe in words the set {2, 4, 6, 8, 10, 12}.

   {2, 4, 6, 8, 10, 12} = {even numbers from 2 to 12 inclusive}

3   Write down in words the given sets:

   a   {1,3,5,7,9}

   b   {Monday, Tuesday, Wednesday, Thursday, Friday}
   Note that these descriptions must be
   very precise, e.g. it is correct to say
   {1, 2, 3, 4, 5} = {first five natural numbers}
   but it is incorrect to say
   {alsation, boxer} = {breeds of dogs}
   because there are many more breeds than the two that are given.

4   Describe a set that includes the given members of the following sets and state another member of each.

   a   Hungary, Poland, Slovakia, Bulgaria   b   10, 20, 30, 40, 50

Write each of the following statements in set notation.

5   John is a member of the set of boys' names.

6   English is a member of the set of school subjects.

7   June is not a day of the week.

8   Monday is not a member of the set of domestic furniture.

State whether the following statements are true or false.

9 32 ∈ { odd numbers }

10 Washington ∈ {American states}

11 Washington ∈ {capital cities}

12 1 ∉ {prime numbers}

Finite, infinite, equal and empty sets

When we can write down all the members of a set, the set is called a *finite set*, e.g. A = {days of the week} is a finite set because there are seven days in a week. If we denote the number of members in the set A by $n(A)$, then $n(A) = 7$.

Similarly if B = {5, 10, 15, 20, 25, 30}, $n(B) = 6$ and if C = {letters in the alphabet}, $n(C) = 26$.

If there is no limit to the number of members in a set, the set is called an *infinite set* e.g. {even numbers} is an infinite set because we can go on adding 2 time and time again.

Two sets are *equal* if they contain exactly the same elements, not necessarily in the same order,

e.g. if A = {prime numbers greater than 2 but less than 9}

and B = {odd numbers between 2 and 8}

then $A = B$, i.e. they are equal sets.

A set that has no members is called an *empty* or *null* set. It is denoted by ∅ or { }.

Exercise 25b

Are the following sets finite or infinite sets?

1 {odd numbers}

2 {the number of leaves on a particular tree}

3 {trees more than 60 m tall}

4 {the decimal numbers between 0 and 1}

Find the number of elements in each of the following sets.

5 A = {vowels}

6 C = {prime numbers less than 20}

If $n(A)$ is the number of elements in set A, find $n(A)$ for each of the following sets.

7 A = {5, 10, 15, 20, 25, 30}

8 A = {the consonants }

9 A = {players in a soccer team}

State whether or not the following sets are equal.

10 $A = \{8, 4, 2, 12\}$, $B = \{2, 4, 6, 8\}$

11 $C = \{$letters of the alphabet except consonants $\}$, $D = \{i, o, u, a, e\}$

12 $X = \{$integers between 2 and 14 that are exactly divisible by 3 or 4$\}$,
$Y = \{3, 4, 6, 8, 9, 12\}$

Determine whether or not the following sets are null sets.

13 {animals that have travelled in space}

15 {prime numbers less than 2 }

14 {multiples of 11 between 12 and 20}

16 {consonants}

Universal sets

Think of the set {pupils in my class}.

With this group of pupils in mind we might well think of several other sets,

i.e. $A = \{$pupils wearing spectacles$\}$

$B = \{$pupils wearing brown shoes$\}$

$C = \{$pupils with long hair$\}$

$D = \{$pupils more than 150 cm tall$\}$

We call the set {pupils in my class} a *universal set* for the sets A, B, C and D.

All the members of A, B, C and D must be found in a universal set, but a universal set may contain other members as well.

We denote a universal set by U or $\mathscr{E}$.

{pupils in my year at school} or {pupils in my school} would also be suitable universal sets for the sets A, B, C and D given above.

Exercise 25c   Suggest a universal set for {5, 10, 15, 20} and {6, 18, 24}

$U = \{$integers$\}$

In questions **1** to **3** suggest a universal set for:

1 {knife, dessert spoon}, {fork, spoon}

2 {10, 20, 30, 40}, {15, 25, 35}

3 {8, 12, 16, 20, 24}, {9, 12, 15, 18, 21, 24}

4 $U = \{$integers from 1 to 20 inclusive$\}$

$A = \{$prime numbers$\}$ $B = \{$multiples of 3 $\}$

Find $n(A)$ and $n(B)$.

5 U = {positive integers less than 16}

A = {factors of 12}   B = {prime numbers}

C = {integers that are exactly divisible by 2 and by 3}

List the sets A, B and C.

6 $U = \{x,$ a whole number, such that $4 \leqslant x \leqslant 20\}$

A = {multiples of 5}   B = {multiples of 7}   C = {multiples of 4}

Find $n(A)$, $n(B)$ and $n(C)$.

Subsets

If all the members of a set B are also members of a set A. then the set B is called a *subset* of the set A. This is written $B \subseteq A$. We use the symbol $\subseteq$ rather than $\subset$ if we don't know whether B could be equal to A.

Subsets that do not contain all the members of A are called *proper subsets*. If B is such a subset we write $B \subset A$.

Exercise **25d**

If A = {David, Edward, Fritz, Harold}, write down all the subsets of A with exactly three members.

The subsets of A with exactly three members are

{David, Edward, Fritz}

{David, Edward, Harold}

{David, Fritz, Harold}

{Edward, Fritz, Harold}

1 If A = {John, Jill, Peter, Audrey, Janet}, write down all the subsets of A with exactly two female members.

2 If N = {positive integers from 1 to 15 inclusive}, list the following subsets of N:

A = {even numbers from 1 to 15 inclusive}

B = {prime numbers less than 15}

C = {multiples of 3 that are less than or equal to 15}

Do sets A and B have any element in common?

3 If A = {even numbers from 2 to 20 inclusive}, list the following subsets of A:

B = {multiples of 3 }

C = {prime numbers}

D = {numbers greater than 12}

Puzzle

During the day, because of the heat, the pendulum of a clock lengthens, causing it to gain half a minute during daylight hours. During the night the pendulum cools, causing it to lose one-third of a minute. The clock shows the correct time at dawn on the first of August. When will it be five minutes fast?

Venn diagrams

In the Venn diagram the universal set (U) is usually represented by a rectangle and the subsets of the universal set by circles within the rectangle.

If U = {families}, A = {families with one car} and B = {families with more than one car} the Venn diagram would be as shown.

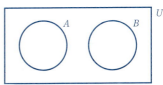

No family can have just one car and, at the same time, more than one car,

i.e. A and B have no members in common.

Two such sets are called *disjoint sets*.

Exercise 25e

1

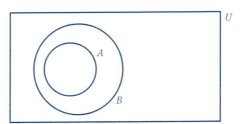

You are given the following information. U = {pupils in my year}

A = {pupils in my class who are my friends} B = {pupils in my class}

a Shade the region that shows the pupils in my class that are not my friends

b Are all my friends in my class?

For each of questions **2** to **5** draw the diagram given below.

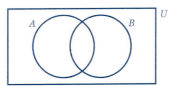

In questions **2** to **5**

U = {pupils who attend my school}

A = {pupils who like coming to my school},

B = {pupils who are my friends}

In each case describe, in words, the shaded area.

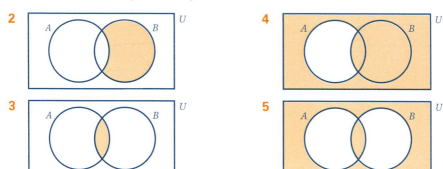

2

3

4

5

Union and intersection of two sets

If we write down the set of all the members that are in either set A or set B we have what we call the *union* of the sets A and B.

The union of A and B is written $A \cup B$.

The set of all the members that are members both of set A and of set B is called the *intersection* of A and B, and is written $A \cap B$.

Exercise 25f

U = {1, 2, 3, 4, 5, 6, 7, 8}

If A = {2, 4, 6, 8} and B = {1, 2, 3, 4, 5} find $A \cup B$ illustrating these sets on a Venn diagram.

$A \cup B$ = {1, 2, 3, 4, 5, 6, 8 }

We could show this on a Venn diagram as follows.

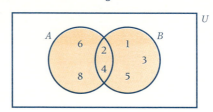

The shaded area represents the set $A \cup B$

In questions **1** to **3** find the union of the two given sets, illustrating your answer on a Venn diagram.

1 U = {girls' names beginning with the letter J}

A = {Janet, Jill, Jamila} B = {Judith, Janet, Jacky }

2 $U = \{$positive integers from 1 to 16 inclusive$\}$

$X = \{4, 8, 12, 16\}$ $\qquad$ $Y = \{2, 6, 10, 14, 16\}$

3 $U = \{$letters of the alphabet$\}$

$P = \{$letters in the word GEOMETRY$\}$

$Q = \{$letters in the word TRIGONOMETRY$\}$

4 Draw suitable Venn diagrams to show the unions of the following sets, and describe these unions in words as simply as possible.

a $U = \{$quadrilaterals$\}$ $\quad$ $A = \{$parallelograms$\}$ $\quad$ $B = \{$trapeziums$\}$

b $U = \{$angles$\}$ $\qquad$ $P = \{$obtuse angles$\}$ $\qquad$ $Q = \{$reflex angles$\}$

$U = \{$integers from 1 to 12 inclusive$\}$

If $A = \{1, 2, 3, 4, 5, 6, 7, 8\}$ and $B = \{1, 2, 3, 5, 7, 11\}$ find $A \cap B$ and show it on a Venn diagram.

$A \cap B = \{1, 2, 3, 5, 7\}$

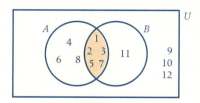

The shaded area represents the set $A \cap B$.

Draw suitable Venn diagrams to show the intersections of the following sets. In each case write down the intersection in set notation.

5 $U = \{$integers from 4 to 12 inclusive$\}$

$X = \{4, 5, 6, 7, 10\}$ $\quad$ $Y = \{5, 7, 11\}$

6 $U = \{$colours of the rainbow$\}$

$A = \{$red, orange, yellow$\}$ $B = \{$blue, red, violet$\}$

7 $U = \{$positive whole numbers$\}$

$C = \{$positive whole numbers that divide exactly into 24$\}$

$D = \{$positive whole numbers that divide exactly into 28 $\}$

8 $U = \{$integers less than 25$\}$

$A = \{$multiples of 3 between 7 and 23$\}$

$B = \{$multiples of 4 between 7 and 23$\}$

Simple problems involving Venn diagrams

Exercise 25g

If U = {girls in my class}

A = {girls who play netball} = {Helen, Bina, Moira, Sara, Lana} and
B = {girls who play tennis} = {Kath, Sara, Helen, Maria}

Illustrate A and B on a Venn diagram. Use this diagram to write down the following sets:

a {girls who play both netball and tennis}

b {girls who play netball but not tennis}

c If $n(U) = 30$ find the number of girls who play neither netball nor tennis.

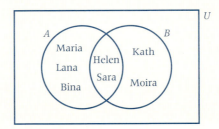

From the Venn diagram

a {girls who play both netball and tennis} = {Helen, Sara}

b {girls who play netball but not tennis} = {Moira, Lana, Bina}

c n(girls who play neither netball nor tennis) = $30 - 7 = 23$

1 U = {the pupils in a class}

X = {pupils who like history}

Y = {pupils who like geography}

List the set of pupils who

a like history but not geography

b like geography but not history

c like both subjects.

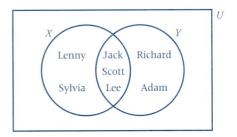

2 U = {boys in my class}

A = {boys who play soccer}

B = {boys who play rugby}

Write down the sets of boys who

a play soccer

b play both games

c play rugby but not soccer.

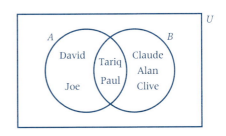

3 $U = \{\text{my friends}\}$

$P = \{\text{friends who wear glasses}\}$

$Q = \{\text{friends who wear brown shoes}\}$

List all my friends who

a wear glasses

b wear glasses but not brown shoes

c wear both glasses and brown shoes.

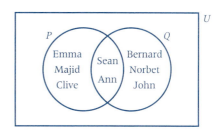

4 $U = \{\text{whole numbers from 1 to 14 inclusive}\}$

$A = \{\text{even numbers between 3 and 13}\}$

$B = \{\text{multiples of 3 between 1 and 14}\}$

Illustrate this information on a Venn diagram and hence write down

a the even numbers between 3 and 13 that are multiples of 3

b $n(A)$ and $n(B)$.

5 $U = \{\text{ letters of the alphabet}\}$

$P = \{\text{different letters in the word SCHOOL}\}$

$Q = \{\text{different letters in the word SQUASH}\}$

Show these on a Venn diagram and hence write down

a $n(P)$

b $n(P \cup Q)$

c $n(P \cap Q)$

> $U = \{\text{months of the year}\}$
>
> $A = \{\text{months of the year beginning with the letter J}\}$
>
> $B = \{\text{months of the year ending with the letter Y}\}$
>
> **a** Find $n(U)$, $n(A)$ and $n(B)$
>
> Hence find
>
> **b** $n(A \cap B)$
>
> **c** $n(A \cup B)$

> **a** $n(U) = 12$ (there are 12 months in a year)
> $A = \{\text{January, June, July}\}$ so $n(A) = 3$
> $B = \{\text{January, February, May, July}\}$ so $n(B) = 4$

b We can illustrate these sets with a Venn diagram using the numbers in each region, rather than the members.

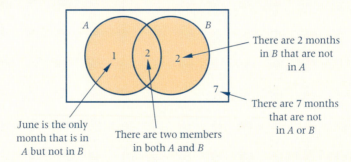

June is the only month that is in A but not in B

There are two members in both A and B

There are 2 months in B that are not in A

There are 7 months that are not in A or B

This shows that $n(A \cap B) = 2$

c $n(A \cup B) = 5$

Alternatively, we know that $A \cup B$ is the set of months in both A and B. However two months, January and July are in both A and B. This means that we cannot find $n(A \cup B)$ just by adding $n(A)$ and $n(B)$, because that includes the two months in $(A \cap B)$ twice.

Hence $\qquad n(A \cup B) = n(A) + n(B) - n(A \cap B)$
$$= 3 + 4 - 2 = 5$$

For any two sets, A and B, $\quad n(A \cup B) = n(A) + n(B) - n(A \cap B)$

6 $U = \{\text{letters of the alphabet}\}$

$P = \{\text{letters used in the word LIBERAL}\}$

$Q = \{\text{letters used in the word LABOUR}\}$

a Find $n(U)$, $n(P)$ and $n(Q)$

b Show these on a Venn diagram.

Hence find **i** $n(P \cap Q)$ **ii** $n(P \cup Q)$ describing each of these sets.

7 $U = \{\text{counting numbers less than 12}\}$

$C = \{\text{prime numbers}\}$ $\qquad D = \{\text{odd numbers }\}$

a Find $n(U)$, $n(C)$ and $n(D)$

b Show these on a Venn diagram

Hence find **i** $n(C \cap D)$ **ii** $n(C \cup D)$

8 $U = \{\text{whole numbers from 1 to 35 inclusive}\}$

$R = \{\text{multiples of 4}\}$ $\qquad S = \{\text{multiples of 6}\}$

a Find $n(U)$, $n(R)$ and $n(S)$

b Find **i** $n(R \cap S)$ **ii** $n(R \cup S)$

9 A and B are two sets such that $n(A) = 8$, $n(B) = 5$ and $n(A \cap B) = 3$. Find $n(A \cup B)$.

Investigation

Ask any 12 members of your class these questions:

Do you swim?   Do you play cricket?   Do you play football?

Now write the following sets

$A = \{$ pupils in my class who swim$\}$

$B = \{$ pupils in my class who play cricket$\}$

$C = \{$ pupils in my class who play football$\}$

Now write each name in the correct place in this Venn diagram.

For example a classmate who swims and plays cricket but does not play football goes in the region that is inside circle A, inside circle B but outside circle C. This is marked with a ×.

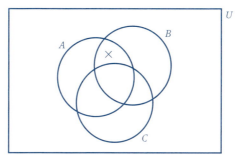

Write down a possible universal set.

Are there any empty sets? If there are, write a sentence to explain what each one means.

IN THIS CHAPTER...

you have seen that:

- an infinite set has no limit on the number of members in it

- in a finite set, all the members can be counted or listed

- the intersection of two sets contains the elements that are in both sets.

- a proper subset of a set A contains some, but not all, of the members of A

- the union of two sets contains all the members of the first set together with the members of the second set that have not already been included

- when two sets have exactly the same members, they are said to be equal.

- a set that has no members is called an empty or null set and is written $\{\}$ or $\varnothing$

26 NUMBER BASES

AT THE END OF THIS CHAPTER...

you should be able to:

1   Use markers to represent groups of fives or powers of fives for a given number.

2   Write in figures the numbers represented by markers under the headings of five and powers of five.

3   Write, in headed columns, numbers to a given base.

4   Write, in base ten, numbers given in other bases.

5   Write a number given in base ten as a number to another given base.

6   Perform operations of addition, subtraction and multiplication in bases other than ten.

7   Determine the base in which given calculations have been done.

Time for another library search!

Find out what you can about hexadecimal numbers and why they are important.

BEFORE YOU START — you need to know:
✓ your multiplication tables – and this means instant recall

KEY WORDS — base, binary, denary system, number base

Denary system (base ten)

We have ten fingers. This is probably why we started to count in tens and developed a system based on ten for recording large numbers. For example

$$3125 = 3 \text{ thousands} + 1 \text{ hundred} + 2 \text{ tens} + 5 \text{ units}$$
$$= 3 \times 10^3 + 1 \times 10^2 + 2 \times 10^1 + 5 \times 10^0$$

Each column is ten times the value of its right-hand neighbour. The base of this number system is ten and it is called the *denary system*.

Base five

If man had started to count using just one hand, we would probably have a system based on five.

Suppose we had eleven stones. Using one hand to count with, we could arrange them like this:

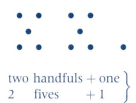

i.e.   two handfuls + one ⎫
or   2 fives   + 1 ⎭

The next logical step is to use a single marker to represent each group of five. We also need to place these markers so that they are not confused with the marker representing the one unit. We do this by having separate columns for the fives and the units, with the column for the fives to the left of the units.

Fives   Units

We can write this number as 21_5 and we call it 'two one to the base five'. We do *not* call it 'twenty-one to the base five', because the word 'twenty' means 'two tens'.

To cope with larger numbers, we can extend this system by adding further columns to the left such that each column is five times the value of its right-hand neighbour. Thus, in the column to the left of the fives column, each marker is worth twenty-five, or 5^2.

twenty-fives fives units

For example

The markers here represent 132_5 and it means

$$(1 \times 5^2) + (3 \times 5) + 2$$

Exercise 26a Write in figures the numbers represented by the markers:

5^3	5^2	5	Units
	●		
	●		●
●	●		●

The number is 1302_5

(The '5' column is empty, so we write zero in this column.)

Write in figures the numbers represented by the markers in questions **1** to **4**:

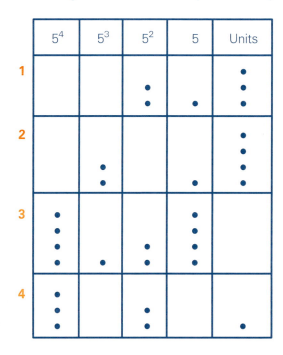

	5^4	5^3	5^2	5	Units
1			●●	●	●●●
2		●●		●	●●●●
3	●●●●	●	●●	●●●	
4	●●●		●●		●

Write 120_5 in headed columns.

5^2	5	Units
1	2	0

Write the following numbers in headed columns.

5 31_5 **7** 410_5 **9** 34_5 **11** 204_5

6 42_5 **8** 231_5 **10** 10_5 **12** 400_5

Write 203_5 as a number to the base 10

$$203_5 = (2 \times 5^2) + (0 \times 5) + 3$$
$$= 50_{10} \qquad + \qquad 3_{10}$$
$$= 53_{10}$$

(Although we do not normally write fifty-three as 53_{10}, it is sensible to do so when dealing with other bases as well.)

Write the following numbers as denary numbers, i.e. to base 10:

13 31_5 **16** 121_5 **19** 32_5 **22** 400_5

14 24_5 **17** 204_5 **20** 20_5 **23** 240_5

15 40_5 **18** 43_5 **21** 4_5 **24** 300_5

Write 38_{10} as a number to the base 5

(To write a number to the base 5 we have to find how many ... 125s, 25s, 5s and units the number contains.)

Method 1 (Starting with the highest value column.)

38 contains no 125s.

$\qquad 38 \div 25 = 1$ remainder 13   i.e. $38 = 1 \times 5^2 + 13$

$\qquad 13 \div 5 = 2$ remainder 3   i.e. $13 = 2 \times 5 + 3$

$\therefore \quad 38 = 1 \times 5^2 + 2 \times 5 + 3$

$\qquad = 123_5$

Method 2 (Starting with the units.)

```
5)38
5) 7    remainder 3 (units)
5) 1    remainder 2 (fives)
   0    remainder 1 (twenty-fives)
```

$\therefore \qquad\qquad\qquad 38 = 123_5$

Write the following numbers in base 5:

25 8_{10} **28** 39_{10} **31** 7_{10} **34** 128_{10}

26 13_{10} **29** 43_{10} **32** 21_{10} **35** 82_{10}

27 10_{10} **30** 150_{10} **33** 30_{10} **36** 100_{10}

Other bases

Any number can be used as a base for a number system.

If the base is 6, we write the number in columns such that each column is 6 times the value of its right-hand neighbour. For example, writing 253_6 in headed columns gives

6^2	6	Units
2	5	3

We see that 253_6 means $(2 \times 6^2) + (5 \times 6) + 3$

Similarly 1011_2 means $(1 \times 2^3) + (0 \times 2^2) + (1 \times 2) + 1$

Exercise 26b Write 425_7 in headed columns and then write it as a denary number.

7^2	7	Units
4	2	5

$\therefore$

$$425_7 = (4 \times 7^2) + (2 \times 7) + 5$$
$$= 196_{10} + 14_{10} + 5_{10}$$
$$= 215_{10}$$

Write the following numbers

a in headed columns   **b** as denary numbers

1 23_4 **5** 57_8 **9** 21_3 **13** 303_4

2 15_7 **6** 204_5 **10** 18_9 **14** 1001_2

3 131_4 **7** 210_3 **11** 24_6 **15** 1211_3

4 101_2 **8** 574_9 **12** 175_8 **16** 1000_6

Write 29_{10} as a number

a to the base 9

b to the base 2

a

$9\,)\,29$
$9\,)\,3\ \ \text{r.}\ \ 2\ (\text{units})$
$\ \ \ \ 0\ \ \text{r.}\ \ 3\ (\text{nines})$

$\therefore$ $29_{10} = 32_9$

b

$2\,)\,29$
$2\,)\,14\ \ \text{r.}\ \ 1\ (\text{unit})$
$2\,)\,7\ \ \text{r.}\ \ 0\ (\text{twos})$
$2\,)\,3\ \ \text{r.}\ \ 1\ (2^2)$
$2\,)\,1\ \ \text{r.}\ \ 1\ (2^3)$
$\ \ \ \ 0\ \ \text{r.}\ \ 1\ (2^4)$

$\therefore$ $29_{10} = 11101_2$

Write the following denary numbers to the base indicated in brackets:

17	9 (4)		**21**	13 (5)		**25**	8 (3)
18	12 (5)		**22**	32 (6)		**26**	15 (6)
19	24 (7)		**23**	53 (8)		**27**	34 (9)
20	7 (2)		**24**	49 (7)		**28**	28 (3)
29	56 (7)		**33**	163 (8)		**37**	43 (6)
30	89 (9)		**34**	640 (4)		**38**	55 (5)
31	45 (2)		**35**	142 (2)		**39**	99 (2)
32	333 (3)		**36**	158 (6)		**40**	394 (7)

Express each of the following numbers as a number to the base indicated in brackets:

41	45_6 (4)	**44**	432_5 (3)	**47**	11011_2 (8)
42	23_4 (6)	**45**	562_8 (4)	**48**	378_9 (3)
43	17_8 (2)	**46**	2120_3 (5)	**49**	3020_4 (8)

Addition, subtraction and multiplication

Numbers with a base other than 10 do not need to be converted to base 10; provided that they have the same base they can be added, subtracted and multiplied in the usual way, as long as we remember which base we are working with.

For example, to find $132_5 + 44_5$ we work in fives, not tens. To aid memory, the numbers can be written in headed columns:

Twenty-fives	Fives	Units	
1	3	2	
①	4 ①	4	+
2	3	1	
	⑧	⑥	

Adding the units gives 6 units:

$$6 \,(\text{units}) = 1\,(\text{five}) + 1\,(\text{unit})$$

We put 1 in the units column and carry the single five to the fives column.

Adding the fives gives 8 fives

$$8\,(\text{fives}) = 5\,(\text{fives}) \qquad + 3\,(\text{fives})$$
$$= 1\,(\text{twenty-five}) \quad + 3\,(\text{fives})$$

We put 3 in the fives column and carry the 1 to the next column.

Adding the numbers in the last column gives 2,

i.e. $132_5 + 44_5 = 231_5$

If the numbers are not to the same base we cannot add them in this way.

For example, 432_7 and 621_8 cannot be added directly.

Exercise 26c   Find $174_8 + 654_8$

8^3	8^2	8	Units	
	1	7	4	
①	6 ①	5 ①	4	+
1	0	5	0	
	⑧	⑬	⑧	

$\therefore$ $174_8 + 654_8 = 1050_8$

Find:

1 $12_5 + 31_5$

2 $11_3 + 2_3$

3 $11_4 + 13_4$

4 $10_2 + 1_2$

5 $24_6 + 35_6$

6 $21_3 + 11_3$

7 $43_8 + 52_8$

8 $101_2 + 11_2$

9 $43_5 + 24_5$

10 $132_4 + 201_4$

11 $345_6 + 402_6$

12 $1101_2 + 111_2$

13 $122_3 + 101_3$

14 $231_4 + 103_4$

15 $635_7 + 62_7$

16 $10010_2 + 1111_2$

> **Tip** Make sure that you are clear about which base you are working in.

Find $132_4 - 13_4$

(There are two methods of doing subtraction and we show both of them here.)

First method

4^2	4	Unit	
1	$\cancel{3}^{②}$	$\cancel{2}^{⑥}$	
	1	3	−
1	1	3	

(We cannot take 3 from 2 so we take one 4 from the fours' column, change it to 4 units and add it to the 2 units.)

Second method

If you use the 'pay back' method of subtraction, the calculation looks like this:

4^2	4	Unit
1	3	$2^{⑥}$
	$1_{①}$	3	—
1	1	3

In either case $\quad 132_4 - 13_4 = 113_4$

Find:

17	$153_6 - 24_6$	**21**	$210_3 - 1_3$	**25**	$231_4 - 32_4$	**29**	$144_6 - 53_6$
18	$110_3 - 2_3$	**22**	$30_5 - 14_5$	**26**	$153_7 - 64_7$	**30**	$1010_2 - 101_2$
19	$32_4 - 23_4$	**23**	$253_8 - 25_8$	**27**	$205_6 - 132_6$	**31**	$724_8 - 56_8$
20	$52_7 - 14_7$	**24**	$10_2 - 1_2$	**28**	$100_2 - 10_2$	**32**	$120_3 - 12_3$

Find $352_6 \times 4_6$

6^3	6^2	6	Units
	3	5	2
②	③	①	4 $\quad\times$
2	3	3	2

$\quad$ $15 = 2 \times 6 + 3$ $\quad$ $21 = 3 \times 6 + 3$ $\quad$ $8 = 6 + 2$

$\therefore \quad\quad\quad\quad\quad 352_6 \times 4_6 = 2332_6$

Find:

33	$4_5 \times 3_5$	**37**	$5_6 \times 4_6$	**41**	$132_4 \times 2_4$
34	$2_4 \times 3_4$	**38**	$13_5 \times 2_5$	**42**	$501_6 \times 5_6$
35	$12_3 \times 2_3$	**39**	$24_7 \times 3_7$	**43**	$202_3 \times 2_3$
36	$20_4 \times 3_4$	**40**	$56_9 \times 3_9$	**44**	$241_5 \times 2_5$

Find:

45	$261_7 + 123_7$	**50**	$232_4 - 103_4$
46	$32_4 \times 2_4$	**51**	$22_5 \times 4_5$
47	$434_5 - 142_5$	**52**	$365_8 + 173_8$
48	$36_7 \times 2_7$	**53**	$121_3 - 112_3$
49	$451_6 + 124_6$	**54**	$34_8 \times 5_8$

Harder examples

Calculate $133_4 \times 32_4$

First place the number in headed columns, then use long multiplication, i.e. multiply by 2_4 and then by 30_4:

4^4	$4^3(=64)$	$4^2(=16)$	fours	units
		1	3	3
			3	2
$\times 2$		3	3	2
		$(2+1=3)$	$(6+1=7=4+3)$	$(6=4+2)$
$\times 30$ 1	1	3	1	0
	$(3+2=5=4+1)$ $(9+2=11=2\times4+3)$		$(9=2\times4+1)$	
$+$ 1	2	3	0	2

$$113_4 \times 32_4 = 12302_4$$

Find:

1 $123_4 \times 23_4$

2 $413_5 \times 24_5$

3 $1001_2 \times 1101_2$

4 $73_8 \times 26_8$

5 $2120_3 \times 212_3$

6 $46_8 \times 35_8$

7 $234_5 \times 423_5$

8 $452_7 \times 324_7$

9 **a** Find $64_8 \times 27_8$ as a number to the base 8.

   b Express 64_8 and 27_8 as denary numbers.

   c Multiply together your two answers for **b**.

   d Change your answer to **c** into a number to the base 8. Does this answer agree with your answer to **a**?

10 **a** Find $476_9 \times 57_9$ as a number to the base 9.

   b Express 476_9 and 57_9 as denary numbers.

   c Multiply together your two answers for **b**.

   d Change your answer to **c** into a number to the base 9. Does this answer agree with your answer to **a**?

11 Find $55_8 \times 43_8$ and use the process described in questions 9 and 10 as a check on your working.

12 Choose a base and make up a long multiplication question of your own. Check your calculation using the process above.

Binary numbers

Numbers with a base of two are called *binary numbers*. We have singled binary numbers out for special attention because of the wide application that they have, especially in the world of computers.

Exercise 26e

1 If you have access to a microcomputer and have done some programming using machine code, or have copied program listings from magazines, you will have seen instructions such as

BIN 1101, 1011, 11001, 100, 1100, 11101

'BIN' means 'binary'.

Convert the binary numbers given above to denary numbers.

2 Basically, computers are very simple; their fundamental computing parts can only be off (0) or on (1), i.e. computers count in binary numbers.

+	0	1
0		
1		

   a Complete the adjacent addition table for binary numbers.

   b Find $1011011_2 + 110101_2$.

3 a Subtract 23_{10} as many times as you can from 138_{10}. Hence write down the value of $138_{10} \div 23_{10}$.

   b Subtract 11_2 from 1111_2 as many times as you can. Hence find $1111_2 \div 11_2$.

4 Complete the following multiplication table for binary numbers.

×	0	1
0		
1		

 Investigation

   a Convert 7_{10}, 5_{10}, 10_{10}, 16_{10}, 19_{10}, 24_{10}, into binary numbers.

   b How many symbols are needed to represent all binary numbers?

   c Write each of the denary numbers 1 to 20 as binary numbers.

   d How can you see when a binary number is even?

   e How can you see when a binary number is odd?

Mixed exercises

Exercise 26f

1 Convert the following denary numbers to base 3 numbers:

   a 5   **b** 8   **c** 12   **d** 31

2 How many different symbols are needed to represent all numbers in base three?

3 Make up a multiplication table for base three numbers.

4 Use repeated subtraction to find the value of $1111_3 \div 101_3$.

5 Convert the following denary numbers to base 5 numbers:

   a 27   **b** 18   **c** 153

6 What do you think 31.2_5 could mean?

<u>7</u> If a number to the base 10 ends in 0, what does the same number to the base 5 end in?

<u>8</u> Is it possible to write a number in base one?

<u>9</u> How many digits are there in 5^3 written in base 5?

<u>10</u> How many digits are there in 3^7 written in base 3?

<u>11</u> **a** Convert 10_2, 10_3, 10_4, 10_5, 10_6, 10_7, 10_8 into denary numbers.

   b Find (i) $1101_2 \times 10_2$ (ii) $121_3 \times 10_3$ (iii) $175_8 \times 10_8$

   c What is the effect of multiplying a number by the base number?

Find the base in which the following calculation has been done: $13 + 5 = 22$

$$\begin{array}{r} 13 \\ 5 + \\ \hline 22 \\ \hline \end{array}$$

From the addition, $5 + 3 = 8$. To leave 2 in the units column, 6 units have been carried to the next column. The total in the next column is 2, so the 6 units have been carried as 1 to the next column.

∴ the base is 6

Find the bases in which the following calculations have been done.

<u>12</u> $15 + 23 = 42$ <u>16</u> $13 - 4 = 4$

<u>13</u> $12 + 13 = 31$ <u>17</u> $21 - 2 = 17$

<u>14</u> $110 + 121 = 1001$ <u>18</u> $13 \times 2 = 31$

<u>15</u> $134 + 213 = 350$ <u>19</u> $21 \times 3 = 103$

<u>20</u> Is the following statement true or false?
   A number written to any base is even if it ends in zero.

Exercise 26g

1 Write the following numbers as denary numbers

   a 12_4 **b** 101_2 **c** 403_6

2 Write the denary number 20 as a number to the base

   a 5   **b** 3   **c** 8

3 Find:

   a $204_5 + 132_5$ **b** $110_3 - 2_3$ **c** $212_8 \times 3_8$

4 There are four possible answers given below to the calculation $213_4 \times 2_4$. Only one answer is correct. Which one is it?

   a 426_4 **b** 21304_4 **c** 1032_4 **d** 221_4

MATHS IS OUT THERE

The JIBARO Indians of the Amazon rain forest express the number 'five' by the phrase 'wehe amukei', meaning 'I have finished one hand', and the number 'ten' by 'mai wehe amukei' meaning 'I have finished both hands'.

IN THIS CHAPTER...

you have seen that:

- numbers can be expressed in any base

- if a number is in base a, the first column from the right is units, the second column gives the number of as, the next column gives the number of a^2s and so on. For example, 165_8 means $1 \times 8^2 + 6 \times 8^1 + 5 \times 8^0$

- you can add, subtract and multiply numbers in a base other than ten in the usual way, but it is sensible to write them in headed columns so that you can keep track of your working.

REVIEW TEST 3: CHAPTERS 18 to 26

In questions **1** to **8**, choose the letter for the correct answer.

1 The mode of the set of cricket scores 54, 30, 60, 54, 18, 62, 30, 54 is

    A 62   **B** 54   **C** 36   **D** 18

2 The scores in six rounds of golf are 73, 75, 80, 84, 73, 73. The median score is

    A 84   **B** 82   **C** 74   **D** 73

Use the diagram below to answer questions 3, 4 and 5.

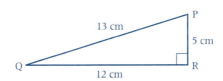

3 Tan $\hat{P}$ =

    A $\frac{5}{13}$ **B** $\frac{12}{5}$ **C** $\frac{12}{13}$ **D** $\frac{13}{12}$

4 Sin $\hat{P}$ =

    A $\frac{5}{13}$ **B** $\frac{5}{12}$ **C** $\frac{12}{13}$ **D** $\frac{13}{12}$

5 Cos $\hat{P}$ =

    A $\frac{5}{13}$ **B** $\frac{5}{12}$ **C** $\frac{12}{13}$ **D** $\frac{13}{12}$

6 The solution of the inequality $8 < 3 - 5x$ is

    A $x > -1$ **B** $x < 1$ **C** $x > 1$ **D** $x < -1$

7 What is the gradient of the line segment joining the points $(-3, 2)$ and $(5, 7)$?

    A $\frac{9}{2}$ **B** $\frac{5}{2}$ **C** $\frac{9}{8}$ **D** $\frac{5}{8}$

8 A line parallel to the x-axis at a distance k away from it has the equation

    A $x = k$ **B** $y = k$ **C** $x + y = k$ **D** $y = x - k$

9 a Find $303_4 + 212_4$

    b Write 45_{10} **i** to base 7

                    ii to base 2

10 State, with brief reasons, whether these two triangles are congruent.

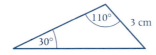 

11 **a** Find the value of t if $(-2, t)$ is on the line $y = 7 - 2x$.

b What is the gradient of the line $2x + 3y = 7$?

c Find the equation of the line with gradient 4 and y-intercept -5.

12 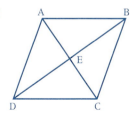 ABCD is a parallelogram in which $A\hat{D}B = B\hat{D}C = 30°$ and $D\hat{A}C = 60°$.

a Prove that triangles ADE and ABE are congruent.

b Prove that triangle ADC is isosceles.

13 **a** Calculate the angle between the diagonal and the longer side of a rectangle with sides 5 cm and 7 cm long.

b A ladder 5 metres long is leaning against a wall at an angle of 30 degrees to the horizontal. Calculate how far up the wall the ladder reaches.

14 Draw a diagram to represent the region defined by the inequalities $1 \leqslant x < 5$, $-2 < y \leqslant 3$.

15 This stem-and-leaf diagram shows the journey to work of a group of factory employees.

Time in minutes   1 | 5 means 15 minutes

```
0 | 7  8  8  9
1 | 0  2  2  5  7  7  8  8
2 | 0  5  6  7  8  8  9
3 | 1  4  4  4  6  7
```

a How many employees are there in the group?

b How long is the longest journey?

c Find the modal time.

d Find the median.

e Find the mean time taken.

16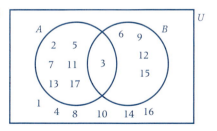

a Describe in words

   i Set A   **ii** Set B

b Find   **i** $n(A \cup B)$ **ii** $n(A \cap B)$

1 The value of $-2 - (-4) + (-3)$ is

   A -9 **B** -7 **C** -1 **D** 5

2 The expression $3(4 - x) - (7 - x)$ simplifies to

   A $5 - 2x$ **B** $5 - 4x$ **C** $-2x$ **D** $5 + 2x$

3 The value of x that satisfies the equation $5 - 2x = 15 - 7x$ is

   A $\frac{20}{9}$ **B** $\frac{1}{2}$ **C** -2 **D** 2

4 The value of $7^4 \times 7^3 \div 7^6$ is

   A $\frac{1}{7}$ **B** 7 **C** 7^2 **D** 1

5 The angles marked p and q are

   A alternate angles

   B corresponding angles

   C vertically opposite angles

   D none of these

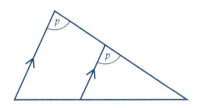

6

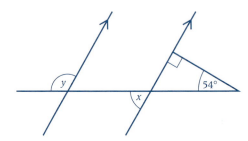

The values of x and y are respectively

   A $54°$ and $126°$ **B** $36°$ and $144°$

   C $36°$ and $164°$ **D** $54°$ and $146°$

7 The largest integer x such that $17 - x > 3x - 11$ is

   A 3 **B** 6 **C** 7 **D** 8

8 The bearing of a ship S, from a lighthouse L, is $235°$. This means that the bearing of the lighthouse from the ship is

   A $125°$ **B** $305°$ **C** $055°$ **D** $215°$

9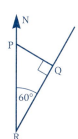

P, Q and R lie on level ground with angle PQR = 90°.

P is due North of R and Q lies on a bearing of 60° from R. The bearing of P from Q is

A 330°   **B** 150°

C 090°   **D** 030°

10 In the right angled-triangle, PQR,

PQ = 6 cm and angle P = 60°.

The length of PR, in cm, is

A 15   **B** 12   **C** 9   **D** 8

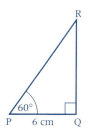

11 In a game, the score occurs with the frequency shown in the table.

x	1	2	4	6	12
Frequency	12	6	3	2	1

The mean score is

A 1   **B** 2.5   **C** 11   **D** 12

12

                                    2 | 7 means 27

```
1 | 0  1  4  4  6  7
2 | 5  5  7  8  8  9  9
3 | 2  4  6  6  6  8
```

The difference between the mode and the median of the data in this stem-and-leaf diagram is

A 2   **B** 4   **C** 6   **D** 8

13 A fair coin and a fair die are tossed. The probability that a 'head' and an 'odd number' be obtained is

A $\frac{1}{12}$ **B** $\frac{1}{4}$ **C** $\frac{1}{2}$ **D** $\frac{2}{3}$

14 Which of the following computations does not have the same result as the others?

A $\sqrt{(2-1)^3}$ **B** $(2-1)^{-2}$ **C** $(1-2)^{-2}$ **D** $(1-2)^3$

15 The lines $y = -3$ and $x = 2$ intersect at the point

A $(-3, 2)$ **B** $(3, -2)$ **C** $(-2, 3)$ **D** $(2, -3)$

16 10% of 20 differs from 0.5 of 2 by

A 1   **B** 2   **C** 4   **D** 5

17 The line ST is parallel to the line

$y = 2x - 1$.

The most likely equation for ST is

A $y = 3x - 3$ **B** $y = 3x + 2$

C $y = x + 2$ **D** $y = 2x + 3$

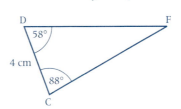

18

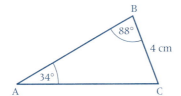

The condition that makes triangles ABC and DEF congruent is

A two sides and the included angle

B three sides

C two angles and a corresponding side

D right angle, hypotenuse and side

19

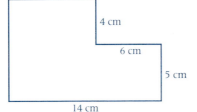

The area of this shape is

A $46 \, \text{cm}^2$

B $94 \, \text{cm}^2$

C $102 \, \text{cm}^2$

D $80 \, \text{cm}^2$

20 If $X = 3a - 4b$ and $Y = a - 5b$ the value of $X - Y$ is

A $2a - 9b$ **B** $4a + b$ **C** $2a - b$ **D** $2a + b$

21 The equation of the straight line which passes through the point (0, 3) with gradient $\frac{1}{2}$ is

A $y = 2x + 3$ **B** $y = 2x + 6$ **C** $2y = x + 3$ **D** $2y = x + 6$

22 The value of x that satisfies the equation $5x - 4 = 2(x + 7)$ is

A 6   **B** 4   **C** 3   **D** 2

23 An item of jewellery appreciates by 5% each year. If its original cost was $2000, its value at the end of two years will be

A $2400   **B** $2205   **C** $2200   **D** $2010

24 The area, in square metres, of a square of side 5 cm is

A 5×10^{-2} **B** 2.5×10^{-3} **C** 2.5×10^{-4} **D** 5×10^{-4}

25 In a certain town of 12 000 people, two-thirds are children and one-half of the remainder are women. A person is chosen at random. The probability that this person is a woman is

A $\frac{2}{3}$ **B** $\frac{1}{2}$ **C** $\frac{1}{4}$ **D** $\frac{1}{6}$

26 $P = \{1, 2, 3, \ldots 10\}$. In the set P, the largest odd number exceeds the largest prime number by

A 0 **B** 1 **C** 2 **D** 3

27 Which one of the following statements defines an infinite set?

A The fifth form students of a certain school.

B Plane figures bounded by three or more straight lines.

C Vehicles that carry passengers.

D Sixth form students who can sing well.

28 Before a sale a store owner raises the price of an item by 10%. In the sale he offers a 10% discount. The cost of the item in the sale is

A the same as before he raised the price.

B less than before he raised the price.

C more than before he raised the price.

29 The order of rotational symmetry of a square is

A 1 **B** 2 **C** 3 **D** 4

30 The square root of 0.000 049 is

A 0.07 **B** 0.007 **C** 0.0022 **D** 0.0007

31 If the formula $P = \dfrac{Q}{5} - 2$ is rearranged to make Q the subject, the expression for Q is

A $5P + 2$ **B** $5P - 10$ **C** $P + 10$ **D** $5P + 10$

32 The statement $9 \times (4 - 3) = 9 \times 4 - 9 \times 3$ is an example of

A the associative law **B** the commutative law

C the distributive law **D** none of these

33 A salesman is given commission of 5% of his weekly sales exceeding $1000. His sales in a certain week are $3500. His commission is

A $20 **B** $125 **C** $175 **D** $50

34 Quadrants of radius 1 cm are cut from each corner of the square ABCD to leave the figure shaded in the diagram.

The area of the shaded region, in cm^2, is

A $4 - \pi$ **B** $1 - \pi$ **C** $8 - 2\pi$ **D** $4 - 2\pi$

35 The range of values for which the inequalities $-3 - x < 2x + 3 \leqslant 9$ are true is

   A $2 < x \leqslant 3$ **B** $0 \leqslant x < 3$ **C** $-2 < x \leqslant 3$ **D** $-6 < x \leqslant 3$

36 If $a = \begin{pmatrix} 7 \\ -3 \end{pmatrix}$ and $b = \begin{pmatrix} -3 \\ -2 \end{pmatrix}$ the value of $3a - 2b$ is

   A $\begin{pmatrix} 15 \\ -13 \end{pmatrix}$ **B** $\begin{pmatrix} 4 \\ -5 \end{pmatrix}$ **C** $\begin{pmatrix} 10 \\ 1 \end{pmatrix}$ **D** $\begin{pmatrix} 27 \\ -5 \end{pmatrix}$

37 The set of inequalities that describe the unshaded region are

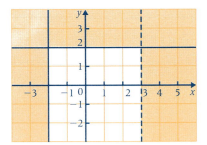

   A $x \geqslant -2, x < 3, y \leqslant 2$

   B $x > -2, x < 3, y < 2$

   C $x > 2, x \leqslant 3, y < 2$

   D $x \geqslant -2, x < 3, y \leqslant 2$

38 In the diagram $AB = BC = BD$.

The transformation that will not transform ABC to BCD is

   A a rotation through $90°$ clockwise about B

   B a reflection in BC

   C rotation of $270°$ anticlockwise about B

   D rotation about C through $90°$ clockwise.

39

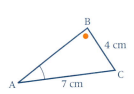
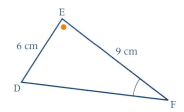

In $\triangle$'s ABC and DEF , $\angle A = \angle F$, $\angle E = \angle B$, $BC = 4\,cm$, $AC = 7\,cm$, $DE = 6\,cm$ and $EF = 9\,cm$. The length of AB is

   A $9\,cm$ **B** $8\,cm$ **C** $7\,cm$ **D** $6\,cm$

40 The vector that translates ABC to DEF is

   A $\begin{pmatrix} -6 \\ 3 \end{pmatrix}$ **B** $\begin{pmatrix} -6 \\ -3 \end{pmatrix}$

   C $\begin{pmatrix} 6 \\ -3 \end{pmatrix}$ **D** $\begin{pmatrix} 6 \\ 3 \end{pmatrix}$

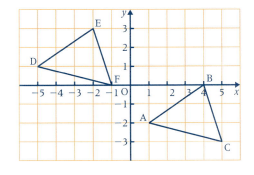

GLOSSARY

acute angle	an angle less than $90°$
alternate angles	equal angles on opposite sides of a transversal, e.g.
amount	the sum of the interest and the principal (the original money invested)
angle of depression	the angle between looking straight ahead and down at an object
angle of elevation	the angle between looking straight ahead and up at an object
angles at a point	a group of angles round a point that make a complete revolution, e.g.
angles on a straight line	a group of angles that together make a straight line, e.g.
annulus	area between two concentric circles
approximation	an estimate of the value of a calculation or quantity
area	the amount of surface covered
arc	part of a curve
associative law	this says that for the same operation brackets can be removed without changing the result, e.g. $2 + (3 + 4)) = (2 + 3) + 4 = 2 + 3 + 4$
axis of symmetry	a line about which a shape is the same on either side
bar chart	a diagram of bars; each represents a quantity. The height of the bar represents the number (frequency) of that quantity
bearing	the direction of one place from another
biased	not equally likely
binary system	a counting system based on 2
bisect	divide into two equal parts
boundary line	a line showing the edge of an area or region
centimetre	a measurement of length
centre (of a circle)	the point that is the same distance from any point on the circumference
centre of enlargement	the point that, with a scale factor, determines the position and size of an image in relation to the object
centre of rotation	the point about which an object is turned to give an image
chord	the straight line joining two points on a curve
circle	a curve made by moving one point at a fixed distance from another
circumcircle	the circle that passes through the vertices of a plane figure
circumference	the edge of a circle
common factor	a number that divides exactly into two or more other numbers
commutative law	the law that says that order does not matter in an operation, e.g. $7 \times 6 = 6 \times 7$
compass direction	a measure of direction using the compass directions N, S, E and W
complementary angles	angles that add up to $90°$
composite number	a number that can be written as the product of two or more prime numbers
compound interest	where the interest for a given period is added to the principal to give the new principal to find the interest for the next period

cone	a pyramid with a circular base
congruent	exactly the same shape and size
construction	drawing a figure exactly
coordinates	an ordered pair of numbers giving the position of a point on a grid
corresponding angles	equal angles on the same side of a transversal, e.g.
cosine of an angle	the ratio of the adjacent side to the hypotenuse in a right-angled triangle
cross-section	the shape formed by a plane cutting a solid
cube	a solid with six faces, each of which is a square, e.g.
cubic unit	a measure of volume
cuboid	a solid with six faces, each of which is a rectangle, e.g.
cylinder	a prism with a uniform circular cross-section
data	a collection of facts
decimal	a fraction expressed by numbers on the right of a point, e.g. 0.2
decimal place	the position of a figure after the decimal point
degree	unit of measure for an angle or for temperature
denary system	a counting system based on 10
denominator	the bottom of a fraction
diagonal	a line from one corner to another in a figure
diameter	a line across a circle going through the centre
difference	the value of the larger number minus the smaller number
digit	one of the figures 0, 1,2,3,4,5,6,7,8,9
dimension	a measurable length in an object
direct proportion	two quantities such that one is always the same multiple of the other
directed numbers	positive and negative numbers collectively
disjoint sets	sets that have no common elements
displacement	the distance and direction of an object from some fixed point
distributive law	the law that says that one operation can be distributed over another, e.g. $3 \times (4 + 5) = 3 \times 4 + 3 \times 5$
enlargement	increasing or decreasing the size of a shape so that all its sides remain in the same proportion
equally likely	events that each have the same chance of happening
equation	two expressions connected by an equals sign
equilateral triangle	a triangle whose sides are all the same length
estimate	an approximate value
event	something that may happen
exchange rate	the number of units of one currency equal to one unit of another
experiment	an action where the outcome is not certain
expression	a collection of algebraic terms connected with plus and minus signs, without an equals sign
factor	a number or letter that divides exactly into another number or algebraic term
fair	all possible outcomes are equally likely
finite set	a set whose members are limited in number
formula	a general rule for expressing one quantity in terms of other quantities

fraction	part of a quantity
frequency	the number of times that a value occurs
gradient	a measure of the slope of a line
gram	a measure of weight
highest common factor	the largest number that divides exactly into two or more other numbers
horizontal	parallel to the surface of the earth
hundredweight	a measure of mass
hypotenuse	the longest side in a right-angled triangle
identity element	a number that preserves the identity of a given number under an operation
image	the resulting shape after a transformation of an object
improper fraction	a fraction whose top is larger than its bottom
incircle	the circle inside a shape that touches all its sides
index (pl. indices)	a superscript to a number that tells you how many of those numbers are multiplied together
inequality	the relationship between two quantities that are not equal
inflation	the increase in prices
intercept	the distance from the origin to where a line crosses an axis
infinite set	a set with an unlimited number of members
integer	a whole number
interest	the money paid for the use of money lent or borrowed
interior angles	angles between a transversal and two parallel lines that add to $180°$, e.g.
intersection of sets	the set of elements common to two or more sets
invariant	a line or point that does not change under a transformation
inverse	the reverse of an operation
isosceles triangle	a triangle with two equal sides
kilogram	a measure of mass
kilometre	a measure of length
kilowatt-hour	a unit of energy
kite	a quadrilateral with two pairs of adjacent sides that are equal, e.g.
like terms	terms that contain the same combination of letters, e.g. $3xy^2$ and $7xy^2$ but not xy^2 and xy
line of symmetry	a line that divides a figure into two identical shapes
line segment	a line with a beginning and an end .
litre	a measure of capacity
lowest common multiple	the lowest number that two or more other numbers divide into
magnitude	the size of a quantity
mass	the quantity of matter in an object
mean	the sum of a set of values divided by the number of values
median	the middle item of a set of items arranged in order of size
member	an item that belongs to a set of items
mile	a measure of length
millilitre	a measure of capacity
millimetre	a measure of length

mirror line	the line in which an object is reflected to give its image
mixed number	the sum of a whole number and a fraction
mixed operations	a calculation involving two or more of addition, subtraction, multiplication and division
mode	the most frequent item in a set
multiple	a particular number multiplied by any other number is a multiple of that particular number
multiplying factor	a number that multiplies to increase or decrease a quantity
natural number	a counting number, i.e. 1,2,3,4,...
net pay	pay received after deductions for tax and other purposes
null set	a set with no members
number line	a line on which numbers can be marked to show their relative size
number pair	a pair of numbers in a particular order, e.g. the coordinates of a point
number pattern	a pattern of numbers with a rule for getting the next number in the pattern
numerator	the top number of a fraction
object	the original shape before a transformation is performed
obtuse	an angle whose size is between 90° and 180°
operation	a way of combining two numbers such as addition, subtraction, multiplication or division
order of rotation	the number of different positions in which an object looks the same when rotated about a fixed point
origin	the point where the x-axis and y-axis cross
outcome	the result of an experiment, e.g. getting a head when a coin is tossed
parallel	two lines that are always the same distance apart
parallelogram	a four sided figure whose opposite sides are parallel
pentagon	a plane figure bounded by five straight lines
perpendicular bisector	a line that bisects a line segment at right angles
percentage	out of a hundred
perimeter	the total distance round the edge of a figure
perpendicular	at right angles to a line or surface
pi (π)	the ratio of the circumference of a circle to its diameter
place value	the position of a digit in a number that shows its value
polyhedron	a solid with many plane faces
possibility space	a table showing all the possible outcomes of two events happening
pound	a measure of mass or the currency of the UK
prime number	a number whose only factors are 1 and itself (1 is not a prime number)
principal	the amount of money invested
prism	a solid whose cross-section is the same everywhere along its length
probability	a measure of the likelihood that an event will happen
probability tree	a diagram showing the different possibilities in which events can happen

product	the result of multiplying two or more numbers together
proper subset	a subset that is not the whole set
proper fraction	a fraction whose numerator is less than its denominator
proportion	the ratio of one quantity to the other
protractor	an instrument for measuring angles
pyramid	a solid with a polygon as a base and triangular sides that meet at a common point
quadrant	a quarter of a circle
quadrilateral	a plane figure bounded by four straight lines
radius	the distance from the centre of a circle to the edge
random	each item has an equal chance of being chosen
rate per cent	the annual (or other time span) interest paid as a percentage of the money lent or borrowed
ratio	the comparison between the sizes of two quantities
reciprocal of a number	that number divided into 1, e.g. the reciprocal of 4 is $\frac{1}{4}$
rectangle	a four sided figure whose angles are each $90°$
recurring decimal	a decimal that never terminates but where the figures form a repeating pattern, e.g. $0.191919\ldots$,
reflex angle	an angle whose size is between $180°$ and $360°$
regular pentagon	a plane figure bounded by five straight lines whose sides are all the same length
relation	a set of ordered pairs
remainder	the amount left over when one number is divided by another
revolution	a complete turn
rhombus	a quadrilateral whose sides are all the same length, e.g. a diamond
right angle	one quarter of a revolution ($90°$)
rotational symmetry	a figure has rotational symmetry' when it can be turned about a point to another position and still look the same
scalar	a quantity that has size but not direction, e.g. speed
scale factor	the factor by which a transformation changes the size of an object
sector	part of a circle enclosed by two radii and an arc
semicircle	half a circle
sequence	an ordered collection of items, e.g. 2, 4, 8, 16,...
set	a collection of items
significant figure	position of a figure in a number, e.g. in 2731 the third significant figure is 3
similar	the same shape, but not the same size
simple interest	the sum paid or charged when a sum of money is lent or borrowed
simultaneous equations	equations that are valid for the same set of values
sine of an angle	the ratio of the opposite side to the hypotenuse in a right-angled triangle
solution	the correct answer to a problem or equation
speed	the rate at which an object covers distance
square	a four-sided figure whose sides are all the same length and each of whose angles is a right angle
square root	the number which when multiplied by itself gives the original number

square unit	a measure of area
standard form	a number between 1 and 10 multiplied by a power of 10
subset	a set whose members are also members of another set
supplementary angles	two angles whose sum is $180°$
symmetry	having congruent parts each side of a line or around a point
table of values	a table giving corresponding values to two or more variables
tangent of an angle	the ratio of the opposite side to the adjacent side in a right-angled triangle
three-figure bearing	a three-figure number giving direction by measuring clockwise from north
tonne	a measure of mass
transformation	a change in position and/or size
translation	a movement in one direction
transversal	a line that crosses two or more parallel lines
trapezium	a four-sided figure with one pair of sides parallel
triangle	a plane figure bounded by three straight lines
unbiased	all possible outcomes are equally likely
uniform	everywhere the same
union	the set containing all the different elements of two or more sets
unit	a standard quantity used to measure, e.g. a metre, a litre, a square inch
universal set	a set that contains all the elements of the sets being considered
unlike terms	terms containing different combinations of letters
variable	a quantity that can vary in value
vector	a quantity that has size and direction, e.g. velocity
Venn diagram	a diagram used to show the elements in two or more sets
vertex (pl. vertices)	corner
volume	a measure of space

ANSWERS

CHAPTER 1

Exercise 1a   page 3

1 2, 3, 6, 1, −5, −3, 5, −3, −5, 5, 0
2 2, −2, 5, −4, 2, 5, −5, 0

3 5 below
4 3 above
5 1 below
6 10 above
7 on *x*-axis
8 4 below
9 3 right
10 5 left
11 2 right
12 7 left
13 on *y*-axis
14 9 left

15 A($-2,3$), B($3,1$), C($2,−2$), D($−3,1$),
E($6,1$), F($−2,−2$), G($−4,−4$), H($1,2$),
I($4,−4$), J($−4,3$)

18 square
19 isosceles triangle
20 rectangle
21 right-angled

Exercise 1b   page 4

1 6
2 8
3 6
4 2
5 2
6 7
7 5
8 7
9 11
10 11

11 ($−1,1$)
12 ($1,−2$)
13 ($−1,3$)
14 ($−6,−1$)
15 ($−5,1$)
16 ($0,−1$)
17 ($3,2$)
18 ($−1,2$)
19 ($−1,3$)
20 ($1,0$)
21 ($4,2$)
22 ($2,−1$)
23 ($−\frac{7}{2},3$)
24 ($−3,−1$)
25 ($−5,−2$)
26 ($4,\frac{3}{2}$)
27 ($−1,3$)
28 ($−1,0$)
29 ($0,0$)
30 ($−1,0$)

Exercise 1c   page 5

1 a ($1,2$), ($3,6$), ($−3,−6$), ($−2,−4$), ($2,4$)
b 10
c 16, 20, −8, 6, 9, −5, $2a$
2 a ($2,2$), ($4,3$), ($6,4$), ($10,6$), ($−4,−1$),
($−8,−3$), ($0,1$)
b *y*-coordinate $= \frac{1}{2}$ (*x*-coordinate) $+ 1$
c 5
d 7, 11, 16, −5, 16, $\frac{1}{2}a + 1$
3 a ($3,−1$), ($5,−3$), ($6,−4$), ($8,−6$), ($−2,4$),
($−4,6$), ($1,1$)
b −5, −8, −10, −18, 9, 11, −8, 10, −10

Exercise 1d   page 7

1 a parallelogram
c no
d both
e no
2 a square
c yes
d both
e yes
3 a trapezium
c no
d neither
e no
4 a rhombus
c no
d both
e yes
5 a rectangle
c yes
d both
e no
6 rectangle, square
7 rhombus, square
8 parallelogram, rectangle, rhombus, square

Exercise 1e   page 8

1 $+10°$
2 $−7°$
3 $−3°$
4 $+5°$
5 $−8°$
6 $0°$
7 2° below
8 3° above
9 4° above
10 10° below
11 8° above
12 freezing point

13 10°
14 12°
15 4°
16 −3°
17 2°
18 −2°
19 1°
20 3°
21 −7°
22 −2°
24 −5 s
25 +5 s
26 +50 c
27 −50 c
28 −1 min
29 +$50
30 −$5
31 +5 paces
32 −5 paces
33 +200 m
34 −5 m
35 −3 °C
36 +21 °C
37 +150 m
38 −3 °C
39 +25 c
40 6 paces in front

Exercise 1f   page 10

1 >
2 >
3 >
4 <
5 >
6 <
7 >
8 >
9 >
10 <
11 <
12 >
13 10, 12
14 −10, −12
15 −2, −4
16 2, 4
17 0, −3
18 5, 8
19 −7, −11
20 16, 32
21 $\frac{1}{6}, \frac{1}{36}$
22 −4, −2
23 −8, −16
24 −2, −3

Exercise 1g   page 11

1 −3
2 3
3 −2
4 −2
5 2
6 7
7 1
8 2
9 −12
10 −1
11 5
12 −2
13 −2
14 −1
15 −4
16 6
17 2
18 −3
19 −3
20 −1
21 3
22 −6
23 −10
24 −5
25 4
26 6
27 3
28 0
29 −3
30 −5
31 1
32 2
33 2
34 −2
35 −1
36 −2
37 1
38 2
39 5
40 16

Exercise 1h   page 13

1 2
2 −3
3 7
4 3
5 −9
6 3
7 −3
8 6
9 −14
10 10
11 −14
12 0
13 0
14 6
15 −6
16 7
17 −3
18 2
19 −4
20 5
21 13
22 13
23 −6
24 8
25 1

Exercise 1i   page 14

1 1
2 −5
3 9
4 8
5 −12
6 7
7 4
8 10
9 15
10 2
11 5
12 −12
13 5
14 −9
15 1
16 9
17 −1
18 0
19 2
20 16
21 5
22 −4
23 −8
24 19
25 −4
26 −4
27 4
28 −3
29 −3
30 −19
31 2
32 3
33 0
34 0
35 −1
36 0
37 9
38 −7
39 −4
40 3
41 −10
42 −3
43 −2
44 1
45 2
46 −12
47 3
48 18
49 −2
50 1
51 2
52 −15
53 −9
54 −6
55 −8

STP Caribbean Mathematics 2

Exercise 1j page 15

1	-3	**7**	-10	**13**	-2
2	-2	**8**	-3	**14**	-2
3	-5	**9**	-5	**15**	-4
4	-4	**10**	-4	**16**	-9
5	-4	**11**	-1	**17**	-4
6	-2	**12**	-2	**18**	-2

Exercise 1k page 17

1	-15	**11**	$+27$	**21**	-6
2	-8	**12**	-16	**22**	$+15$
3	$+14$	**13**	-35	**23**	-18
4	$+4$	**14**	$+24$	**24**	$+20$
5	-42	**15**	-15	**25**	-24
6	$+12$	**16**	-45	**26**	-24
7	-18	**17**	-24	**27**	$+45$
8	$+16$	**18**	$+8$	**28**	-20
9	-5	**19**	$+3$	**29**	-28
10	$+18$	**20**	-8	**30**	$+36$

Exercise 1l page 18

1	$-6x + 30$	**16**	$-3x - 2$
2	$-15c - 15$	**17**	$16 - 24x$
3	$-10e + 6$	**18**	$-6y + 12x$
4	$-3x + 4$	**19**	$20x - 5$
5	$-16 + 40x$	**20**	$-5 + 20x$
6	$-7x - 28$	**21**	$24 + 30x$
7	$-6d + 6$	**22**	$-24 - 30x$
8	$-8 - 4x$	**23**	$24 - 30x$
9	$-14 + 21x$	**24**	$-24 + 30x$
10	$-4 + 5x$	**25**	$-5a - 5b$
11	$12x + 36$	**26**	$6x + 4y + 2$
12	$10 + 15x$	**27**	$-25 - 10x$
13	$6x - 18$	**28**	$4x - 4y$
14	$-14 - 7x$	**29**	$-4c + 5$
15	$-6x + 2$	**30**	$18x - 9$

Exercise 1m page 18

1	$25x + 12$	**16**	$12x - 14$
2	$27 - 6c$	**17**	$4x - 12$
3	$14m - 20$	**18**	$9x + 19$
4	$3 - 6x$	**19**	$x - 21$
5	$6x - 4$	**20**	$31x - 11$
6	$13 - 8g$	**21**	$14x + 11$
7	$x - 2$	**22**	$-6x - 19$
8	$4f + 12$	**23**	$14x - 19$
9	$4s - 3$	**24**	$-6x + 11$
10	$19x - 3$	**25**	$15x - 9$
11	$17x - 1$	**26**	$11x + 7$
12	$9x - 18$	**27**	$-7 - 15x$
13	$9x + 1$	**28**	$2x + 21$
14	$15 - 5x$	**29**	$2x + 15$
15	$12x + 8$	**30**	$5x - 2$

Exercise 1n page 19

1	-2	**11**	7	**21**	3	**31**	1	**41**	$\frac{2}{3}$
2	-5	**12**	$-\frac{3}{4}$	**22**	2	**32**	2	**42**	3
3	-1	**13**	6	**23**	2	**33**	$\frac{1}{2}$	**43**	$-\frac{1}{2}$
4	-1	**14**	5	**24**	1	**34**	2	**44**	$\frac{3}{10}$
5	-2	**15**	7	**25**	6	**35**	2	**45**	-1
6	-4	**16**	2	**26**	-4	**36**	-2	**46**	3
7	4	**17**	1	**27**	3	**37**	1	**47**	$2\frac{1}{2}$
8	1	**18**	3	**28**	-3	**38**	0	**48**	1
9	3	**19**	1	**29**	$1\frac{1}{3}$	**39**	2	**49**	$\frac{1}{4}$
10	5	**20**	2	**30**	1	**40**	-2	**50**	2

Exercise 1p page 21

1 $-5°$

2 a $<$ **b** $>$

3	2	**6**	4	**9**	-24	**12**	-14
4	-5	**7**	0	**10**	-12	**13**	-24
5	-2	**8**	5	**11**	10	**14**	7

15 a $-6 + 15x$ **b** $22 - 7x$

Exercise 1q page 22

1 a $(2, 1)$ **b** $(-1, -1)$ **c** $(6, -3)$

2 a a parallelogram

  b i $(3, 1)$ **ii** $(1, -\frac{1}{2})$

3 a $1 - 3b$ **b** $11x - 13$

4 a $s = -5$ **b** $x = 3$ **c** $x = -3$

CHAPTER 2

Exercise 2a page 25

1 Associative for multiplication
2 Associative for addition
3 Commutative
4 Distributive
5 Commutative
6 Associative for multiplication
7 Distributive
8 Distributive
9 Commutative
10 Distributive
11 Identity for addition
12 Identity for multiplication
13 Inverse of 10 under addition
14 -9
15 divide by 9

16	$+5$	**19**	B	**22**	D
17	B	**20**	A	**23**	C
18	C	**21**	A	**24**	C

25 a true   **b** false   **c** true
    d true   **e** sometimes   **f** usually false
    g true

Exercise 2b page 27

1 a 0.75 **b** 0.6 **c** 0.3
    d 0.15 **e** 0.875 **f** 0.24

2 0.47

3 a $\frac{3}{50}$ **b** $\frac{1}{250}$ **c** $15\frac{1}{2}$
    d $2\frac{1}{100}$ **e** $3\frac{1}{4}$

4 $\frac{43}{50}$

5 $\frac{1}{20}$

6 a 30% **b** 20% **c** 70%
    d 3.5% **e** 92.5%

7 a 132% **b** 150% **c** 240%
    d 105% **e** 255.5%

8 a 0.45 **b** 0.6 **c** 0.95
    d 0.055 **e** 0.125

9 a $\frac{2}{5}$ **b** $\frac{13}{20}$ **c** $\frac{27}{50}$ **d** $\frac{1}{4}$

10 a 40% **b** 15% **c** 42% **d** 37.5%

11 $\frac{3}{5}$ **13** $\frac{19}{20}$ **15** 85% **17** 60%

12 $\frac{7}{20}$ **14** $\frac{8}{25}$ **16** 34% **18** 12.5%

19

Fraction	Percentage	Decimal
$\frac{3}{5}$	60%	0.6
$\frac{4}{5}$	80%	0.8
$\frac{3}{4}$	75%	0.75
$\frac{7}{10}$	70%	0.7
$\frac{11}{20}$	55%	0.55
$\frac{11}{25}$	44%	0.44

20 a $\frac{1}{20}$ **b** 5%
21 a 42% **b** 0.42
22 a $\frac{7}{25}$ **b** 60% **c** 12%
23 a $\frac{3}{5}$ **b** 35% **c** 0.05 **d** 12:1

Exercise 2c   page 29

1 9
2 4
3 100
4 125
5 1000
6 81
7 128
8 10
9 64
10 10 000
11 1 000 000
12 27
13 7200
14 893
15 65 000
16 3820
17 27.5
18 537 000
19 46.3
20 503.2
21 709
22 69.78

Exercise 2d   page 30

1 3^7 **3** 9^{10} **5** b^5 **7** 12^9 **9** 4^{16}
2 7^8 **4** 2^{11} **6** 5^8 **8** p^{14} **10** r^8

Exercise 2e   page 30

1 4^2 **5** q^4 **9** 9^1 **13** 2^1 **17** 4^1
2 7^6 **6** 15^4 **10** p^1 **14** a^{12} **18** a^1
3 5^1 **7** 6^5 **11** 6^{11} **15** c^3 **19** 3^8
4 10^5 **8** b^2 **12** 3^3 **16** 2^9 **20** b^9

Exercise 2f   page 32

1 $\frac{1}{4}$ **5** $\frac{1}{7}$ **9** $\frac{1}{9}$ **13** $\frac{1}{15}$ **17** $\frac{1}{100}$
2 $\frac{1}{27}$ **6** $\frac{1}{16}$ **10** $\frac{1}{4}$ **14** $\frac{1}{6}$ **18** $\frac{1}{8}$
3 $\frac{1}{16}$ **7** $\frac{1}{81}$ **11** $\frac{1}{64}$ **15** $\frac{1}{49}$ **19** $\frac{1}{10}$
4 $\frac{1}{3}$ **8** $\frac{1}{5}$ **12** $\frac{1}{36}$ **16** $\frac{1}{125}$ **20** $\frac{1}{64}$

21 0.0034
22 0.26
23 0.062
24 0.008 21
25 0.000 538
26 0.000 046 7
27 0.3063
28 0.028 05
29 0.005 173
30 3.004
31 5^{-2}
32 3^{-3}
33 6^{-3}
34 2^2
35 a^{-2}
36 10^{-3}
37 b^{-4}
38 4^5
39 c^1
40 2^{a-b}

Exercise 2g   page 33

1 4 **4** $\frac{1}{3}$ **7** 81 **10** $\frac{1}{36}$
2 $\frac{1}{25}$ **5** 1 **8** 1 **11** $\frac{1}{1000}$
3 64 **6** 125 **9** 4 **12** $\frac{1}{49}$
13 2410
14 0.7032
15 497.1
16 0.007 805
17 59 200
18 0.1074
19 783.4
20 3050
21 5.99
22 0.000 386 01

23 2^7 **28** 5^2 **33** 2^9 **38** a^{10}
24 4^3 **29** 3^0 **34** a^{12} **39** 3^{-2}
25 3^2 **30** 6^0 **35** 3^4 **40** b^0
26 a^7 **31** 4^4 **36** 7^0 **41** 5^{-5}
27 a^4 **32** 5^{-6} **37** 4^5 **42** a^0

Exercise 2h   page 34

1 3780
2 0.001 26
3 5 300 000
4 740 000 000 000 000
5 0.000 13
6 0.000 003 67
7 30 400
8 0.000 850 3
9 4 250 000 000 000
10 0.000 000 064 3

Exercise 2i   page 35

1 2.5×10^3
2 6.3×10^2
3 1.53×10^4
4 2.6×10^5
5 9.9×10^3
6 3.907×10^4
7 4.5×10^6
8 5.3×10^8
9 4×10^4
10 8×10^{10}
11 2.603×10^4
12 5.47×10^5
13 3.06×10^4
14 4.06×10^6
15 7.04×10^2
16 2.6×10^{-2}
17 4.8×10^{-3}
18 5.3×10^{-2}
19 1.8×10^{-5}
20 5.2×10^{-1}
21 7.9×10^{-1}
22 6.9×10^{-3}
23 7.5×10^{-6}
24 4×10^{-10}
25 6.84×10^{-1}
26 9.07×10^{-1}
27 8.05×10^{-2}
28 8.808×10^{-2}
29 7.044×10^{-4}
30 7.3×10^{-11}
31 7.93×10^1
32 5.27×10^{-3}
33 8.06×10^4
34 9.906×10^{-1}
35 7.05×10^{-2}
36 6.05×10^1
37 3.005×10^{-3}
38 6.0005×10^{-1}
39 7.08×10^6
40 5.608×10^5
41 5.3×10^{12}
42 5.02×10^{-8}
43 $7.008 09 \times 10^{-3}$
44 7.08×10^5
45 4.05×10^1
46 8.892×10^1
47 5.06×10^{-5}
48 5.7×10^{-8}
49 5.03×10^8
50 9.9×10^7
51 8.4×10^1
52 3.51×10^2
53 9×10^{-2}
54 7.05×10^{-3}
55 3.6×10^1
56 5.09×10^2
57 2.68×10^5
58 3.07×10^1
59 5.05×10^{-3}
60 8.8×10^{-6}

Exercise 2j   page 36

1 1550, 1500, 2000
2 8740, 8700, 9000
3 2750, 2800, 3000
4 36 840, 36 800, 37 000
5 68 410, 68 400, 68 000
6 5730, 5700, 6000
7 4070, 4100, 4000
8 7510, 7500, 8000
9 53 800, 53 800, 54 000
10 6010, 6000, 6000
11 4980, 5000, 5000
12 8700, 8700, 9000
13 54, 45
14 45 499, 44 500
15 1549, 1450
16 $2 500 000
17 1950

Exercise 2k   page 38

1 2.76, 2.8, 3
2 7.37, 7.4, 7
3 16.99, 17.0, 17
4 23.76, 23.8, 24
5 9.86, 9.9, 10
6 3.90, 3.9, 4
7 8.94, 8.9, 9
8 73.65, 73.6, 74
9 6.90, 6.9, 7
10 55.58, 55.6, 56
11 5.1
12 0.009
13 7.90
14 34.8
15 0.0078
16 0.975

17 5.551
18 285.6

19 6.7
20 10.00

Exercise 2l   page 39

1 3
2 8

3 6
4 8

5 7
6 8

7 0
8 0

9 0
10 8

Exercise 2m   page 39

1 60 000
2 4000
3 4 000 000
4 600 000
5 80 000
6 500
7 50 000
8 4000
9 700 000
10 900
11 30
12 1000
13 4700
14 57 000
15 60 000
16 890 000
17 7000
18 10 000

19 73 000
20 440
21 50 000
22 54 000
23 480
24 600
25 0.008 46
26 0.826
27 5.84
28 78.5
29 46.8
30 0.007 85
31 7.51
32 370
33 0.990
34 54.0
35 47
36 0.006 845

37 600 000
38 500
39 7.82
40 5000
41 37.9
42 7000
43 0.0709
44 0.07
45 3.3
46 1.7
47 13
48 13
49 14
50 29
51 24
52 0.23
53 0.026
54 0.000 43

Exercise 2n   page 41

1 100
2 36
3 0.35
4 20
5 180 000
6 0.8
7 0.48
8 3.6
9 1.3
10 3 500 000

11 600
12 4.5
13 2
14 0.7
15 17
16 0.003
17 0.0056
18 80
19 90 000
20 1.5

21 10
22 0.36
23 10
24 2
25 32
26 1.2
27 15
28 0.25
29 0.12
30 140

Exercise 2p   page 42

1 7.08
2 7.55
3 7.02
4 8.54
5 9.19
6 7.71
7 7.49
. 9.15
9 1.61
10 1.56
11 3.80
12 1.50
13 2.94
14 1.54
15 1.44
16. 1330
17 8370
18 6580
19 15.5
20 6.65
61 6340
62 0.006 08
63 34.8
64 484 000
65. 0.361
66 0.0203
67 0.000 123

21 172
22 14.7
23 11.2
24 1170
25 12 600
26 36.8
27 1950
28 38.0
29 1350
30 14 400
31 2.70
32 0.0196
33 0.0549
34 526
35 4.65
36 0.0481
37 1.79
38 0.005 15
39 3.97
40 0.548
68 631
69 0.000 000 096 1
70 4950
71 0.174
72. 16.7
73 0.000 146
74 13.4

41 0.121
42 0.0825
43 0.393
44 0.103
45 0.139
46 124
47 55.8
48 91.7
49 186
50 957
51 49.0
52 11 200
53 83.6
54 2.28
55 0.672
56 9.83
57 0.693
58 0.742
59 0.128
60 10 300

Exercise 2q   page 43

1 $\frac{1}{16}$
2 $1/b^3$
3 1

4 3.64×10^4
5 5.07×10^{-3}
6 60 000

7 0.0614
8 3.71
9 2.88

Exercise 2r   page 44

1 216
2 2^{-2}
3 $\frac{1}{5}$
10 a $\frac{7}{20}$

4 a^7
5 6.5×10^8
6 46 000
b 0.35

7 21 500
8 1350
9 0.699

Exercise 2s   page 44

1 5
2 $1/a$
3 1
10 a 62.5%

4 5.708×10^{-3}
5 10 000
6 0.0508
b 0.625

7 9
8 9.89
9 4.70

CHAPTER 3

Exercise 3b   page 49

1 g
2 e

3 d
4 e

5 f
6 f

7 d
8 g

Exercise 3d   page 52

1 60°
2 110°
3 75°

4 60°
5 60°
6 80°

7 110°
8 120°
9 30°

10 130°
11 130°

Exercise 3e   page 53

1 50°, 55°
2 130°, 130°, 50°
3 60°, 60°, 60°, 120°, 60°
4 50°, 80°, 50°
5 70°, 80°, 30°
6 115°, 115°
7 140°, 40°, 40°
8 70°, 110°, 70°, 70°
9 50°, 45°, 50°
10 55°, 125°, 55°
11 110°, 70°, 130°, 130°
12 40°, 100°
13 80°
14 90°, 90°, 50°
15 120°
16 40°
17 70°
18 60°
19 135°
20 55°
21 55°
22 120°
23 120°
24 45°

Exercise 3f   page 56

1 e
2 e
3 d

4 d
5 d
6 g

7 g
8 e
9 d

Exercise 3g   page 57

1 50°, 130°
2 130°, 50°
3 50°, 70°
4 260°, 40°, 60°
5 70°, 70°, 70°
6 45°, 90°

7 55°, 65°
8 60°
9 45°
10 30°
11 90°

Exercise 3h   page 59

1 e, g
2 e, d
3 e, g
4 e, d
5 h, f
6 d, g
7 70°, 110°, 180°
8 130°, 50°, 180°
9 40°, 40°, 80°
10 120°, 60°, 180°

Exercise 3i   page 61

1 120°
2 130°, 50°
3 85°
4 40°, 100°, 60°
5 55°, 125°

6 40°
7 80°, 80°
8 130°, 130°, 50°
9 80°, 100°, 80°, 100°
10 70°, 110°

Exercise 3j page 62

1 65°   **4** 110°   **7** 45°
2 140°   **5** 70°   **8** parallel
3 55°   **6** 70°

Exercise 3k page 62

1 80°   **3** 110°   **5** 50°   **7** 40°
2 60°   **4** 40°   **6** 40°

CHAPTER 4

Exercise 4a page 67

1 60°   **2** 75°   **3** 100°   **4** 130°
5 $d = 60°$, $e = 120°$ **7** $p = 130°$, $q = 50°$
6 70°   **8** $s = 70°$, $t = 110°$
9 $l = 60°$, $m = 100°$, $n = 20°$
10 $d = 30°$, $e = 75°$, $f = 105°$

Exercise 4c page 70

11 90°   **12** 45°   **14** they are parallel

Exercise 4d page 72

2 they are equal   **6** at the midpoint of AB
3 AB and CD   **7** at the midpoint of CD
4 coincident   **8** 90°
5 coincident   **9** each is 90°

Exercise 4e page 74

4 The perpendicular bisector of LM passes through N.
5 The perpendicular bisector of PR does not pass through Q.
6 The perpendicular bisector of the chord AB which passes through the centre C.

CHAPTER 5

Exercise 5a page 80

11 500 m   **12** 2.29 m

Exercise 5b page 83

1 23 m   **4** 38 m   **7** 55 m   **10** 46 m
2 22 m   **5** 70 m   **8** 58 m   **11** 91 m
3 50 m   **6** 32 m   **9** 9 m

Exercise 5c page 85

1 86 m   **5** 339 m   **9** 528 m   **13** 433 m
2 77 m   **6** 824 m   **10** 54 m   **14** 134 m
3 71 m   **7** 112 m   **11** 1170 m   **15** 582 m
4 81 m   **8** 923 m   **12** 8660 m   **16** 280 m

Exercise 5e page 89

1 87 m   **3** 274 m   **5** 1227 m
2 163 m   **4** 51 m

Exercise 5f page 90

1 860 cm

2

3

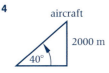

4

5

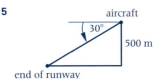

Exercise 5g page 90

1 94 m

2

3

4

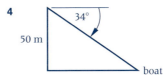

5

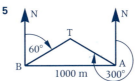

$T\widehat{B}A = T\widehat{A}B$ so $\triangle TAB$ is isosceles
$\therefore AT = BT$

Exercise 5h page 91

1 354 m

2

3

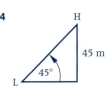

4

5 **a** i 043° **ii** 132° **iii** 223°
   c 18.7 km
   d i 345° **ii** 165°
6 **a** i 078° **ii** 143° **iii** 323° **iv** 258°
   b 65°, 16°, 99°
   c 16.4 km
7 **a** 19.1 km   **b** 13.1 km

STP Caribbean Mathematics 2

CHAPTER 6

Exercise 6a   page 94

1 4	**9** 2	**17** 3	**25** 1	**33** $\frac{2}{3}$
2 4	**10** 3	**18** 8	**26** 1	**34** 2
3 12	**11** 4	**19** -2	**27** 4	**35** 3
4 2	**12** 5	**20** 1	**28** 1	**36** 2
5 3	**13** 3	**21** 2	**29** $4\frac{1}{2}$	**37** $5\frac{1}{2}$
6 4	**14** 4	**22** 4	**30** $1\frac{1}{2}$	**38** $\frac{4}{5}$
7 1	**15** 1	**23** 2	**31** 2	**39** 1
8 3	**16** 1	**24** 1	**32** 1	**40** 2

Exercise 6b   page 96

1 $6x + 24$	**17** $28x + 27$	**33** 1
2 $6x + 3$	**18** $18x - 44$	**34** 2
3 $4x - 12$	**19** $4x + 25$	**35** $\frac{1}{2}$
4 $6x - 10$	**20** $-30x + 47$	**36** $1\frac{1}{2}$
5 $12 - 8x$	**21** $10x + 5$	**37** $1\frac{1}{4}$
6 $20x + 10$	**22** $-x + 12$	**38** 1
7 $6 - 9x$	**23** $17x - 33$	**39** -1
8 $35 - 28x$	**24** $-4x + 14$	**40** -4
9 $10x - 14$	**25** $-10x + 14$	**41** $1\frac{1}{2}$
10 $42 + 12x$	**26** $36x + 26$	**42** $-\frac{2}{3}$
11 $8x + 18$	**27** $-7x + 32$	**43.** 3
12 $26x + 13$	**28** $x + 22$	**44** 1
13 $34x - 13$	**29** $21x - 19$	**45** 3
14 $8x + 4$	**30** $-6x + 2$	**46** $\frac{1}{2}$
15 $21x + 5$	**31** 2	
16 $13x + 19$	**32** $1\frac{1}{2}$	

Exercise 6c   page 98

1 $\frac{x}{2}$	**6** $2x$	**11** $\frac{5x}{8}$	**16** $\frac{3x}{10}$
2 $\frac{x}{6}$	**7** $\frac{3x}{10}$	**12** $\frac{x}{18}$	**17** $\frac{2x}{3}$
3 $\frac{3x}{2}$	**8** $\frac{3x}{2}$	**13** $2x$	**18** $9x$
4 $\frac{2x}{3}$	**9** $6x$	**14** $\frac{x}{2}$	**19** $9x$
5 $\frac{4x}{5}$	**10** $\frac{x^2}{6}$	**15** $\frac{4x}{5}$	**20** $\frac{2x^2}{3}$

Exercise 6d   page 98

1 15	**11** $\frac{3}{2}$	**21** $6\frac{3}{4}$	**31** $1\frac{6}{7}$	**41** $\frac{1}{30}$
2 8	**12** $\frac{5}{9}$	**22** $3\frac{1}{2}$	**32** $\frac{23}{27}$	**42** 7
3 48	**13** $1\frac{1}{3}$	**23** $12\frac{2}{3}$	**33** $\frac{18}{23}$	**43** $2\frac{17}{26}$
4 12	**14** $1\frac{1}{20}$	**24** 20	**34** $\frac{1}{2}$	**44** 2
5 $3\frac{5}{9}$	**15** $\frac{5}{12}$	**25** $-1\frac{1}{4}$	**35** $1\frac{1}{7}$	**45** $-\frac{1}{4}$
6 $22\frac{1}{2}$	**16** $\frac{7}{10}$	**26** $1\frac{5}{7}$	**36** $1\frac{2}{3}$	**46** $\frac{28}{33}$
7 14	**17** $2\frac{1}{4}$	**27** $\frac{3}{4}$	**37** $1\frac{1}{6}$	
8 $8\frac{1}{3}$	**18** $13\frac{3}{4}$	**28** $1\frac{1}{3}$	**38** $5\frac{5}{6}$	
9 $\frac{1}{6}$	**19** $3\frac{6}{13}$	**29** 11	**39** $1\frac{1}{22}$	
10 $\frac{3}{20}$	**20** $1\frac{11}{17}$	**30** $\frac{1}{14}$	**40** $1\frac{18}{23}$	

Exercise 6e   page 100

1 $150	**4** 12	**7** 9	**9** 12
2 40	**5** 24 cm	**8** 3	**10** $1000
3 30 cm	**6** 5 cm		

Exercise 6f   page 102

1 -8	**11** $17x - 20$	**21** 2
2 15	**12** $-15x - 30$	**22** 13
3 -24	**13** $25 - 14x$	**23** $\frac{13}{11}$
4 -3	**14** $15x - 20$	**24** $\frac{13}{16}$
5 2	**15** $-9x + 6$	**25** 10
6 28	**16** 2	**26** 7
7 -4	**17** $25x - 46$	**27** -4
8 $\frac{1}{3}$	**18** $5x - 17$	**28** $\frac{7}{16}$
9 3	**19** 1	
10 $-2x + 17$	**20** 3	

Exercise 6g   page 103

1 $2l + 2w$	**11** $N = 10n$
2 $2l + d$	**12** $C = nx$
3 $3l$	**13** $L = l - d$
4 $5l$	**14** $p = 6l$
5 $2l + s + d$	**15** $A = 2l^2$
6 $W = x + y$	**16** $N = S - T$
7 $P = 2l + 2b$	**17** $W = T + S$
8 $T = N + M$	**18** $S = N - L - R$
9 $T = N - L$	**19** $r = p - q$ or $r = q - p$
10 $A = l^2$	**20** $W = Kn$
21 $d = b - a$	

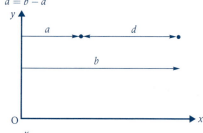

22 $q = \dfrac{x}{5}$

23 $L = \dfrac{ny}{100}$

24 $A = 100lb$

25 $T = t + \dfrac{s}{60}$

Exercise 6h   page 106

1 10	**5** 20	**9** 25	**13** 5
2 100	**6** 200	**10** $7\frac{1}{2}$	**14** 33
3 30	**7** 24	**11** -1	**15** 50
4 2	**8** 15	**12** -12	**16** 19
17 16	**21** 15	**25** $1\frac{3}{4}$	**29** 31
18 2	**22** 200	**26** -21	**30** -3
19 105	**23** $3\frac{1}{3}$	**27** 0	
20 $3\frac{1}{3}$	**24** 7	**28** $\frac{5}{24}$	

Exercise 6i   page 108

1 a 48	b -18	c 6	d 5
2 a 4	b 20	c 8	d -12
3 a 52	b 20	c 96	d -4
4 a 5	b 3	c 38	d -24
5 a $1\frac{1}{4}$	b $4\frac{7}{8}$	c $12\frac{5}{6}$	d $\frac{5}{24}$
6 a 15	b -1.1	c -15.9	d 0.38

7 $C = 50n$,   600 c or $6

8 $L = \dfrac{n}{2}$, $5

9 $V = lbd$,   1200 cm³

10 $P = 2a + 2b$,   70 cm

11 $P = 6x$,   6 cm

12 $P = L - Nr$,   5 m

13 $P = 3a$,   24 cm

14 $W = Na + p$,   45

15 $A = 2lw + 2lh + 2hw$,   6200 cm²

Exercise 6j   page 110

1. $T = N - G$
2. $x = \dfrac{z}{y}$
3. $d = St$
4. $X = L + Y$
5. $a = S - 2b$
6. $u = v - t$
7. $d = S + t$
8. $z = P - 2y$
9. $T = \dfrac{C}{R}$
10. $a = L - b - c$
11. $a = P - b$
12. $T = N - R$
13. $c = b - a - d$
14. $u = v - rt$
15. $n = \dfrac{N}{r}$
16. $y = x + z$
17. $c = P - ab$
18. $m = Ln$
19. $u = v - at$
20. $y = s - ax$

Exercise 6k   page 111

1. formula
2. equation
3. equation
4. expression
5. formula
6. expression
7. expression
8. equation
9. equation
10. formula
11. formula
12. expression

Exercise 6l   page 111

1. $2\frac{1}{2}$
2. $3\frac{1}{3}$
3. $6x - 24$
4. $6x$
5. 12
6. $\frac{1}{3}$
7. $5x - 3$
8. $P = 4l + f + g$
9. -3
10. $N = R + D$

Exercise 6m   page 112

1. $\frac{1}{3}$
2. $\frac{2}{3}$
3. $6x$
4. $\dfrac{15x}{2}$
5. $1\frac{1}{2}$
6. $-8x + 10$
7. $5\frac{9}{10}$
8. 15
9. $N = a + b + c$
10. $N = n + ab$

Exercise 6n   page 112

1. $\dfrac{15x}{4}$
2. $\frac{1}{3}$
3. 2
4. $\dfrac{11x}{12}$
5. $\dfrac{x}{4}$
6. $x + 6$
7. $-\frac{7}{20}$
8. 10
9. $P = 6a$
10. $p = \dfrac{L}{3q}$

CHAPTER 7

Exercise 7b   page 117

1. a $\frac{1}{6}$ b $\frac{1}{9}$ c $\frac{1}{6}$
2. a $\frac{4}{25}$ b $\frac{16}{25}$
3. $\frac{1}{6}$
4. a $\frac{1}{16}$ b $\frac{1}{8}$ c $\frac{3}{16}$ d $\frac{5}{8}$
5. $\frac{1}{4}$
6. a $\frac{5}{36}$ b $\frac{1}{18}$ c $\frac{1}{18}$ d $\frac{19}{36}$
7. $\frac{7}{10}$
8. a $\frac{3}{10}$ b $\frac{3}{20}$
9. a $\frac{1}{4}$ b $\frac{1}{16}$ c $\frac{3}{4}$ d $\frac{1}{4}$

Exercise 7c   page 120

1. a $\frac{1}{6}$ b $\frac{1}{6}$ c $\frac{1}{3}$
2. a $\frac{5}{9}$ b $\frac{4}{9}$
3. a $\frac{3}{10}$ b $\frac{7}{10}$
4. a $\frac{5}{7}$ b $\frac{4}{7}$
5. a $\frac{3}{4}$ b $\frac{3}{8}$
6. a $\frac{2}{3}$ b $\frac{3}{4}$
7. 2
8. 80
9. 2
10. 80

Exercise 7d   page 123

1. a $\frac{2}{5}$
   b

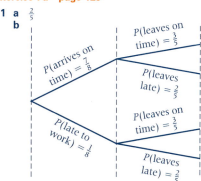

   Getting to work   Leaving work

2. a $\frac{3}{4}$ b $\frac{1}{16}$ c $\frac{9}{16}$
3. a $\frac{2}{5}$ b $\frac{2}{15}$
4. a i $\frac{1}{6}$ ii $\frac{5}{6}$
   b i $\frac{1}{36}$ ii $\frac{5}{36}$ iii $\frac{5}{36}$
   c $\frac{5}{18}$
5. a $\frac{7}{15}$ b $\frac{3}{5}$ c $\frac{2}{15}$
6. a $\frac{1}{3}$ b $\frac{1}{2}$ c $\frac{1}{6}$

Exercise 7e   page 125

1. $\frac{8}{15}$
2. $\frac{7}{20}$
3. a $\frac{5}{8}$ b $\frac{1}{3}$ c $\frac{1}{24}$
4.

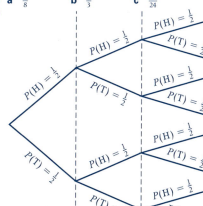

   a $\frac{1}{8}$ b $\frac{1}{8}$ c $\frac{3}{8}$

Exercise 7f   page 130

5 $\frac{1}{2}$ **6** $\frac{1}{4}$

CHAPTER 8

Exercise 8a   Page 130

1 $11.32, £8.68   **4** $15.72, $4.28
2 $15.36, $4.64   **5** $15.04, $4.96
3 $12.00, $8.00

6 $	**7** $	**8** $
47.50	23.20	3.50
16.20	3.60	6.24
51.60	3.40	4.50
115.30	30.20	2.64
		16.88

9 $	**10** $
7.50	12.50
10.50	10.35
8.40	3.18
15.30	2.37
41.70	5.89
	34.29

11 $123.45   **12** $7200

Exercise 8b   page 132

1 $200
2 $192.50
3 $215.60
4 $237.60
5 $292.30
6 a 7 hours 50 min
  b 37 hours 50 min, $72.42
7 $43\frac{3}{4}$ hours, $229.25
8 a $7\frac{3}{4}$ **b** $38\frac{3}{4}$, $175.15
9 a $117.80 **b** $164.30 **c** $173.60
10 $100.94
11 $157.25
12 $153.26
13 $234.36
14 $188.37
15 $133.94
16 a $7\frac{1}{2}$ hours **b** $37\frac{1}{2}$ hours **c** $106.50
  d $1\frac{1}{2}$ hours **e** $112.89

Exercise 8c   page 135

1 $540 **3** $211 **5** $382 **7** $402.50
2 $368 **4** $524 **6** $500 **8** $321.60
9 a 1125 **b** 500 **c** 625 **d** $481.25
10

	a	b i	b ii	c
Ms Arnold	186	100	86	$211.10
Mr Beynon	158	80	78	$181.30
Miss Capstick	194	100	94	$221.90
Mr Davis	225	100	125	$263.75
Mr Edmunds	191	100	91	$217.85

Exercise 8d   page 137

1 a $3000   **b** $17 000
2 a $6300 **b** i $28 7000 ii $551.92
3

M Davis	$82	$738
P Evans	$147.60	$1082.40
G Brown	$317.70	$1447.30
A Khan	$545.50	$1636.50

4 a $4.56   **b** $42.56
5 $1880
6 $23.50
7 $2467.50
8 $753.25
9 $376, $390.40
10 a $1265   **b** no
  c The manager added a $2\frac{1}{2}$% to the original sale price. He should have added $17\frac{1}{2}$% to the pre-sales tax price.

Exercise 8e   page 138

1 a $126 **b** $201.60 **c** $375.30
  d $500 **e** $163.35
2 $440 **8** $42 **14** $48 400
3 7 years **9** $76.32 **15** $6272
4 6.5% **10** $103.88 **16** $76.04
5 $300 **11** $191.77 **17** $1093.50
6 4% **12** $143.99 **18** $12 800
7 $406 **13** $229.40

Exercise 8f   page 140

1 $65.50 **5** $168.77 **9** $90.20
2 $84.32 **6** $100.75 **10** $345.65
3 $105.80 **7** $116
4 $133.99 **8** $116.40

Exercise 8g   page 141

1 3 **15** 16 **23** 4 hours
2 $\frac{1}{10}$ **16** 1 **24** $\frac{1}{2}$ hour
3 $1\frac{1}{2}$ **17** 12 **25** 10 hours
4 1.2 **18** 3 **26** $2\frac{7}{9}$ hours
5 0.06 **19** 1.8 **27** 3 c
6 0.02 **20** 0.144 **28** 12 c
7 8 **21** 0.56 **29** 3.024 c
8 2 **22** 0.84 **30** $1\frac{1}{2}$ c

Exercise 8h   page 143

1 $45 **5** $126.69 **9** $108
2 $54 **6** $111.24 **10** £113
3 $125 **7** $101.83
4 $118 **8** $145

Exercise 8i   page 144

1 $276.30
2 $728.75
3 £220
4 $137.80
5 a $200 **b** $181.82
6 a $173 **b** $173
7 a $100 **b** $91
8 a $250 **b** $250
9 a $1200 **b** $1167
10 a $400 **b** $349
11 a $17 **b** $16
12 a $400 **b** $448
13 a $600 **b** $594
14 $727.27
15 $23 169.81
16 £825
17 $8.96
18 $84 174

CHAPTER 9

Exercise 9a   page 147

1 scalar   3 scalar   5 vector
2 vector   4 scalar

Exercise 9b   page 148

1 $\begin{pmatrix} 3 \\ 2 \end{pmatrix}$ 3 $\begin{pmatrix} 4 \\ 0 \end{pmatrix}$ 5 $\begin{pmatrix} -3 \\ 4 \end{pmatrix}$

2 $\begin{pmatrix} 4 \\ 1 \end{pmatrix}$ 4 $\begin{pmatrix} -2 \\ 2 \end{pmatrix}$ 6 $\begin{pmatrix} -5 \\ -3 \end{pmatrix}$

7 $\mathbf{g} = \begin{pmatrix} 5 \\ 0 \end{pmatrix}$ $\mathbf{h} = \begin{pmatrix} -4 \\ 0 \end{pmatrix}$ $\mathbf{i} = \begin{pmatrix} 6 \\ 2 \end{pmatrix}$

$\mathbf{j} = \begin{pmatrix} -6 \\ 7 \end{pmatrix}$ $\mathbf{k} = \begin{pmatrix} -6 \\ -2 \end{pmatrix}$ $\mathbf{l} = \begin{pmatrix} 3 \\ -1 \end{pmatrix}$

$\mathbf{m} = \begin{pmatrix} 0 \\ -4 \end{pmatrix}$ $\mathbf{n} = \begin{pmatrix} 4 \\ 2 \end{pmatrix}$

17 Both pairs are parallel.

Exercise 9c   page 150

1 (7, 4)   9 (2, 0)
2 (1, −2)   10 (7, −4)
3 (−3, 7)   11 (−9, −1)
4 (1, −5)   12 (−7, −3)
5 (8, 1)   13 (−6, −1)
6 (8, 0)   14 {−2, −3}
7 (−1, 0)   15 (3, −2)
8 (−9, −8)   16 (−2, −3)
17 (1, −3)   20 (−1, −10)
18 (1, 5)   21 (−6, −6)
19 (−7, 4)   22 (−1, 10)

Exercise 9d   page 151

1 $\begin{pmatrix} 6 \\ 2 \end{pmatrix}$ 5 $\begin{pmatrix} -5 \\ -6 \end{pmatrix}$ 9 $\begin{pmatrix} 0 \\ -12 \end{pmatrix}$

2 $\begin{pmatrix} 5 \\ -1 \end{pmatrix}$ 6 $\begin{pmatrix} 2 \\ -2 \end{pmatrix}$ 10 $\begin{pmatrix} 2 \\ 8 \end{pmatrix}$

3 $\begin{pmatrix} -6 \\ -1 \end{pmatrix}$ 7 $\begin{pmatrix} -2 \\ -2 \end{pmatrix}$

4 $\begin{pmatrix} 6 \\ 5 \end{pmatrix}$ 8 $\begin{pmatrix} -4 \\ -5 \end{pmatrix}$

Exercise 9e   page 152

1 a $\mathbf{b} = 2\mathbf{a}$ c $\mathbf{d} = 3\mathbf{a}$ e $\mathbf{b} = 2\mathbf{e}$
   b $\mathbf{c} = -\mathbf{a}$ d $\mathbf{e} = -\mathbf{a}$ f $\mathbf{d} = -3\mathbf{c}$

2 $\mathbf{a} = \begin{pmatrix} 4 \\ -2 \end{pmatrix}$ $\mathbf{b} = \begin{pmatrix} -2 \\ -3 \end{pmatrix}$ $\mathbf{c} = \begin{pmatrix} -4 \\ -6 \end{pmatrix}$ $\mathbf{d} = \begin{pmatrix} 2 \\ 3 \end{pmatrix}$

$\mathbf{e} = \begin{pmatrix} 8 \\ -4 \end{pmatrix}$ $\mathbf{f} = \begin{pmatrix} -4 \\ 2 \end{pmatrix}$ $\mathbf{g} = \begin{pmatrix} 6 \\ 9 \end{pmatrix}$ $\mathbf{h} = \begin{pmatrix} -8 \\ 4 \end{pmatrix}$

$\mathbf{e} = 2\mathbf{a}$, $\mathbf{f} = -\mathbf{a}$, $\mathbf{h} = -2\mathbf{a}$, $\mathbf{c} = 2\mathbf{b}$, $\mathbf{d} = -\mathbf{b}$
$\mathbf{g} = -3\mathbf{b}$, $\mathbf{h} = -\mathbf{e}$, $\mathbf{g} = 3\mathbf{d}$, $\mathbf{h} = 2\mathbf{f}$, ...

3 $\begin{pmatrix} 8 \\ 12 \end{pmatrix}$, $\begin{pmatrix} -4 \\ -6 \end{pmatrix}$, $\begin{pmatrix} 2 \\ 3 \end{pmatrix}$

4 $\begin{pmatrix} 2 \\ -4 \end{pmatrix}$, $\begin{pmatrix} -4 \\ 8 \end{pmatrix}$, $\begin{pmatrix} 4 \\ -8 \end{pmatrix}$

5 $\begin{pmatrix} 10 \\ -8 \end{pmatrix}$, $\begin{pmatrix} -5 \\ 4 \end{pmatrix}$, $\begin{pmatrix} 15 \\ -12 \end{pmatrix}$

6 $\begin{pmatrix} 3 \\ 6 \end{pmatrix}$, $\begin{pmatrix} -6 \\ -12 \end{pmatrix}$, $\begin{pmatrix} 6 \\ 12 \end{pmatrix}$

7 $\begin{pmatrix} 10 \\ 2 \end{pmatrix}$, $\begin{pmatrix} -5 \\ -1 \end{pmatrix}$, $\begin{pmatrix} 15 \\ 3 \end{pmatrix}$, $\begin{pmatrix} -20 \\ -4 \end{pmatrix}$

8 $\begin{pmatrix} -6 \\ 0 \end{pmatrix}$, $\begin{pmatrix} 4 \\ 0 \end{pmatrix}$, $\begin{pmatrix} -10 \\ 0 \end{pmatrix}$, $\begin{pmatrix} 8 \\ 0 \end{pmatrix}$

9 $\begin{pmatrix} -6 \\ 4 \end{pmatrix}$, $\begin{pmatrix} 18 \\ -12 \end{pmatrix}$, $\begin{pmatrix} 3 \\ -2 \end{pmatrix}$, $\begin{pmatrix} -12 \\ 8 \end{pmatrix}$

10 $\begin{pmatrix} -18 \\ -60 \end{pmatrix}$, $\begin{pmatrix} 24 \\ 80 \end{pmatrix}$, $\begin{pmatrix} -3 \\ -10 \end{pmatrix}$, $\begin{pmatrix} 30 \\ 100 \end{pmatrix}$

Exercise 9f   page 154

1 $\begin{pmatrix} 7 \\ -1 \end{pmatrix}$ 6 $\begin{pmatrix} 4 \\ 3 \end{pmatrix}$ 11 $\begin{pmatrix} 6 \\ 9 \end{pmatrix}$ 16 $\begin{pmatrix} -2 \\ 10 \end{pmatrix}$

2 $\begin{pmatrix} -8 \\ 2 \end{pmatrix}$ 7 $\begin{pmatrix} -6 \\ -6 \end{pmatrix}$ 12 $\begin{pmatrix} 7 \\ 11 \end{pmatrix}$ 17 $\begin{pmatrix} -2 \\ -4 \end{pmatrix}$

3 $\begin{pmatrix} 7 \\ -4 \end{pmatrix}$ 8 $\begin{pmatrix} 3 \\ 6 \end{pmatrix}$ 13 $\begin{pmatrix} 7 \\ 10 \end{pmatrix}$ 18 $\begin{pmatrix} -5 \\ -2 \end{pmatrix}$

4 $\begin{pmatrix} 2 \\ 6 \end{pmatrix}$ 9 $\begin{pmatrix} 7 \\ 8 \end{pmatrix}$ 14 $\begin{pmatrix} 10 \\ 0 \end{pmatrix}$ 19 $\begin{pmatrix} -8 \\ 5 \end{pmatrix}$

5 $\begin{pmatrix} 10 \\ 0 \end{pmatrix}$ 10 $\begin{pmatrix} 6 \\ -4 \end{pmatrix}$ 15 $\begin{pmatrix} -1 \\ 11 \end{pmatrix}$ 20 $\begin{pmatrix} 0 \\ 0 \end{pmatrix}$

Exercise 9g   page 156

1 a $\begin{pmatrix} 7 \\ 5 \end{pmatrix}$ d $\begin{pmatrix} 8 \\ 6 \end{pmatrix}$ g $\begin{pmatrix} 10 \\ 9 \end{pmatrix}$

   b $\begin{pmatrix} 7 \\ 5 \end{pmatrix}$ e $\begin{pmatrix} 4 \\ 6 \end{pmatrix}$ h $\begin{pmatrix} 10 \\ 9 \end{pmatrix}$

   c $\begin{pmatrix} 8 \\ 6 \end{pmatrix}$ f $\begin{pmatrix} 6 \\ 9 \end{pmatrix}$

2 a $\begin{pmatrix} 3 \\ 2 \end{pmatrix}$ c $\begin{pmatrix} 0 \\ -5 \end{pmatrix}$ e $\begin{pmatrix} -6 \\ 12 \end{pmatrix}$

   b $\begin{pmatrix} 3 \\ 2 \end{pmatrix}$ d $\begin{pmatrix} 0 \\ -5 \end{pmatrix}$ f $\begin{pmatrix} -20 \\ -12 \end{pmatrix}$

3 a $\begin{pmatrix} 5 \\ 10 \end{pmatrix}$ b $\begin{pmatrix} 18 \\ 24 \end{pmatrix}$ c $\begin{pmatrix} 12 \\ 24 \end{pmatrix}$

4 a $\begin{pmatrix} -19 \\ -1 \end{pmatrix}$ b $\begin{pmatrix} 4 \\ -11 \end{pmatrix}$

Exercise 9h   page 157

1 $\begin{pmatrix} 5 \\ 3 \end{pmatrix}$ 7 $\begin{pmatrix} 11 \\ 9 \end{pmatrix}$ 13 $\begin{pmatrix} 5 \\ 10 \end{pmatrix}$

2 $\begin{pmatrix} 0 \\ 6 \end{pmatrix}$ 8 $\begin{pmatrix} 5 \\ 8 \end{pmatrix}$ 14 $\begin{pmatrix} 4 \\ -5 \end{pmatrix}$

3 $\begin{pmatrix} 2 \\ 4 \end{pmatrix}$ 9 $\begin{pmatrix} -3 \\ -2 \end{pmatrix}$ 15 $\begin{pmatrix} 4 \\ -1 \end{pmatrix}$

4 $\begin{pmatrix} -5 \\ 1 \end{pmatrix}$ 10 $\begin{pmatrix} -7 \\ 3 \end{pmatrix}$ 16 $\begin{pmatrix} 2 \\ 3 \end{pmatrix}$

5 $\begin{pmatrix} 2 \\ 1 \end{pmatrix}$ 11 $\begin{pmatrix} 2 \\ 4 \end{pmatrix}$ 17 $\begin{pmatrix} -3 \\ 11 \end{pmatrix}$

6 $\begin{pmatrix} 3 \\ 2 \end{pmatrix}$ 12 $\begin{pmatrix} 1 \\ -1 \end{pmatrix}$ 18 $\begin{pmatrix} -11 \\ 7 \end{pmatrix}$

19 a $\begin{pmatrix} 1 \\ 2 \end{pmatrix}$ b $\begin{pmatrix} -1 \\ -2 \end{pmatrix}$

20 a $\begin{pmatrix} -6 \\ -4 \end{pmatrix}$ b $\begin{pmatrix} -3 \\ -3 \end{pmatrix}$ c $\begin{pmatrix} 3 \\ 3 \end{pmatrix}$

STP Caribbean Mathematics 2

21 a $\begin{pmatrix} 8 \\ 2 \end{pmatrix}$ **c** $\begin{pmatrix} 0 \\ -22 \end{pmatrix}$ **e** $\begin{pmatrix} 0 \\ 3 \end{pmatrix}$

 b $\begin{pmatrix} 9 \\ 19 \end{pmatrix}$ **d** $\begin{pmatrix} 10 \\ 11 \end{pmatrix}$

REVIEW TEST 1   page 159

1 A **3** C **5** D **7** A **9** A
2 A **4** D **6** B **8** A **10** A
11 C
12 a 3×10^3 **b** 70%, \$150
13 a $a = 80°$, $b = 75°$, $c = 25°$, $d = 75°$, $e = 105°$,
  $f = 100°$
15 a $x = 1.5$ **b** $x = 5$
16 (1, 1) (1, 2) (1, 3) (1, 4)
  (2, 1) (2, 2) (2, 3) (2, 4)
  (3, 1) (3, 2) (3, 3) (3, 4)
  (4, 1) (4, 2) (4, 3) (4, 4)

  There are 16 outcomes, $\frac{3}{16}$

17 \$1000.00
18 a $\frac{2}{3}$ **b** $\frac{1}{9}$ **c** $\frac{4}{9}$
19 \$124.80
20 \$632.50
21 a $\begin{pmatrix} 2 \\ -5 \end{pmatrix}$ **b** $\begin{pmatrix} 8 \\ -1 \end{pmatrix}$ **c** $\begin{pmatrix} 13 \\ -13 \end{pmatrix}$

CHAPTER 10

Exercise 10a   page 162

1 a, **b** and **c**

2

5

3

6

4

7

Exercise 10b   page 163

1

3
None

2

4

5

6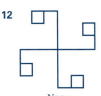

12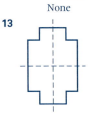
None

13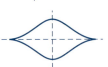

7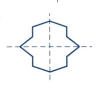

14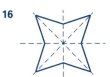

8

15

9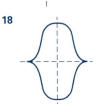

16

10

17

11

18

Exercise 10c   page 164

1

3

2

4

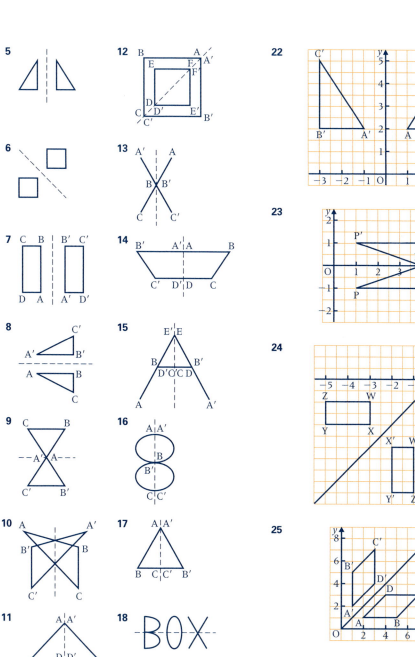

19 Q9: A and A′, Q11: A, A′; D D′, Q12: A, A′; F, F′; D, D′; C, C′, Q13: B, B′; E, E′, Q14: A, A′ and D, D′, Q15: E, E′ and C, C′, Q16: A, A′; B B′; C, C′, Q17: A, A′; C, C′.
They all lie on the axis of symmetry.

20 Equal distances; perpendicular lines.

21 Equal distances; perpendicular lines.

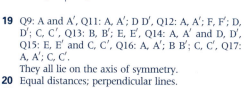

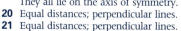

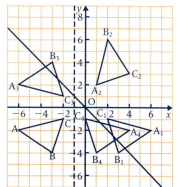

1

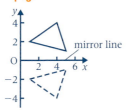

2

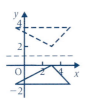

3

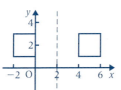

4

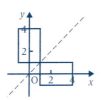

5

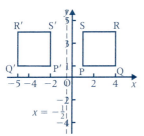

$x = -\frac{1}{2}$

6

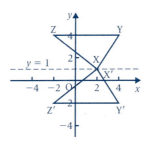

X, X' are invariant points

7

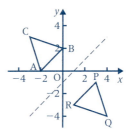

R is the image of A. There are no invariant points.

8

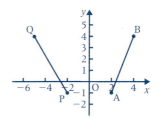

If there is a mirror line it has to be the perpendicular bisector of AP. But this line does not pass through the midpoint of QB, so PQ is not the reflection of AB.

9

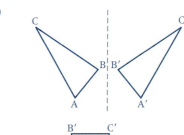

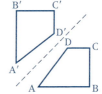

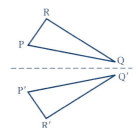

10

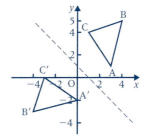

11

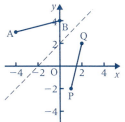

12

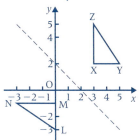

Exercise 10e   page 170

1 Yes

2

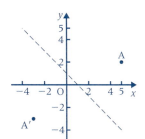

3

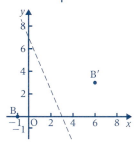

Exercise 10f   page 171

1 a and c

2 Translation e and b
Reflection a and c
Neither d

3 Translation 2; reflection 1; neither 3 and 4.

Exercise 10g   page 172

1

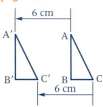

2

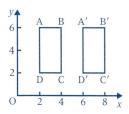

3

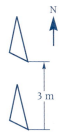

4

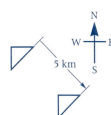

5

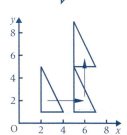

Exercise 10h   page 172

1 (7, 3)	**5** (1, 3)	**9** (9, −6)
2 (6, 9)	**6** (6, −7)	**10** (2, 0)
3 (2, 7)	**7** (−2, 2)	
4 (1, 5)	**8** (−4, −2)	

11 $\begin{pmatrix} 4 \\ 1 \end{pmatrix}$ **13** $\begin{pmatrix} 4 \\ 3 \end{pmatrix}$ **15** $\begin{pmatrix} -2 \\ -2 \end{pmatrix}$

12 $\begin{pmatrix} -1 \\ 1 \end{pmatrix}$ **14** $\begin{pmatrix} 4 \\ 0 \end{pmatrix}$ **16** $\begin{pmatrix} 1 \\ 1 \end{pmatrix}$

17 (5, 6)   **18** (−2, 3)   **19** (−4, −5)

Exercise 10i   page 174

1 $\overrightarrow{AA'} = \begin{pmatrix} -5 \\ 1 \end{pmatrix}$, $\overrightarrow{BB'} = \begin{pmatrix} -5 \\ 1 \end{pmatrix}$, $\overrightarrow{CC'} = \begin{pmatrix} -5 \\ 1 \end{pmatrix}$
Yes, Yes.

2 $\overrightarrow{LL'} = \begin{pmatrix} 4 \\ 2 \end{pmatrix}$, $\overrightarrow{MM'} = \begin{pmatrix} 4 \\ 2 \end{pmatrix}$, $\overrightarrow{NN'} = \begin{pmatrix} 4 \\ 3 \end{pmatrix}$,
No, No.

3 $\begin{pmatrix} 6 \\ 3 \end{pmatrix}$ **4 a** $\begin{pmatrix} -5 \\ 4 \end{pmatrix}$ **b** $\begin{pmatrix} 5 \\ -4 \end{pmatrix}$

5 **a** $\begin{pmatrix} 0 \\ -4 \end{pmatrix}$ **b** $\begin{pmatrix} -6 \\ 0 \end{pmatrix}$ **c** $\begin{pmatrix} 5 \\ 5 \end{pmatrix}$ **d** $\begin{pmatrix} 0 \\ 0 \end{pmatrix}$

6

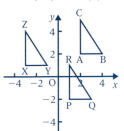

a $\begin{pmatrix} -1 \\ -4 \end{pmatrix}$ **b** $\begin{pmatrix} 1 \\ 4 \end{pmatrix}$ **c** $\begin{pmatrix} -4 \\ 3 \end{pmatrix}$ **d** $\begin{pmatrix} 0 \\ 0 \end{pmatrix}$

7

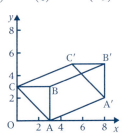

Yes, $\begin{pmatrix} 5 \\ 2 \end{pmatrix}$, parallelogram – the opposite sides are parallel. AA′C′C, BB′C′C

8 **a**

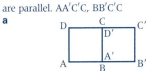

b

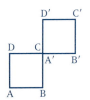

9

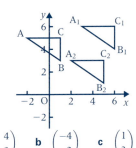

a $\begin{pmatrix} 4 \\ -2 \end{pmatrix}$ **b** $\begin{pmatrix} -4 \\ 2 \end{pmatrix}$ **c** $\begin{pmatrix} 1 \\ 3 \end{pmatrix}$

Chapter 11

Exercise 11a   page 178

1 **a** $\frac{1}{4}$ **b** $\frac{1}{2}$ **c** $\frac{1}{3}$

2 **a**, **b** and **c**

Exercise 11b   page 179

1 4, 2, 3

2 **a** 6   **b** 2

3

4

5

6

7

8

9 90°, 120°, 180°, 90°, 120°, 180°

Exercise 11c   page 180

1 rotational   **4** line   **7** both
2 rotational   **5** both   **8** both
3 line   **6** both   **9** rotational

Exercise 11d   page 182

1 90° clockwise   **6** (1, 0), 90° anticlockwise
2 90° clockwise   **7** (1, 0), 180°
3 180° either way   **8** (2, 0), 180°
4 90° clockwise   **9** (2, 1), 90° clockwise
5 origin, 180°   **10** (3, 1), 180°

11

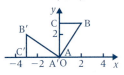

12

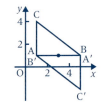

13
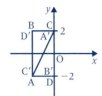

14
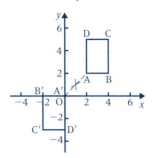

15
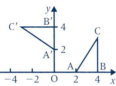

16
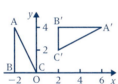

17
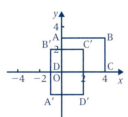

18
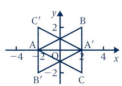

19
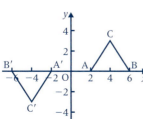

a a semicircle   **b** OC = OC′, OB = OB′

20 Because for 180° the direction of rotation does not matter.

21

Exercise 11e   page 186

1 c (0, 4)   **e** 90° clockwise
2 c (−2, −2)   **e** 90° clockwise
3 c (−1, 3)   **e** 90° anticlockwise

Exercise 11f   page 187

1 90° anticlockwise   **2** 90° clockwise

Exercise 11g   page 188

Simple models may again prove useful.

1 Translation given by $\begin{pmatrix} -2 \\ 2 \end{pmatrix}$

2 Reflection in $x = 0$
3 Reflection in $x = \frac{1}{2}$

4 Translation given by $\begin{pmatrix} -2 \\ 0 \end{pmatrix}$

5 Reflection in $y = -x$
6 Rotation through 90° anticlockwise about (−1, −1)
7 Rotation through 90° anticlockwise about (0, 1)
8 Rotation through 180° about (0, 2)

9 Rotation through 180° about $\left(\frac{5}{2}, \frac{3}{2} \right)$

10 Reflection in $y = x + 1$
11 Reflection in BC, rotation about B through 90° clockwise
12 Reflection in y-axis, rotation about O through 180°, translation parallel to x axis.
13 (1) Reflection in OB
    (2) Translation parallel to AB
    (3) Rotation about B through 120° clockwise
    (4) Rotation about O through 120° clockwise
14 (1) Reflection in BE
    (2) Translation parallel to AB
    (3) Rotation about B through 90° clockwise
    (4) Rotation about the midpoint of BE, through 180°
    (5) Rotation about E through 90° anticlockwise

15 Translation given by the vector $\begin{pmatrix} -8 \\ 0 \end{pmatrix}$

16 Translation given by the vector $\begin{pmatrix} 4 \\ 0 \end{pmatrix}$

17

centre of the turning circle

STP Caribbean Mathematics 2

18 Rotations about different vertices, reflections, translations

19

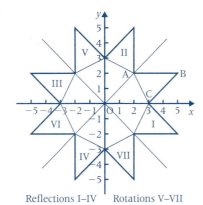

Reflections I–IV   Rotations V–VII

20 a Reflection in the line $y = -x$
   b Yes

CHAPTER 12

Exercise 12a   page 194

1 12 cm
2 10 m
3 30 mm
4 7 cm
5 2 km
6 9.2 cm
7 approx 3.14
8 approx 3.14

Exercise 12b   page 195

1 14.5 m
2 28.9 cm
3 18.2 cm
4 333 mm
5 54.7 m
6 1570 mm
7 226 cm
8 30.2 m
9 11.3 m
10 0.0880 km
11 44.0 cm
12 176 mm
13 8.80 m
14 220 mm
15 35.2 cm
16 970 mm
17 88 cm
18 24 m
19 1300 mm
20 220 cm
21 1600 mm
22 2000 cm
23 29 m

Exercise 12c   page 197

1 10.3 cm
2 10.7 cm
3 18.3 cm
4 20.5 cm
5 27.9 cm
6 33.6 cm
7 94.3 cm
8 62.8 mm
9 20.6 cm
10 45.1 cm

Exercise 12d   page 198

1 78.5 mm
2 62.8 mm, 88.0 mm
3 4.4
4 194 cm
5 176 cm
6 176 cm, 200
7 12.6 cm
8 94.3 cm

9 62.8 m
10 6.28 secs, 9.55 revolutions
11 3140 cm
12 12.6 m
13 70.7
14 94.3 m

Exercise 12e   page 201

1 7.00 cm
2 19.3 mm
3 87.5 m
4 43.8 cm
5 73.5 mm
6 132 cm
7 5.76 mm
8 62.2 m
9 92.6 cm
10 13.9 m
11 16.5 m
12 59.8 m
13 31.8 cm
14 20.0 m
15 4.93 cm
16 9.55 cm each
17 3.82 cm, 45.8 cm
18 37.7 cm
19 4.77 cm
20 9.55 cm
21 9.55 cm, 29.1 cm

Exercise 12f   page 203

1 50.2 cm^2
2 201 m^2
3 78.5 m^2
4 78.5 mm^2
5 38.5 cm^2
6 11 300 cm^2
7 45.3 m^2
8 9.62 km^2
9 20 100 m^2
10 25.1 cm^2
11 51.3 m^2
12 58.9 cm^2
13 118 mm^2
14 451 mm^2
15 374 cm^2
16 457 cm^2
17 714 m^2
18 943 cm^2
19 3540 cm^2
20 193 cm^2

Exercise 12g   page 205

1

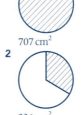

707 cm^2

2

236 cm^2

3 491 mm^2

4

26.2 cm^2

5 No
6 21.5 cm^2
7 8, 110 cm^2
8 11 700 cm^2
9 2

Exercise 12h   page 207

1 17.6 mm
2 9.55 m
3 37.7 cm
4 26.4 m^2
5 491 cm^2
6 28.6 mm
7 7.95 cm^2

Exercise 12i   page 207

1 62.8 m
2 452 cm^2
3 57.3 cm
4 50.2 m^2
5 89.2 mm
6 40.9 cm
7 87.5 cm^2

Exercise 12j   page 208

1 12.6 km^2
2 308 mm
3 14 m
4 154 cm^2
5 32.2 cm^2
6 18.1 m^2

CHAPTER 13

Exercise 13a   page 210

1

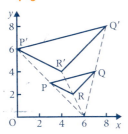

Centre of enlargement is (6, 0)

2

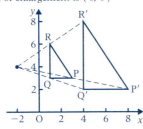

Centre of enlargement is (−2, 4)

3

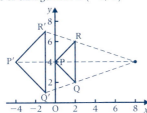

Centre of enlargement is (8, 4)

4 In 1 PQ∥P′Q′, PR∥P′R′, RQ∥R′Q′

In 2 PQ∥P′Q′, PR∥P′R′, RQ∥R′Q′
In 3 PQ∥P′Q′, PR∥P′R′, RQ∥R′Q′

5

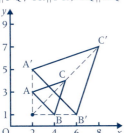

a   Centre of enlargement is (2, 1)

6

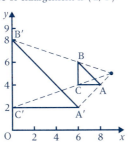

Centre of enlargement is (9, 5)

7

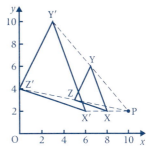

Centre of enlargement is (10, 2)

Exercise 13b   page 212

1

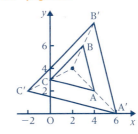

Centre of enlargement is (2, 4)

2 Centre of enlargement is (2, 2)

3

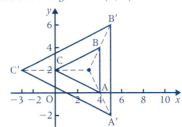

Centre of enlargement is (3, 2)

4

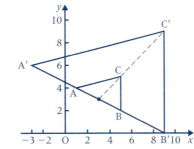

Centre of enlargement is (3, 3)

Exercise 13c   page 214

1

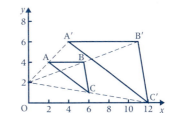

2

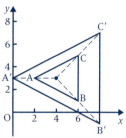

3

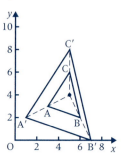

4

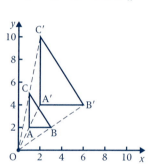

5

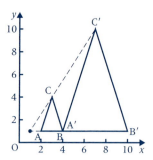

6

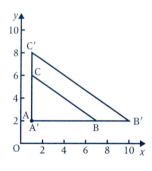

9

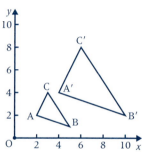

10

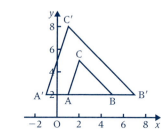

Exercise 13d   page 216

1 $(6, 3), \frac{1}{3}$　　　　**3** $(3\frac{1}{2}, 4), \frac{1}{3}$

2 $(-1, 0), \frac{1}{2}$　　　　**4** $(1, 2), \frac{1}{2}$

5

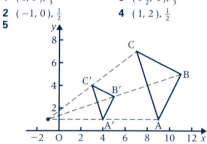

6

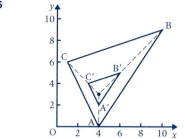

Exercise 13e   page 218

1 $(5, 6), -2$

2 $(0, 1), -3$

3

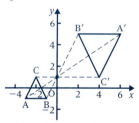

Centre $(0, 1)$, scale factor -2

4

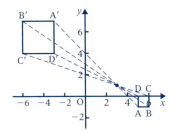

Centre (3, 1), scale factor −3

5

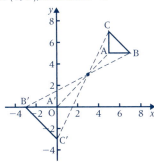

Centre (3, 3), scale factor $-1\frac{1}{2}$

6

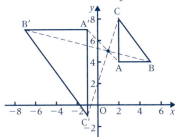

Centre (1, 5), scale factor −2

7

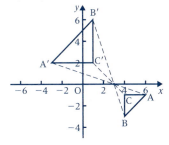

Centre (3, 0), scale factor −2

8

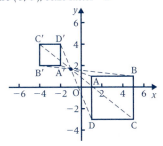

Centre (−1, $1\frac{2}{3}$), scale factor $-\frac{1}{2}$

9

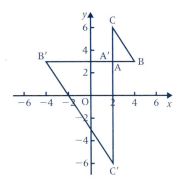

Centre (2, 3), scale factor −3

10 a Centre (0, 0), scale factor −1
    b Rotation about 0 through 180°

11

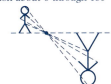

12

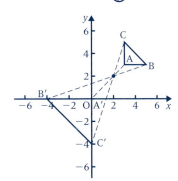

13

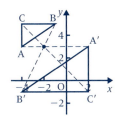

14

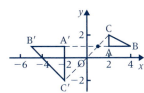

CHAPTER 14

Exercise 14a page 224

1 yes	**5** yes	**9** no
2 no	**6** yes	**10** no
3 yes	**7** yes	**11** A and D
4 no	**8** no	

Exercise 14b   page 226

1 a yes
 b AC = 4.1 cm, CB = 3.2 cm, A′C′ = 8.2 cm, C′B′ = 6.4 cm
 c each is 2
 d all are equal to 2
2 a yes
 b AC = 8.6 cm, CB = 7.7 cm, A′C′ = 5.7 cm, C′B′ = 5.1 cm
 c each is 0.67 or $\frac{2}{3}$
 d all equal to 0.67
3 a yes
 b AC = 7.9 cm, CB = 6.4 cm, A′C′ = 3.9 cm, C′B′ = 3.2 cm
 c each is 0.5 or $\frac{1}{2}$
 d all equal to 0.5
4 a yes
 b AC = 10.1 cm, CB = 6.6 cm, A′C′ = 7.6 cm, C′B′ = 4.9 cm
 c each is 0.75 or $\frac{3}{4}$
 d all equal 0.75
5 a yes
 b AC = 6.1 cm, CB = 9.2 cm, A′C′ = 9.2 cm, C′B′ = 13.8 cm
 c each is 1.5 or $\frac{3}{2}$
 d all equal 1.5
6 80°, 52°, yes
7 72°, 72°, yes
8 70°, 70°, yes
9 93°, 52°, no

Exercise 14c   page 228

1 yes, $\dfrac{AB}{PQ} = \dfrac{BC}{QR} = \dfrac{AC}{PR}$
2 yes, $\dfrac{AB}{PR} = \dfrac{BC}{RQ} = \dfrac{AC}{PQ}$
3 no
4 yes, $\dfrac{AC}{QP} = \dfrac{CB}{PR} = \dfrac{AB}{QR}$
5 yes, $\dfrac{AB}{PQ} = \dfrac{BC}{QR} = \dfrac{AC}{PR}$
6 yes, $\dfrac{AB}{RP} = \dfrac{BC}{PQ} = \dfrac{AC}{RQ}$
7 yes, $\dfrac{AB}{RQ} = \dfrac{BC}{QP} = \dfrac{AC}{RP}$
8 no

Exercise 14d   page 230

1 yes, 2.5 cm
2 yes, 7.2 cm
3 no
4 yes, 6.3 cm
5 7.5 cm
6 7.5 cm
7 $8\frac{1}{3}$ cm
8 $4\frac{1}{2}$ cm
9 b 4 cm
10 b CD = 9 cm, DE = 10.5 cm
11 b 5 cm
12 b DE = 18 cm, AE = 13.5 cm, CE = 4.5 cm

Exercise 14e   page 234

1 8 cm
2 6 cm
3 10 cm
4 30 cm
5 24 cm

Exercise 14f   page 236

1 yes, $\widehat{P}$
2 yes, $\widehat{Q}$
3 no
4 yes, $\widehat{P}$
5 no
6 yes, $\widehat{Q}$
7 yes, $\widehat{B} = \widehat{D}$, $\widehat{C} = \widehat{E}$, they are parallel

Exercise 14g   page 238

1 yes, CB = 3.6 cm
2 no
3 yes, RQ = 35 cm
4 yes, RQ = 7.2 cm
5 yes, AC = $10\frac{2}{3}$ cm
6 5.1 cm
7 3 cm

Exercise 14h   page 241

1 yes, 4 cm
2 yes, 2.4 cm
3 yes, 83°
4 no
5 yes, 34°
6 yes, 32°
7 yes, $3\frac{1}{2}$ cm
8 yes, 18 cm
9 b AC = 3.15 cm, CE = 1.05 cm
10 b 143 cm
11 c yes
12 10 m
13 19.2 m
14 60 cm

CHAPTER 15

Exercise 15a   page 245

1 216 cm³
2 432 m³
3 180 000 cm³
4 105.4 cm³
5 1600 mm³
6 58.5 cm³
7 403.2 mm³
8 49.68 m³
9 0.000 008 cm³
10 0.3968 cm³
11 112.5 cm3
12 189 cm³
13 129.6 cm³
14 133.28 m³
15 144.6 cm³
16 230 400 cm³

Exercise 15b   page 248

1 720 cm³
2 2160 cm³
3 1120 cm³
4 720 cm³
5 1242 cm³
6 128 cm³
7 660 cm³
8 192 cm³
9 2400 cm³
10 2880 cm³
11 315 cm³
12 450 cm³
13 690 cm³
14 624 cm³
15 864 cm³
16 720 cm³
17 5.184 m³
18 1344 cm³
19 624 m³

Exercise 15c   page 252

1 126 cm³
2 113 cm³
3 314 cm³
4 59.4 cm³
5 3.14 cm³
6 15.1 m³
7 37.7 cm³
8 50.9 cm³
9 4520 cm³
10 1390 cm³
11 322 cm³
12 407 cm³
13 330 cm³
14 652 cm³
15 70 800 cm³
16 2810 cm³
17 941 mm³
18 825 cm³
19 1.60 m³
20 44.0 cm³

Exercise 15d   page 253

1 1000 cm³
2 402 cm³
3 34.5 cm³
4 203 cm³
5 628 cm³
6 2160 cm³

CHAPTER 16

Exercise 16a   page 256

1 9
2 25
3 81
4 900
5 0.16
6 2500
7 90 000
8 0.0004
9 250 000
10 100
11 0.09
12 4 000 000
13 0.000 016
14 1
15 0.0009
16 900
17 10 000
18 16
19 0.09
20 64
21 1600
22 1 000 000
23 4900
24 0.0009
25 8100
26 0.0064
27 40 000

Exercise 16b   page 256

1	60.84	**18**	134.6
2	1444	**19**	58 080
3	6273	**20**	0.6790
4	0.1681	**21**	0.7726
5	0.0256	**22**	0.001 310
6	0.001 024	**23**	5242
7	2323	**24**	14.29
8	127.7	**25**	0.0603
9	2632	**26**	0.005 184
10	96.04	**27**	201.6
11	146.4	**28**	20 160
12	8.644	**29**	0.020 16
13	1.040	**30**	94.67
14	185.0	**31**	193.2
15	289	**32**	0.005 285
16	1.232	**33**	**c** 4.8, 3.2, 9.6, 7.3
17	51.98	**34**	**c** 30, 70, 164, 185

Exercise 16c   page 258

1	5.76 cm^2	**4**	1.12 m^2	**7**	0.003 84 m^2
2	92.2 m^2	**5**	296 cm^2	**8**	105 000 km^2
3	1050 cm^2	**6**	2700 mm^2	**9**	0.0961 cm^2

Exercise 16d   page 258

1	3	**7**	7	**13**	70	**19**	0.3
2	5	**8**	8	**14**	700	**20**	0.4
3	2	**9**	1	**15**	0.2	**21**	0.02
4	9	**10**	90	**16**	20	**22**	500
5	10	**11**	0.9	**17**	50	**23**	2000
6	6	**12**	0.8	**18**	100	**24**	0.004

Exercise 16e   page 259

1	4. – – –	**6**	3. – – –	**11**	0.4 – – –
2	3. – – –	**7**	9. – – –	**12**	9. – – –
3	6. – – –	**8**	4. – – –	**13**	3. – – –
4	6. – – –	**9**	2. – – –	**14**	0.7 – – –
5	1. – – –	**10**	0.2 – – –	**15**	2. – – –

Exercise 16f   page 259

1	30	**8**	200	**15**	2000
2	200	**9**	60	**16**	60
3	20	**10**	100	**17**	20
4	80	**11**	600	**18**	3
5	20	**12**	10	**19**	1
6	100	**13**	20	**20**	6
7	50	**14**	200		

Exercise 16g   page 260

1	6.20	**8**	2.39	**15**	101
2	4.45	**9**	25.5	**16**	642
3	20.7	**10**	8.06	**17**	27.0
4	65.0	**11**	3.35	**18**	85.3
5	5.66	**12**	7.62	**19**	7.81
6	3.13	**13**	4.90	**20**	2700
7	8.19	**14**	4.36		

21 37.4, 250, 25.0, 84.9, 26.8, 118, 57.1, 204, 64.5, 629, 19.9, 122, 27.5, 275, 2750, 64.2, 27.0, 3.91, 1.92, 6.28, 19.9

Exercise 16h   page 261

1	0.205	**4**	0.748	**7**	0.775
2	0.648	**5**	0.0118	**8**	0.527
3	0.118	**6**	0.707	**9**	0.167

10	0.0527	**14**	0.831	**18**	0.566
11	0.548	**15**	0.208	**19**	0.228
12	0.416	**16**	0.0980	**20**	0.866
13	0.447	**17**	0.912	**21**	0.008 54

Exercise 16i   page 262

1	9.22 cm	**5**	0.245 m	**9**	0.0922 km
2	11.0 cm	**6**	3.89 cm	**10**	7.68 cm
3	22.4 m	**7**	27.4 mm	**11**	15.5 m
4	5.66 m	**8**	290 km	**12**	7.81 cm

CHAPTER 17

Exercise 17a   page 264

1	**a**	90 km	**b**	2 hours	**c**	45 km
2	**a**	30 km	**b**	3 hours	**c**	10 km
3	**a**	107 km	**b**	3.2 hours	**c**	33.4 km
4	**a**	50 miles	**b**	2 hours	**c**	25 miles

Exercise 17b   page 266

The scales in some of these answers have been halved.

1

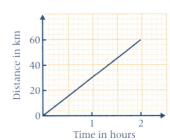

2

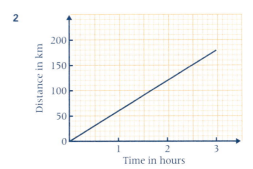

3

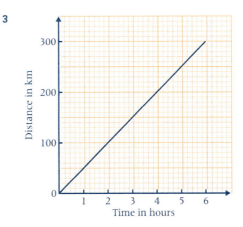

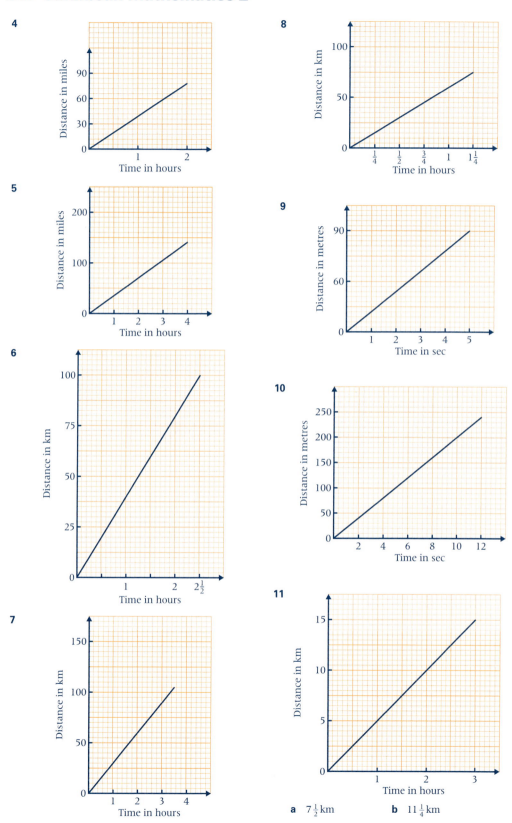

4

5

6

7

8

9

10

11

a $7\frac{1}{2}$ km   b $11\frac{1}{4}$ km

12

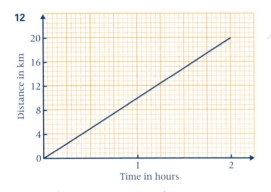

| **a** | $7\frac{1}{2}$ km | **b** | $12\frac{1}{2}$ km |

13

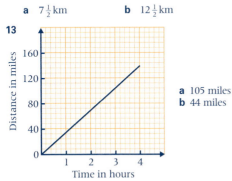

a 105 miles
b 44 miles

14

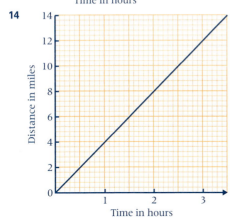

	a	2 miles	**b**	14 miles		
15	**a**	800 km	**b**	1100 km		
16	**a**	48 km	**b**	84 km	**c**	54 km
17	**a**	1200 miles	**b**	1650 miles		
18	**a**	90 km	**b**	135 km		
19	**a**	9 miles	**b**	15 miles		
20	**a**	52.5 m	**b**	89.25 m		
21	**a**	32 miles	**b**	38 miles		
22	**a**	4 km	**b**	$2\frac{2}{3}$ km	**c**	10 km
23	**a**	37 miles	**b**	185 miles		
24	**a**	500 m	**b**	850 m		
25	**a**	1755 miles	**b**	4185 miles		
26	**a**	30	**b**	72		

Exercise 17c   page 268

1	**a**	2 hours	**b**	3 hours
2	**a**	5 hours	**b**	$3\frac{1}{4}$ hours
3	**a**	$\frac{1}{2}$ hour	**b**	$1\frac{1}{4}$ hours

4	**a**	$2\frac{1}{2}$ hours	**b**	$5\frac{1}{3}$ hours
5	**a**	$1\frac{1}{2}$ hours	**b**	5 hours
6	**a**	$1\frac{1}{2}$ hours	**b**	$4\frac{1}{2}$ hours
7	**a**	25 sec	**b**	200 sec
8	**a**	24 min	**b**	54 min
9	**a**	216 hours = 9 days	**b**	$5\frac{1}{4}$ days
10	**a**	$1\frac{1}{4}$ hours	**b**	$2\frac{3}{4}$ hours
11	**a**	$2\frac{1}{2}$ hours	**b**	5 hours 20 min
12	**a**	$\frac{3}{4}$ hour	**b**	$3\frac{1}{4}$ hours

Exercise 17d   page 269

1	80 km/h	**17**	12 km/h
2	60 km/h	**18**	8 km/h
3	60 m.p.h.	**19**	18 km/h
4	120 m.p.h.	**20**	18 km/h
5	20 m/s	**21**	54 m.p.h.
6	45 m/s	**22**	54 m.p.h.
7	50 km/h	**23**	60 m.p.h.
8	65 km/h	**24**	105 m.p.h.
9	35 m.p.h.	**25**	74.33 km/h
10	8 m.p.h.	**26**	138.5 km/h
11	36 m/s	**27**	693.3 km/h
12	17 m/s	**28**	482 km/h
13	80 km/h	**29**	162.86 km/h
14	90 km/h	**30**	253.62 km/h
15	64 km/h	**31**	102.97 km/h
16	120 km/h		

Exercise 17e   page 272

1	9 km/h	**4**	7 m.p.h.	**7**	3 knots
2	10 m.p.h.	**5**	75 km/h		
3	7 m.p.h.	**6**	200 km/h		

Exercise 17f   page 272

1	**a**	**i** 1215	**ii** 1348	**iii** 1445
	b	$2\frac{1}{2}$ hours		
	c	**i** $1\frac{1}{4}$ hours	**ii** $1\frac{1}{4}$ hours	
	d	64 km/h		
2	**a**	**i** 90 km	**ii** 50 km	
	b	5 hours	**d**	28 km
	c	28 km/h	**e**	**i** 42 km **ii** 48 km
3	**a**	45 km	**d**	1 hour
	b	$1\frac{1}{2}$ hours	**e**	45 km/h
	c	30 km/h	**f**	32 km/h
4	**a**	**i** B	**ii** B	
	b	**i** 80 km/h	**ii** 64 km/h	
	c	$\frac{1}{2}$ hour		
	d	$2\frac{3}{4}$ hour		
	e	58.2 km/h (counting the stop)		

Exercise 17g   page 276

1	**a**	150 km/h	**d**	1 hour
	b	2 hours	**e**	1330; $2\frac{1}{2}$ hours
	c	75 km/h	**f**	60 km/h
2	**a**	First 34 m.p.h.; second 40 m.p.h.		
	b	at 11.15, 40 miles from Kingston		
	c	60 miles		
3	**a**	56 miles	**c**	56 m.p.h.
	b	45 minutes	**d**	36 m.p.h.

4 a i 0830   **ii** 1330
  b 5 h
  c $1\frac{1}{2}$ hours
  d 4 km/h
  e 7 hours
5 a 80 km/h, 1430
  b 100 km/h, 1354
  c at 1410, 153 miles from A
  d 52 miles
6 a Betty, Chris, Audrey
  b 10 km/h
  c 15 km/h
  d 20 km/h
  e at 2.30 p.m., after 25 km
  f Audrey 10 km, Betty 9 km, Chris 15 km
  g $2\frac{1}{2}$ km
7 a at 3.23 p.m., $9\frac{1}{2}$ miles from Jane's home
  b 3.8 miles
8 a at 3.14 p.m., 60 miles from A
  b 6 miles from B
  c 26 miles from A

Exercise 17h   page 281

1 a 20 km   **b** $2\frac{1}{2}$ hours   **c** 8 km/h
2 a 35 km   **b** $1\frac{1}{4}$ hours
3 a 14 hours   **b** 57 hours
4 80 km/h
5 420 m.p.h.
6 5 km/h
7 a 15 km   **d** 10 km/h
  b $1\frac{1}{2}$ hours   **e** 45 km/h
  c 10 min

Exercise 17i   page 282

1 a 175 km   **c** the bus stopped
  b $\frac{3}{4}$ hour   **d** 120 km/h
2

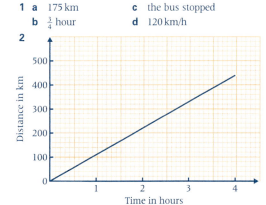

3 a 900 m   **b** 1575 m, 54 km/h
4 a 3 hours   **b** $1\frac{3}{4}$ hours
5 200 km/h by 5.6 m/s or 20 km/h
6 12 km/h
7 48 m.p.h.

REVIEW TEST 2   page 284

1 D   **3** A   **5** B   **7** D   **9** B
2 A   **4** B   **6** C   **8** B
10 a $(2, -5), (3, -8), (7, -1)$
  b $(-2, 5), (-3, 8), (-7, 1)$
  c $(5, 2), (8, 3), (1, 7)$

11 yes, 3.6 cm
12 49.82 cm^2
13 480 cm^3
14 a i 1870   **ii** 0.0210
  b $(1, 3), (5, 0), (-2, -5)$
15 missing values are 60, 80 160, 200
  a 140 cm   **b** 4.5 seconds
16 4 km/h

CHAPTER 18

Exercise 18a   page 287

1 b $26\frac{1}{2}°$ **c** 0.5
2 b $26\frac{1}{2}°$ **c** 0.5
3 b $26\frac{1}{2}°$ **c** 0.5
4 b $26\frac{1}{2}°$ **c** 0.5
5 b $26\frac{1}{2}°$ **c** 0.5
6 yes
7 b 37°   **c** 0.75
8 b 37°   **c** 0.75
9 b 31°   **c** 0.6
10 b 31°   **c** 0.6
11 b 50°   **c** 1.2
12 b 50°   **c** 1.2
13 $\dfrac{B_1 C_1}{AB_1} = \dfrac{B_2 C_2}{AB_2} = \dfrac{B_3 C_3}{AB_3}$

14

	Angle A	$\dfrac{BC}{AB}$
1	$26\frac{1}{2}°$	0.5
2	$26\frac{1}{2}°$	0.5
3	$26\frac{1}{2}°$	0.5
4	$26\frac{1}{2}°$	0.5
5	$26\frac{1}{2}°$	0.5
6	37°	0.75
7	37°	0.75
8	31°	0.6
9	31°	0.6
10	50°	1.2
11	50°	1.2

Exercise 18b   page 290

1 0.364   **7** 0.344   **13** 0.0699
2 0.532   **8** 0.213   **14** 0.754
3 3.08   **9** 0.384   **15** 0.9661
4 1.33   **10** 1.00   **16** 57.3
5 1.66   **11** 1.80   **17** 1.28
6 0.158   **12** 2.75   **18** 0.700
19

Angle	Tangent of angle
26.5°	0.5
37°	0.754
31°	0.601
50°	1.19

Exercise 18c   page 290

1 0.277   **7** 0.591   **13** 0.913
2 0.568   **8** 0.285   **14** 2.94
3 0.202   **9** 0.180   **15** 1.17
4 1.74   **10** 0.0664   **16** 1.65
5 2.86   **11** 1.15   **17** 2.17
6 1.05   **12** 0.642   **18** 1.98

Exercise 18d   page 291

1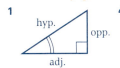

4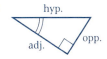

2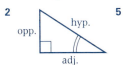

5

3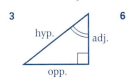

6

Exercise 18e   page 291

1 5.40 cm	**7** 7.77 cm	**13** 6.43 cm
2 5.81 cm	**8** 3.12 cm	**14** 5.22 cm
3 0.975 cm	**9** 7.00 cm	**15** 3.00 m
4 4.55 cm	**10** 5.40 cm	**16** 17.8 cm
5 1.43 cm	**11** 4.50 cm	
6 5.38 cm	**12** 7.05 cm	

Exercise 18f   page 294

1 5.77 cm	**5** 9.99 cm	**9** 17.9 cm
2 4.60 cm	**6** 14.1 cm	**10** 126 cm
3 3.68 cm	**7** 34.5 cm	
4 5.60 cm	**8** 35.0 cm	

Exercise 18g   page 295

1 14.3 cm	**7** 3.23 cm
2 17.9 cm	**8** 30.8 cm
3 8.16 cm	**9** 5.66 m
4 10.1 cm	**10** 1.40 m
5 5.10 m	**11 a** 16° **b** 17.2 m
6 69.9 m	

Exercise 18h   page 297

1 65.6°	**8** 31.8°	**15** 48.7°
2 19.8°	**9** 34.0°	**16** 74.4°
3 22.3°	**10** 44.8°	**17** 48.1°
4 76.3°	**11** 69.4°	**18** 59.5°
5 54.5°	**12** 18.4°	**19** 45.3°
6 17.2°	**13** 34.9°	**20** 50.4°
7 9.1°	**14** 39.0°	

Exercise 18i   page 297

1 31.0°	**9** 66.0°	**17** 49.4°
2 38.7°	**10** 56.3°	**18** 59.0°
3 26.6°	**11** 6.8°	**19** 23.2°
4 21.8°	**12** 67.4°	**20** 12.5°
5 35.0°	**13** 18.4°	**21** 35.5°
6 8.5°	**14** 8.1°	**22** 66.8°
7 51.3°	**15** 9.5°	**23** 24.0°
8 20.6°	**16** 39.8°	**24** 53.1°

Exercise 18j   page 298

1 38.7°	**5** 26.6°	**9** 33.7°
2 33.7°	**6** 22.8°	**10** 57.5°
3 55.0°	**7** 8.8°	**11** 36.9°
4 50.2°	**8** 59.0°	**12** 24.4°

13 26.6°	**17** 30.3°	**21** 56.3°
14 29.7°	**18** 51.3°	**22** 52.1°
15 33.7°	**19** 42.5°	
16 51.3°	**20** 41.2°	

Exercise 18k   page 300

1 31.0°	**5** 10.2 km
2 26.6°	**6** 26.6°, 45.0°, 18.4°
3 59.0°, 59.0°, 62.0°	**7** 3.08 m
4 56.3°	

8

9 75.6°, 104.4°, 75.6°. 104.4°
10 CÂB = 24.6°, 130.8°
11 15.4 cm

CHAPTER 19

Exercise 19a   page 304

1 0.438	**6** 0.951	**11** 56.5°	**16** 4.0°
2 0.995	**7** 0.289	**12** 24.4°	**17** 40.3°
3 0.429	**8** 0.073	**13** 39.7°	**18** 20.9°
4 0.603	**9** 0.886	**14** 44.7°	**19** 25.3°
5 0.981	**10** 0.946	**15** 69.6°	**20** 15.1°

Exercise 19b   page 305

1 8.83 cm	**9** 36.9°
2 6.22	**10** 30.0°
3 1.95 cm	**11** 42.2°
4 1.07 cm	**12** 45.6°
5 9.54 cm	**13** 2.06 cm
6 4.85 cm	**14** 6.64 cm
7 44.4°	**15** Â = 36.9°, Ĉ = 53.1°
8 23.6°	**16** 28.2°

Exercise 19c   page 307

1 0.515	**8** 0.971	**15** 34.9°
2 0.669	**9** 0.954	**16** 76.1°
3 0.998	**10** 0.954	**17** 20.3°
4 0.708	**11** 64.2°	**18** 42.4°
5 0.498	**12** 44.4°	**19** 51.1°
6 0.391	**13** 45.6°	**20** 32.5°
7 0.139	**14** 19.4°	

Exercise 19d   page 308

1 8.48 cm	**6** 0.799 cm	**11** 41.4°
2 2.68 cm	**7** 53.1°	**12** 63.3°
3 5.07 cm	**8** 41.4°	**13** 66.4°
4 3.22 cm	**9** 38.7°	**14** 56.9°
5 2.78 cm	**10** 60.0°	

Exercise 19e   page 310

1 tan A	**5** tan X	**9** sin N
2 cos A	**6** cos M	**10** cos F
3 sin Q	**7** tan A	
4 sin P	**8** cos P	

11 81.9°, 31.0°, 48.6°, 33.1°, 59.0°, 68.0°, 2.44 cm,
4.90 cm, 6.43 cm, 0.647 cm, 30.9 cm, 13.9 cm

12 36.9°   **15** 61.0°   **18** 1.09 cm
13 49.5°   **16** 3.06 cm   **19** 320 cm
14 41.8°   **17** 0.282 cm

Exercise 19f   page 312

1 44.4°, 45.6°   **7** 13.4 cm
2 4.50 cm   **8** 41.8°
3 71.9°, 18.1°   **9** 45.6°
4 7.61 cm   **10** BC = 2.69 cm
5 12.2 cm   **11** 5.56 cm
6 15.9 m   **12** $\widehat{A} = 54°, \widehat{C} = 36°$

Exercise 19g   page 314

1 48.6°   **7** 48.2°, 83.6°
2 5.15 cm   **8** $\widehat{A} = 65.4°, 65.4°, 49.2°$
3 51.3°   **9** 7.18°
4 53.1°   **10** 9.59°
5 1.69 m   **11** 5.74°
6 7.45 cm   **12** 2.87°
13 a 040°   **b** 090°   **c** 270°   **d** 3.83 km
14 a 6.82 km   **b** 7.31 km
15 a east   **b** 1.16 km   **c** 1.24 km north
16 a 10.1 km **b** 3 km   **c** 7.1 km   **d** 11.7 km

Exercise 19h   page 316

1 a 0.643
  b 0.643; equal
2 a 0.8
  b 0.8; 90°
3 0.3
4 0.8
5 45°, isosceles, 1

Exercise 19i   page 317

1 0.9925   **4** 30.0°   **7** 6.75 cm
2 58.5°   **5** 6.25 cm
3 0.8829   **6** 53.1°

Exercise 19j   page 317

1 0.906   **4** 21.4°   **6** 30.0°
2 68.6°   **5** 12.3 cm   **7** 7.14 cm
3 1.00

CHAPTER 20

Exercise 20a   page 320

1 a 36°C   **b** 78°C   **c** 77°F   **d** 176°F
2 a £112   **b** £67   **c** $174   **d** $109
3 a EC $432 **b** EC $891 **c** US $163 **d** US $363
4 constant speed
  a 12 km   **b** 21 km
  c 1 hour 40 minutes   **d** $3\frac{1}{2}$ hours
5 constant speed
  a 825 km   **b** 2475 km
  c 1 hour 49 minutes   **d** 4 hours 33 minutes
6 a 54%, 77%   **b** $32\frac{1}{2}$, 52
7 a $43.75   **b** $84   **c** $117.25
  d $114.30   **e** $251.43
8 a 34 mpg   **b** 22 km/l   **c** 64 mpg
  d 8 km/l (to nearest unit)
9 a 39 m/s   **b** 166 km/h **c** 65 km/h
  d 49 m/s (to nearest unit)
10 a 9.5 cm   **b** 5.8 cm   **c** 6.5 cm   **d** 9.2 cm

Exercise 20b   page 324

1 a 1290 g   **b** 7 mm
2 a i $8\frac{3}{4}$ s   **ii** $15\frac{1}{2}$ s
  b i 136 km/h   **ii** 191 km/h
3 a i 290 g   **ii** 930 g
  b i 65 days   **ii** 150 days
  c 240 g
  d 20 g
4 a 84 m/s when $t = 4.55$
  b i 81 m/s   **ii** 61.5 m/s
  c 2.25 s and 6.6 s
5 a 19 knots, £16.40
  b 14.5 knots and 24.2 knots
  c i £17.57   **ii** £17.04
6 a i 386 g   **ii** 1340 g
  b i 3.82 cm   **ii** 5.5 cm
7 a 28.8°C, 27.7°C
  b 1120 a.m., 7.40 p.m.
8 a 1745   **b** December
9 a i 1.7 cm   **ii** 10 cm
  b i 1.3 cm   **ii** 8.6 cm

CHAPTER 21

Exercise 21a   page 330

1 a 2   **b** 3   **c** 7   **d** 12
2 a −1   **b** −6   **c** −8   **d** −20
3 a $-3\frac{1}{2}$ **b** $4\frac{1}{2}$ **c** −6.1   **d** 8.3
4 a −7   **b** 2   **c** $-5\frac{1}{2}$ **d** 4.2
5 a 10   **b** −8   **c** 7   **d** −5.2
6 a −1   **b** 3   **c** −2   **d** $\frac{4}{3}$
7 a 3   **b** −6   **c** $\frac{1}{4}$ **d** −4.1
8 a −2   **b** 4   **c** $-\frac{3}{2}$ **d** $\frac{3}{4}$
9 $a = -5, b = 3, c = -4$
10 $a = -2, b = 8, c = 18$
11 $y = 3x$ **13** $y = -\frac{1}{3}x$
12 $y = -2x$ **14** $y = \frac{2}{3}x$
15 $(-2, -4), (6, 12)$
16 $(-2, 6), (1, -3), (8, -24)$
17 a above $(2, 2), (-2, 1), (-4.2, -2)$
  b below $(3, 0)$

Exercise 21b   page 332

1–6

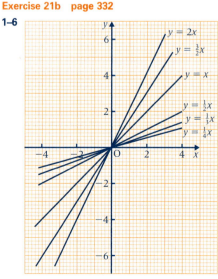

7–12

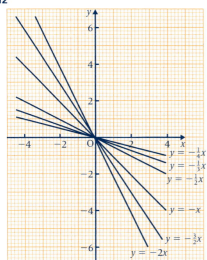

Exercise 21c   page 334

1 a 2   **b** 2   **c** 2
2 a −4   **b** −4   **c** −4
3 a 3   **b** 3   **c** 3
4 a −4   **b** −4   **c** −4
5 2.5
6 −0.5
7 a +   **c** +   **e** −
  b −   **d** −   **f** +

Exercise 21d   page 336

1 $y = 5x$

2 $y = 5x$

3 $y = \frac{1}{2}x$

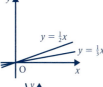

4 $y = -3x$

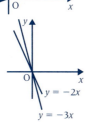

5 $y = 10x$

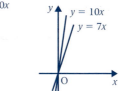

6 $y = -\frac{1}{2}x$

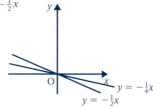

7 $y = -6x$

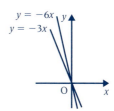

8 $y = 0.75x$

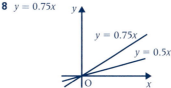

9 acute   **13** acute   **17** obtuse
10 obtuse   **14** acute   **18** obtuse
11 obtuse   **15** acute   **19** obtuse
12 acute   **16** acute   **20** obtuse
21 approximately $\frac{1}{3}$, 1, $-\frac{2}{3}$, 0

Exercise 21e   page 338

1 gradient 3, y-intercept 1, **a** −5 **b** 7
2 gradient −3, y-intercept 4, **a** 7 **b** −5
3 gradient $\frac{1}{2}$, y-intercept 4, **a** 3 **b** 4
4 gradient 1, y-intercept −3, **a** 7 **b** −2
5 gradient $\frac{3}{4}$, y-intercept 3, **a** 4 **b** 2
6 gradient 2, y-intercept −2
7 gradient −2, y-intercept 4
8 gradient 3, y-intercept −4
9 gradient $\frac{1}{2}$, y-intercept 3
10 gradient $-\frac{3}{2}$, y-intercept 3
11 gradient 2, y-intercept 5
12 gradient −2, y-intercept −7
13 gradient −3, y-intercept +2

Exercise 21f   page 340

1 $m = 4, c = 7$ **5** $m = 7, c = 6$
2 $m = \frac{1}{2}, c = -4$ **6** $m = \frac{2}{5}, c = -3$
3 $m = 3, c = -2$ **7** $m = \frac{3}{4}, c = 7$
4 $m = -4, c = 5$ **8** $m = -3, c = 4$

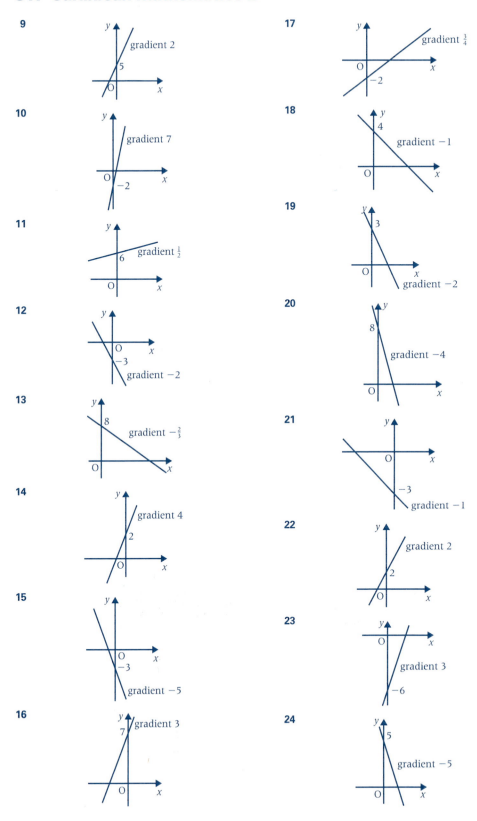

9 gradient 2, 5

10 gradient 7, −2

11 gradient ½, 6

12 −3, gradient −2

13 8, gradient −⅔

14 gradient 4, 2

15 −3, gradient −5

16 gradient 3, 7

17 gradient ¾, −2

18 4, gradient −1

19 3, gradient −2

20 8, gradient −4

21 −3, gradient −1

22 gradient 2, 2

23 gradient 3, −6

24 5, gradient −5

25

26

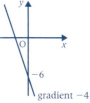

4

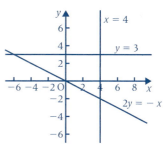

$(4, 3), (4, -2), (-6, 3)$
A right-angled triangle

5

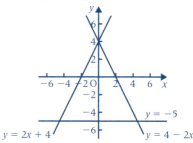

$(0, 4), (4.5, -5), (-4.5, -5)$
An isosceles triangle

Exercise 21g   page 341

1 They are parallel. Their m values are equal.
2 They are parallel. Their m values are equal.

3 Yes	**7** Yes	**11** Yes	**15** No
4 Yes	**8** Yes	**12** Yes	**16** Yes
5 No	**9** No	**13** No	
6 No	**10** Yes	**14** Yes	

Exercise 21h   page 343

1

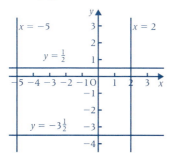

2

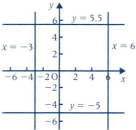

3

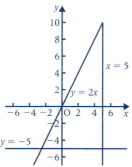

$(5, 10), (5, -5), (-2.5, -5)$
A right-angled triangle

Exercise 21i   page 344

1 $x = 1.5, y = 4.5$
2 $x = 1\frac{1}{3}, y = 3\frac{2}{3}$
3 $x = 1\frac{1}{3}, y = 5\frac{1}{2}$
4 $x = -\frac{1}{2}, y = 1\frac{1}{2}$
5 $x = -\frac{1}{2}, y = 2$
6 $x = 1\frac{1}{2}, y = 3\frac{1}{2}$
7 $x = 2\frac{2}{5}, y = \frac{9}{10}$
8 $x = -\frac{2}{5}, y = 1\frac{3}{5}$
9 $x = 2.4, y = 1.2$
10 $x = \frac{1}{3}, y = 1\frac{2}{3}$

Exercise 21j   page 345

1–4 Each pair of equations gives parallel lines. Parallel lines do not intersect, so there are no solutions.

Exercise 21k   page 345

1 a 2   **b** -4 **c** $\frac{2}{3}$
2 $a = -4, b = \frac{1}{3}, c = -1.5$
3 a $+$ **b** $-$ **c** $+$
4

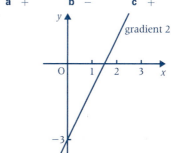

5 a obtuse **c** obtuse
  b acute **d** obtuse
6 $(-1, -6)$

Exercise 21l page 346

1 a 10 **b** 15 **c** $\frac{5}{2}$
2 $a = -5$, $b = 3$, $c = -4$
3 a + **b** − **c** −
4 a gradient 4, y-intercept -7
  b gradient $\frac{5}{2}$, y-intercept 1
  c gradient 3, y-intercept 2
  d gradient $-\frac{1}{3}$, y-intercept -4
5 a Yes **b** No
6

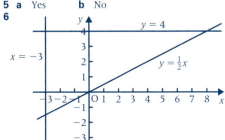

$(-3, 4)$, $(8, 4)$, $(-3, -1\frac{1}{2})$

Exercise 21m page 346

1 a 11 **b** -10 **c** -31
2 $a = -4, b = 11, c = 5$
3

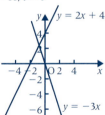

4

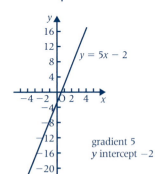

gradient 5
y intercept -2

5 a $y = 2x - 4$
  b $2y = x + 10$
  c $y = -4x - 3$

6

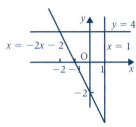

$(1, 4)$, $(1, -4)$, $(-3, 4)$
7 They all pass through the point $(0, 4)$.
8 They all pass through the point $(-1, -6)$.

CHAPTER 22

Exercise 22a page 349

1 ○────── 7
2 ──────○ 4
3 ○────── -2
4 ○────── 0
5 ──────○ -2
6 ○────── $\frac{1}{2}$
7 ──────○ 5
8 ──────○ 0
9 ──────○ 1.5

10 a 2, 3, 4, 6, 7 **b** 2, 5, 7, 8, 9
  c 2, 3, 7, 9 **d** 2, 3, 4, 6, 7
  e 2, 3, 4, 7, 9
12 a $5 > 3$; Yes **b** $1 > -1$; Yes
  c $-2 > -4$; Yes **d** $7 > 5$; Yes
13 a $0 > -1$; Yes **b** $-4 > -5$; Yes
  c $-7 > -8$; Yes **d** $2 > 1$; Yes
14 a $1 < 6$; Yes **b** $-3 < 2$; Yes
  c $-6 < -1$; Yes **d** $3 < 8$; Yes

Exercise 22b page 350

1 $x < 12$ ──────○ 12
2 $x < 2$ ──────○ 2
3 $x > 5$ ○────── 5
4 $x > 2$ ○────── 2
5 $x < -2$ ──────○ -2
6 $x < 3$ ──────○ 3
7 $x < -3$ ──────○ -3
8 $x < -7$ ──────○ -7
9 $x < -5$ ──────○ -5
10 $x < -2$ ──────○ -2
11 $x > -1$ ○────── -1
12 $x < 3$ ──────○ 3
13 $x > 0$ ○────── 0

14 $x > -3$ — diagram, point at -3

15 $x < -3$ — diagram, point at -3

16 $x < 1$ — diagram, point at 1

17 $x > -2$ — diagram, point at -2

18 $x < -5$ — diagram, point at -5

19 $x < 5$ — diagram, point at 5

20 $x < 1$ — diagram, point at 1

21 $x < -1$ — diagram, point at -1

22 $x > 0$ — diagram, point at 0

23 $x > 7$ — diagram, point at 7

24 $x > -5$ — diagram, point at -5

25 $x > -3$ — diagram, point at -3

26 $x < 13$ — diagram, point at 13

27 $x > 12$ — diagram, point at 12

28
a	$24 < 72$	**b**	$3 < 9$
c	$6 < 18$	**d**	$2 < 6$
e	$-24 < -72$	**f**	$-4 < -12$
a Yes	**b** Yes	**c** Yes	
d Yes	**e** No	**f** No	

29
a	$72 > -24$	**b**	$9 > -3$
c	$18 > -6$	**d**	$6 > -2$
e	$-72 > 24$	**f**	$-12 > 4$
a Yes	**b** Yes	**c** Yes	
d Yes	**e** No	**f** No	

30
a	$-36 < -12$	**b**	$-4\frac{1}{2} < -1\frac{1}{2}$
c	$-9 < -3$	**d**	$-3 < -1$
e	$36 < 12$	**f**	$9 < 2$
a Yes	**b** Yes	**c** Yes	
d Yes	**e** No	**f** No	

32 Only when you are multiplying by a positive number.

Exercise 22c   page 351

1 $x < 3$ — diagram, point at 3

2 $x > 1$ — diagram, point at 1

3 $x > 2$ — diagram, point at 2

4 $x < 1$ — diagram, point at 1

5 $x < \frac{1}{2}$ — diagram, point at $\frac{1}{2}$

6 $x > 1\frac{1}{3}$ — diagram, point at $1\frac{1}{3}$

7 $x < 2\frac{1}{4}$ — diagram, point at $2\frac{1}{4}$

8 $x > 1\frac{1}{2}$ — diagram, point at $1\frac{1}{2}$

9 $x \leqslant 1$ — diagram, point at 1

10 $x \leqslant 4$ — diagram, point at 4

11 $x \geqslant -2$ — diagram, point at -2

12 $x \geqslant 1$ — diagram, point at 1

13 $x < -1$ — diagram, point at -1

14 $x \leqslant 2$ — diagram, point at 2

15 $x > 1$ — diagram, point at 1

16 $x \geqslant 1\frac{1}{3}$ — diagram, point at $1\frac{1}{3}$

17 $x \geqslant 0$ — diagram, point at 0

18 $x \leqslant 1$ — diagram, point at 1

19 $x < 1$ — diagram, point at 1

20 $x < -3$ — diagram, point at -3

21 **a** $x > 3$ **b** $2 \leqslant x \leqslant 3$
   c No values of x

22 **a** $0 \leqslant x \leqslant 1$ **b** $x \leqslant 0$
   c No values of x

23 **a** $-2 < x \leqslant 4$ **b** No values of x
   c $x < -2$

24 **a** $-3 < x; -1$ **b** $x < -3$
   c No values of x

25 $x < 12; \ x > -1; \ -1 < x < 12$

26 $x \leqslant -1; \ x \geqslant 3;$ No values of x

27 $x \leqslant 7; \ x \geqslant -2; \ -2 \leqslant x \leqslant 7$

28 $x > 1; \ x < 2; \ 1 < x < 2$

29 $x > 2; \ x < 3; \ 2 < x < 3$

30 $x < 2; \ x > -1; \ -1 < x < 2$

31 $x \geqslant -1; \ x < 2; \ -1 \leqslant x < 2$

32 $x > \frac{1}{2}; \ x \leqslant 3; \ \frac{1}{2} < x \leqslant 3$

33 $2 < x < 5$

34 $-3 \leqslant x \leqslant 2$

35 $x < -2$

36 $x < 0$

37 $x \geqslant 1$

38 $-4 < x < 2$

39 $x < -3$

40 $x < -1$

41 $1\frac{4}{5} < x < 3$

42 $\frac{1}{2} < x < 1$

Exercise 22d   page 355

1

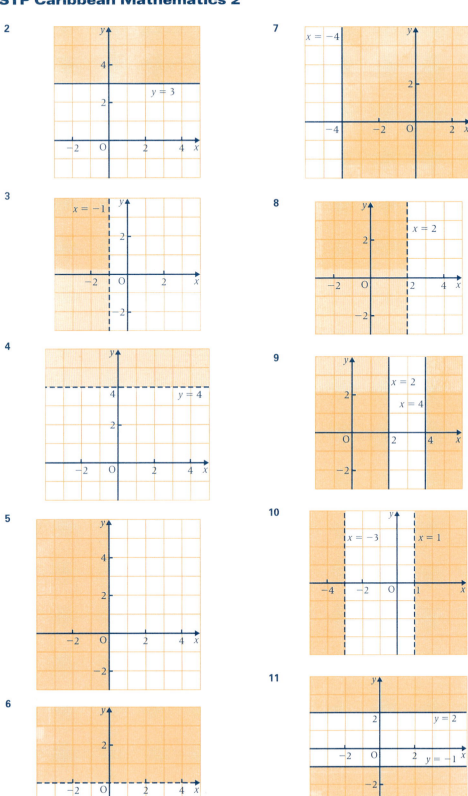

2

$y = 3$

7

$x = -4$

3

$x = -1$

8

$x = 2$

4

$y = 4$

9

$x = 2$

$x = 4$

5

10

$x = -3$ $x = 1$

6

11

$y = 2$

$y = -1$

12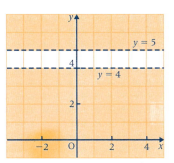

13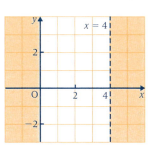

14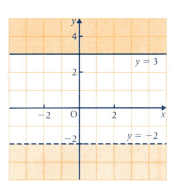

15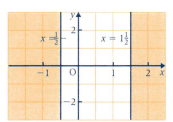

16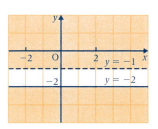

17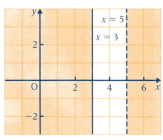

18 10: No   11: No   12: No
19 $x \leqslant 2$ **21** $x < -1$ **23** $-1 \leqslant x < 2$
20 $y < 3$ **22** $-2 \leqslant y \leqslant 2$ **24** $-\frac{1}{2} < y < 2\frac{1}{2}$
25 20: Yes 21: Yes 22: No 23: Yes 24: No 25: No
26 $-3 \leqslant x \leqslant 1$ **28** $2 \leqslant y < 3$
27 $-4 \leqslant y \leqslant -1$ **29** $3 \leqslant x \leqslant 6$
30 26: Yes 27: No 28: Yes 29: No

Exercise 22e   page 358

1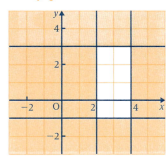

2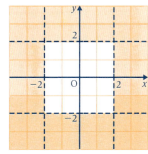

3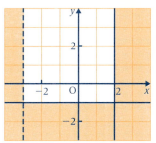

4

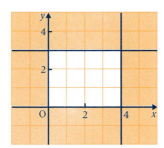

9 $-2 \leqslant x \leqslant 3, \quad -1 \leqslant y \leqslant 2$
10 $-2 < x \leqslant 2, \quad -2 \leqslant y \leqslant 1$
11 9: Yes 10: No
12 $-2 \leqslant x \leqslant 1, \quad y \geqslant -1$
13 $x < 0, \quad y > 0$
14 $-2 < x < 2, \quad -2 < y < 2$
15 $1 < x < 3, \quad 1 < y < 3$

CHAPTER 23

Exercise 23a   page 361

1 $70°$	**3** $60°$	**5** $45°$
2 $110°$	**4** $70°$	**6** $66°$

5

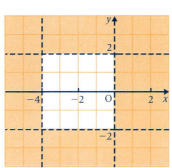

Exercise 23b   page 364

1 Yes	**3** Yes	**5** No	**7** Yes
2 No	**4** No	**6** Yes	**8** Yes

Exercise 23c   page 366

1 Reflection in x-axis; Yes
2 Enlargement, scale factor 2, centre $(0, 0)$; No
3 Translation $\begin{pmatrix} -3 \\ -2 \end{pmatrix}$; Yes
4 Reflection in y-axis; Yes
5 Enlargement, scale factor $\frac{1}{2}$, centre $(-4, 0)$; No
6 Rotation of $90°$ clockwise about $(5, 2)$; Yes

6

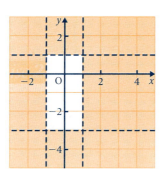

Exercise 23d   page 367

1 $B\widehat{A}C = 38.7°$, $A\widehat{B}C = 51.3°$, $A\widehat{C}B = 90°$; No
2 Yes
3 $B\widehat{A}C = 106°$, $A\widehat{B}C = 39.9°$, $A\widehat{C}B = 34.1°$; No
4 Yes
5 The length of one side

Exercise 23e   page 368

1 Yes; SSS	**5** Yes
2 No	**6** $\triangle$ ABC and $\triangle$ ADC
3 No	**7** Yes
4 No	**8** Yes

7

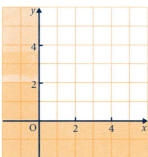

Exercise 23f   page 370

5 Two are: $\triangle$ ABC and $\triangle$ LMN
6 Two

Exercise 23g   page 370

1 Yes; AAS	**6** Yes; AAS
2 No; similar	**7** Yes; AAS
3 Yes; AAS	**8** No
4 Yes; AAS	**9** Yes
5 No	**10** Yes

8

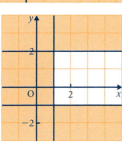

Exercise 23h   page 372

1 Yes; AC $= 4.4$ cm, $\widehat{A} = 34.3°$, $\widehat{C} = 115.7°$
2 No
3 Yes; PR $= 7.2$ cm, $\widehat{R} = 46°$, $\widehat{P} = 74°$
4 No
5 Yes; DF $= 7.8$ cm, $\widehat{D} = 50°$, $\widehat{F} = 40°$

Exercise 23i   page 373

1 Yes
2 No; there are two possibilities
3 Yes

4 No; there are two possibilities
5 Yes
6 Yes

Exercise 23j   page 373

1 Yes; SAS
2 Not necessarily
3 Yes; SHR
4 Yes; SAS
5 Yes; SAS
6 Not necessarily
7 Yes; SHR
8 Not necessarily
9 Yes; SHR
10 Yes

Exercise 23k   page 375

1 No
2 Yes; AAS
3 Yes; SSS
4 Yes; AAS
5 No
6 Yes; SAS
8 Yes; ASS
9 Yes; SHR
10 Yes; SSS
11 Yes; ASS
12 Yes; SAS

Exercise 23l   page 376

9 △BDF and △CDE

Exercise 23n   page 382

1 AC bisects both angles; Yes
2 Both are right angles
3 No
4 They are equal
5 Yes; No; Yes, of AC; they are all right angles
6 No
7 They are equal

Exercise 23p   page 385

1 5 cm
2 60°
4 8 cm
5 5 cm
6 9.5 cm
7 6 cm
8 5.7 cm

CHAPTER 24

Exercise 24a   page 390

1 a 19   **b** 8
2 a tens   **b** 27   **c** 14
d

marks		1 | 4 means 14
1	4 4 5	
2	0 4 5 6 7 7 7 9	
3	0 0 1 2 3 3 4 4 4 4 4 5 5 6 8 9 9	

3 a

Time in minutes		2 | 5 means 25
1	2 4 5 6 8 8 9 7 9	
2	0 0 1 1 5 7 9 7	
3	0 2 5	

Time in minutes		2 | 5 means 25
1	2 4 5 6 7 8 8 9 9	
2	0 0 1 1 5 7 7 9	
3	0 2 5	

b 9   **c** 13   **d** 2

4 a 10
b

Price of books				5 | 70 means $5.70
4	50	90	39	75
5	40	70	95	80
6	20	45		

5 a 28
b

Number of minutes		6 | 1 means 61
3	7	
4	2 1 7 5	
5	9 0 0 9 0 5 2 9 9	
6	1 9 6 0 1 5 8 1 8	
7	4 2 0 4 2	

c

Number of minutes		6 | 1 means 61
3	7	
4	1 2 5 7	
5	0 0 0 2 5 9 9 9 9	
6	0 1 1 1 5 6 8 8 9	
7	0 2 2 4 4	

d 9

Exercise 24b   page 393

	Mean	Mode	Median
1	4.43	2	4
2	14.1	12, 13 and 14	13.5
3	1.84	1.6	1.65
4	1.54	1.3 and 1.8	1.5

5 mean 119.2   median 124
The median, one very low mark brings down the mean.
6 mean $7150   mode $5000   median $5000
The mean. This suggests workers are paid a lot more than they are.
7 mean $180   mode $200   median $175
The median because it is the lowest.
8 mean $21.22   median $18.10
9 a 310   **b** 101   **c** 209.5   **d** 193
10 a 27   **b** 39   **c** 33   **d** 30
11 a 16   **b** 17.36   **c** 16
12 a 38   **b** 38   **c** 36.6

Exercise 24c   page 396

1 2.00 (to 3 s.f.)
2 3.8
3 a 4.28   **b** 5
4 a 3.64   **b** 6
5 a 1.57   **b** 1

Exercise 24d   page 397

1 2
2 1
3 3.5
4 a 3.5   **b** 3   **c** 3.48
5 a 2   **b** 1   **c** 2.2

CHAPTER 25

Exercise 25a   page 401

1 **a** { teachers in my school }
   b { books I have read }
3 **a** odd numbers up to 9
   b the days of the week from Monday to Friday
4 **a** { European countries }, France
   b { multiples of 10 }, 60
5 John $\in$ { boys' names }
6 English $\in$ { school subjects }
7 June $\notin$ { days of the week }
8 Monday $\notin$ { domestic furniture }
9 false   **10** true   **11** true   **12** true

Exercise 25b   page 402

1 infinite   **7** 6   **13** no
2 infinite   **8** 21   **14** yes
3 finite   **9** 11   **15** yes
4 infinite   **10** no   **16** no
5 5   **11** yes
6 8   **12** yes

Exercise 25c   page 403

1 cutlery
2 whole numbers less than 50
3 whole numbers less than 25
4 $n(A) = 8$, $n(B) = 6$
   $B = \{ 3, 6, 9, 12, 15, 18 \}$
5 $A = \{ 1, 2, 3, 4, 6, 12 \}$ $B = \{ 2, 3, 5, 7, 11, 13 \}$
   $C = \{ 6, 12 \}$
6 $n(A) = 4$, $n(B) = 2$, $n(C) = 5$

Exercise 25d   page 404

1 { Audrey, Janet }
2 $A = \{ 1, 3, 5, 7, 9, 11, 13, 15 \}$,
   $B = \{ 2, 3, 5, 7, 11, 13 \}$, $C = \{ 3, 6, 9, 12, 15 \}$
   yes, 3, 5, 7, 11, 13
3 $B = \{ 6, 12, 18 \}$, $C = \{ 2 \}$,
   $D = \{ 13, 14, 15, 16, 17, 18, 19, 20 \}$

Exercise 25e   page 405

1 Your own answers.
2 my friends who do not like coming to my school
3 my friends who like coming to my school
4 all pupils at my school except those who are not my friends and like coming to school
5 My friends who like coming to school and the pupils who are not my friends who do not like coming to school.

Exercise 25f   page 406

1
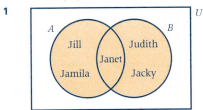

$A \cup B = \{ \text{Janet, Jill, Jenny, Judith, Jacky} \}$

2
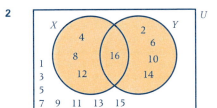

$X \cup Y = \{ 2, 4, 6, 8, 10, 12, 14, 16 \}$

3
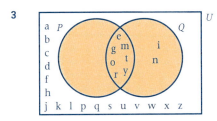

$P \cup Q = \{ \text{e, g, i, m, n, o, t, r, y} \}$

4 **a**
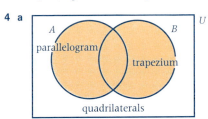

$A \cup B = \{ \text{all parallelograms and trapeziums} \}$

b
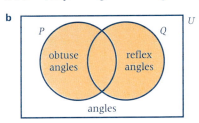

$P \cup Q = \{ \text{angles that are either obtuse or reflex} \}$

5
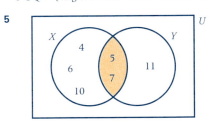

$X \cap Y = \{ 5, 7 \}$

6
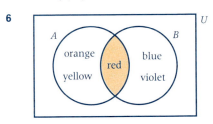

$A \cap B = \{ \text{red} \}$

7

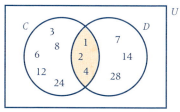

$C \cap D = \{1, 2, 4\}$

8

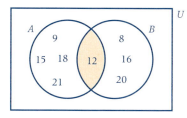

$A \cap B = \{12\}$

Exercise 25g   page 408

1 a Lenny, Sylvia
  b Adam, Richard
  c Jack, Scott, Lee
2 a David, Joe, Terry, Paul
  b Terry, Paul
  c Claude, Alan, Clive
3 a Emma, Majid, Clive, Sean, Ann
  b Emma, Majid, Clive
  c Sean, Ann

4

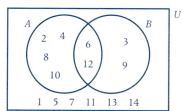

a 6, 12   **b** $n(A) = 6, n(B) = 4$

5

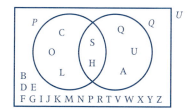

a 5   **b** 8   **c** 2

6 a 26, 6, 6
  b

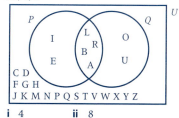

  i 4   **ii** 8

7 a 11, 5, 6
  b

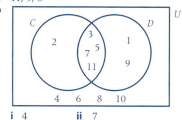

  i 4   **ii** 7
8 a 35, 8, 5
  b i 2   **ii** 11
9 10

CHAPTER 26

Exercise 26a   page 414

1 213_5 **2** 2014_5 **3** 41240_5 **4** 30201_5

	5^3	5^2	5	unit
5			3	1
6			4	2
7		4	1	0
8		2	3	1
9			3	4
10			1	0
11		2	0	4
12		4	0	0

13 16_{10} **19** 17_{10} **25** 13_5 **31** 12_5
14 14_{10} **20** 10_{10} **26** 23_5 **32** 41_5
15 20_{10} **21** 4_{10} **27** 20_5 **33** 110_5
16 36_{10} **22** 100_{10} **28** 124_5 **34** 1003_5
17 54_{10} **23** 70_{10} **29** 133_5 **35** 312_5
18 23_{10} **24** 75_{10} **30** 1100_5 **36** 400_5

Exercise 26b   page 416

1 a

4	1
2	3

  b 11_{10}

2 a

7	1
1	5

  b 12_{10}

3 a

4^2	4	1
1	3	1

  b 29_{10}

4 a

2^2	2	1
1	0	1

  b 5_{10}

5 a

8	1
5	7

  b 47_{10}

6 a

5^2	5	1
2	0	4

  b 29_{10}

7 a

3^2	3	1
2	1	0

  b 12_{10}

8 a

9^2	9	1
5	7	4

  b 472_{10}

9 a

3	1
2	1

b 7_{10}

10 a

9	1
1	8

b 17_{10}

11 a

6	1
2	4

b 16_{10}

12 a

8^2	8	1
1	7	5

b 125_{10}

13 a

4^2	4	1
3	0	3

b 67_{10}

14 a

2^3	2^2	2	1
1	0	0	1

b 9_{10}

15 a

3^3	3^2	3	1
1	2	1	1

b 49_{10}

16 a

6^3	6^2	6	1
1	0	0	0

b 216_{10}

17 21_4
18 22_5
19 33_7
20 111_2
21 23_5
22 52_6
23 65_8
24 100_7
25 22_3
26 23_6
27 37_9
28 1001_3
29 110_7
30 108_9
31 101101_2
32 110100_3
33 243_8
34 22000_4
35 10001110_2
36 422_6
37 111_6
38 210_5
39 1100011_2
40 1102_7
41 131_4
42 15_6
43 1111_2
44 11100_3
45 11302_4
46 234_5
47 33_8
48 102122_3
49 310_8

Exercise 26c   page 418

1 43_5
2 20_3
3 30_4
4 11_2
5 103_6
6 102_3
7 115_8
8 1000_2
9 122_5
10 333_4
11 1151_6
12 10100_2
13 1000_3
14 1000_4
15 1030_7
16 100001_2
17 125_6
18 101_3
19 3_4
20 35_7
21 202_3
22 11_5
23 226_8
24 1_2
25 133_4
26 56_7
27 33_6
28 10_2
29 51_6
30 101_2
31 646_8
32 101_3
33 22_5
34 12_4
35 101_3
36 120_4
37 32_6
38 31_5
39 105_7
40 180_9
41 330_4
42 4105_6
43 1111_3
44 1032_5
45 414_7
46 130_4
47 242_5
48 105_7
49 1015_6
50 123_4
51 143_5
52 560_8
53 2_3
54 214_8

Exercise 26d   page 420

1 10221_4
2 22022_5
3 1110101_2
4 2422_8
5 2011210_3
6 2116_8
7 222142_5
8 220041_7
9 **a** 2254_8 **c** 1196_{10}
   b 52, 23   **d** 2254_8
10 **a** 31026_9 **c** 20436
   b 393, 52   **d** 31026_9
11 3047_8

Exercise 26e   page 421

1 13, 11, 25, 4, 12, 29

2 **a**

+	0	1
0	0	1
1	1	10

b 10010000_2

3 **a** 6   **b** 101_2

4

×	0	1
0	0	0
1	0	1

Exercise 26f   page 422

1 **a** 12_3 **c** 110_3
   b 22_3 **d** 1011_3
2 Three

3

×	0	1	2
0	0	0	0
1	0	1	2
2	0	2	11

4 11_3
5 **a** 102_5 **b** 33_5 **c** 1103_5
6 $31_5 + \frac{2}{5}$ or 16.4_{10}
7 0
8 No
9 Four, $5^3 = 1000_5$
10 Eight, $3^7 = 10000000_3$
11 **a** 2, 3, 4, 5, 6, 7, 8
   b i 11010_2 **ii** 1210_3 **iii** 1750_8
   c The figures move one column to the left; i.e. the effect is the same as multiplying a denary number by ten.
12 6   15 7   18 5
13 4   16 5   19 6
14 3   17 9   20 False

Exercise 26g   page 423

1 **a** 6   **b** 5   **c** 147
2 **a** 40_5 **b** 202_3 **c** 24_8
3 **a** 341_5 **b** 101_3 **c** 636_8
4 part **c**

REVIEW TEST 3   page 424

1 B   3 B   5 A   7 D
2 C   4 C   6 D   8 B
9 **a** 1121   **b i** 63_7 **ii** 101101
10 3rd angle in 1st△ is 40° so triangles are congruent AAS
11 **a** 11   **b** $-\frac{2}{3}$ **c** $y = 4x - 5$
13 **a** 35.5°   **b** 2.5 m

14

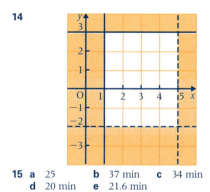

15 a 25 **b** 37 min **c** 34 min
 d 20 min **e** 21.6 min

16 a i A = { prime numbers less than 18 }
 ii B = { multiples of 3 less than 18 }
 b i 11 **ii** 1
17 a 28 **b** 3 **c** 7 **d** 15

Multiple Choice Answers page 426

1	C	**11**	B	**21**	D	**31**	D
2	A	**12**	D	**22**	A	**32**	C
3	D	**13**	B	**23**	B	**33**	B
4	B	**14**	D	**24**	B	**34**	A
5	B	**15**	D	**25**	D	**35**	C
6	B	**16**	A	**26**	C	**36**	D
7	B	**17**	D	**27**	B	**37**	A
8	C	**18**	C	**28**	B	**38**	D
9	A	**19**	C	**29**	D	**39**	D
10	B	**20**	D	**30**	B	**40**	A

INDEX